MATHEMATICS RESEARCH DEVELOPMENTS

FRACTIONAL CALCULUS IN ANALYSIS, DYNAMICS AND OPTIMAL CONTROL

MATHEMATICS RESEARCH DEVELOPMENTS

Additional books in this series can be found on Nova's website under the Series tab.

Additional e-books in this series can be found on Nova's website under the e-book tab.

MATHEMATICS RESEARCH DEVELOPMENTS

FRACTIONAL CALCULUS IN ANALYSIS, DYNAMICS AND OPTIMAL CONTROL

JACKY CRESSON
EDITOR

nova publishers
New York

For permission to use material from this book please contact us:
Telephone 631-231-7269; Fax 631-231-8175
Web Site: http://www.novapublishers.com

LIBRARY OF CONGRESS CATALOGING-IN-PUBLICATION DATA

Fractional calculus in analysis, dynamics, and optimal control / editor, Jacky Cresson (Laboratory of Applied Mathematics, University of Pau, Pau Cedex, France).
pages cm
Includes index.
ISBN 978-1-62948-635-2 (hardcover)
1. Fractional calculus. I. Cresson, Jacky, editor of compilation.
QA314.C74 2014
515'.83--dc23

2013043481

Published by Nova Science Publishers, Inc. † New York

CONTENTS

PREFACE

Fractional calculus has received a big deal of attention in recent years due to its ability to capture many particular features of Physics, Biology or Economy. Many Textbooks on the subject are devoted to the theory of fractional calculus or deal with specific applications in different fields. Mathematical monographs about the subject are most of the time dedicated to fractional versions of classical mathematical fields like fractional differential equations, fractional partial differential equations or fractional calculus of variations, etc. However, the fractional version of many problems is not always supported by any mathematical consideration apart from the generalization of known results. As a consequence, most of the results in these monographs can not be used to solve classical problems of mathematics and are of interest only for people working with fractional calculus.

A non exhaustive list of classical mathematical domains where fractional calculus can be used to obtain new results is:

- Dynamical systems and in particular chaotic Hamiltonian systems
- Analysis of non differentiable functions
- Variational formulations of PDEs or ODEs

None of the previous fields is defined by or concerned with fractional calculus at the beginning.

The first item was developed by Georges Zaslavsky and Rudolph Hilfer at the beginning of the 1990. We can find a synthese of Zaslavsky's work in his book *Hamiltonian Chaos and fractional dynamics* (Oxford University Press, 2005). We refer to the article R. Hilfer untitled *Foundation of fractional dynamics* (Fractals 3(3),549-556,1995) for a good introduction to his point of view. Despite some progress, the tools and rigorous foundations of fractional dynamics to study Hamiltonian chaos are far from being satisfying.

The two others items have appeared after the work of K. Kolvankar and A.D. Gangal around 1997 about the local fractional calculus and its relation to analysis of irregular curves (see K. Kolwankar and A. D. Gangal, Hlder exponents of irregular signals and local fractional derivatives, Pramana J. phys. 48, No. 1 (1997), 4968) and in 1996 after the seminal work of Fred Riewe about a fractional Lagrangian framework for dissipative systems (see F. Riewe, *Nonconservative Lagrangian and Hailtonian mechanics*, Phys. Rev. E (3), 53(2), 1890-1899, 1996).

The aim of this book is to provide examples where fractional calculus can be used to solve classical problems of analysis or at least to provide a new point of view and to present new results in the theory of fractional optimal control. All the contributions are self contain and all the proofs are given in details.

The content of this book is as follows:

The first Chapter authored by Da Prata and Pinto deals with analysis and geometry of non differentiable functions using the so-called local fractional derivative. The author reviewed the initial notion defined by K. Kolvankar and A.D. Gangall and the definition given by F. Ben Adda and J. Cresson. They provide a full comparison of these two notions and they relate precisely the order of differentiation to the fractal dimension of the underlying curve as well as its Hölder regularity. They also extend the notion of local differentiation.

The second Chapter authored by Cresson reviews recent works on the use of a fractional calculus of variations to find Lagrangian variational formulations for dissipative PDEs. This Chapter also gives a complete introduction to the formalism of embedding which is used to obtain these results.

The third Chapter authored by Bourdin studies the discrete/continuous fractional optimal control by proving in particular a fractional Pontryagin's theorem.

The fourth Chapter authored by Pierre Inizan deals with an abstract model associated to chaotic Hamiltonian systems allowing to prove the emergence of fractional dynamics for the asymptotic dynamical behaviour of the systems in the neighbourhood of fractal traps. The construction uses a combination of G. Zaslavsky and R. Hilfer ideas.

The last Chapter authored by Poosed, Almeida and Torres deals more specifically with numerical simulation of fractional optimal control problem.

All the Chapters contain many examples as well as numerical simulations.

We hope that the reader will find in this book many reasons to study fractional calculus as well as new ideas and directions.

2 March 2013

Jacky Cresson
Laboratory of Applied Mathematics
University of Pau Avenue University
P. O. Box 1155
64013 Pau Cedex, France
jacky.cresson@univ-pau.fr

In: Fractional Calculus in Analysis, Dynamics ... ISBN: 978-1-62948-635-2
Editor: Jacky Cresson, pp. 1-63

Chapter 1

LOCAL FRACTIONAL DERIVATIVES

N. C. Dias [*] ***and J. N. Prata*** [†]
Departamento de Matemática
Universidade Lusófona de Humanidades e Tecnologias
Lisboa, Portugal
and
Grupo de Física-Matemática
Univeridade de Lisboa
Lisboa, Portugal

PACS: 05.45-a, 52.35.Mw, 96.50.Fm

Keywords: Fractional calculus, fractals

AMS Subject Classification: 53D, 37C, 65P

1. Introduction

In the course of history, mathematical models of natural phenomena were based on assumptions of regularity and determinism. It was believed that physical phenomena can, by and large, be represented by analytic functions and the dynamics of physical systems can be adequately described by differential equations. These beliefs were largely supported by the successes of physical theories such as Newton's mechanics, electromagnetic theory, acoustics, heat transport, along with diffusion, quantum mechanics and relativity in the 20th century. However, it became clear in more recent years that such phenomena as phase transitions, turbulence, and the rheology of polymeric materials could not be explained using such an approach [27, 42]. In Mandelbrot's own words [25] "clouds are not spheres, mountains are not cones, coastlines are not circles, and bark is not smooth, nor does lightning travel in a straight line".

In his renowned essay [25] Mandelbrot advocated a departure from this tradition of smoothness, regularity and determinism, to one where roughness, irregularity and random-

[*] E-mail address: ncdias@meo.pt
[†] E-mail address: joao.prata@mail.telepac.pt

ness are prevalent. It is not just the fact that natural phenomena are more accurately described by the latter characteristics, but also from a mathematical standpoint, sets and structures which were reputed as exceptional or aberrant should in a sense be the rule. We will recapitulate two notable theorems which prove that, from a topological point of view, the sets of continuous functions and of infinitely differentiable functions are much more profligate at providing functions which are nowhere differentiable or nowhere analytic than the opposite.

A paramount example of randomness and non-differentiability is Brownian motion. The typical path of the Brownian particle is nowhere differentiable [27]. Unlike ordinary Newtonian motion, one cannot write down differential equations for such processes. There is an old answer to this problem, called fractional calculus. Fractional derivatives generalize the notion of derivative of order $n \in \mathbb{N}$ of a function $\frac{d^n f}{dx^n}$ to non-integer orders $\frac{d^\alpha f}{dx^\alpha}$, $\alpha \in \mathbb{R}^+$. One of the truly remarkable properties of fractional derivatives is the fact that a non-differentiable function f may nevertheless admit a fractional derivative of some order $0 < \alpha < 1$. It is by now widely acknowledged that there are numerous situations were fractional differential equations are very effective at describing the situation at hand [42, 44]. It is not just the fact that non-differentiable (fractal) functions may emerge as solutions to fractional differential equations, also the highly non-local nature of fractional derivatives is very useful at addressing systems with long range interactions or prolonged memory effects.

On the other hand, this non-locality prevents the same kind of local description of the geometry of the graph of a function that one has from traditional calculus of differentiable functions. This prompted the search for local versions of fractional derivatives. The first attempts were put forward by Kolwankar and Gangal [17, 18, 19] and several authors then worked on the main mathematical properties of such derivatives [4, 7, 9] and applications [21]. The local fractional derivatives are themselves (as are the non-differentiable functions on which they act) fractal-like. There is a critical order of differentiability below which the derivative vanishes, and above which it diverges. This is strongly reminiscent of the Hausdorff-dimension as one computes the Hausdorff measure of a set below and above its dimension. And indeed the two concepts are inextricably linked. The critical order of differentiability is related to the fractal dimension of the graph of the function.

A structure theorem due to Chen, Yan and K. Zhang [7] proves that if the local fractional derivative of some order $0 < \alpha < 1$ of a function exists in an interval, then it is zero, except possibly in a set of vanishing Lebesgue measure. The interest in local fractional derivatives, then becomes restricted to the following situations.

(i) The function is multi-fractal, and hence the critical order of differentiability varies in the interval. This weakens the conditions of the Chen-Yan-Zhang Theorem and permits the fractional derivative (of varying order) to be finite and non-zero. In this context, various techniques in the framework of wavelet theory have been developed [15].

(ii) The set of zero Lebesgue measure on which the fractional derivative is non-vanishing cannot be discarded as it has non-vanishing and finite Hausdorff measure of dimension $0 < s < 1$ (e.g. the devil staircase function for the Cantor set for $s = \frac{\log 2}{\log 3}$ [40]). This situation could be a physical model for a particle interacting at a set of instants which is a fractal set (e.g. Cantor's set). In such a case the particle's trajectory is certainly not a straight line, which is what one obtains if one neglects the set of vanishing Lebesgue measure where the interactions occur. For this situation a certain type of fractal calculus developed by Parvate and Gangal is more useful [32].

Certain fractal sets display a vanishing or divergent Hausdorff measure for all dimensions $s > 0$. This is because the functions that one uses to "resolve" these sets - the powers x^s - are not tailored for such sets. It is well known that other functions (called dimension functions) may be more appropriate. By the same token, the fact that (in accordance with the Chen-Yan-Zhang Theorem) the local fractional derivatives vanish almost everywhere may in certain cases just be an indication that fractional derivatives of order α are not appropriate and a fine-tuning using orders other than the powers may be more effective. As we will see this is what happens for instance in the case of the Takagi function [39].

In this work we will follow this point of view. Our strategy to circumvent the Chen-Yan-Zhang Theorem is then to develop a local fractional calculus of orders other than the positive reals. We will develop the main properties of this type of calculus, which we dubbed **ψ-calculus**. The fractal nature of the ψ-local fractional derivative manifests itself again. Indeed, if it vanishes nowhere in an interval, then the set of points where it is positive and where it is negative are both dense everywhere in that interval. Moreover, if one draws the graph of the points where the ψ-derivative is positive (or negative) versus the value of the ψ-derivative one obtains a dust-like struture.

Here is a brief outlook of the paper.

In the next section we dwell on the concept of *negligible set* from a topological point of view. We revise the categories of Baire and Baire's Theorem. We apply Baire's Theorem in the proofs of the Banach-Mazurkiewicz Theorem and of the Cater Theorem. The Banach-Mazurkiewicz Theorem says that the set of functions that admit a finite (right) derivative at a point in some interval is a *negligible* subset of the space of continuous functions on that interval. We consider the Weierstrass and the Takagi functions as examples of functions which are continuous but nowhere differentiable in an interval. The Cater Theorem, on the other hand, states that the set of functions which are analytic somewhere in an interval are a *negligible* subset of the space of infinitely differentiable functions on that interval. We also give an example by Cater of a function which is infinitely differentiable but nowhere analytic in an interval.

In section 3 we develop the concept of *dimension function*, a generalization of the topological and the fractal notions of dimension. As an application, we discuss various types of Hölder continuity with respect to dimension functions. The Takagi function constitutes an example of a function displaying a Hölder continuity with respect to one of such functions. The Hausdorff measure and Hausdorff dimension are then presented in connection with the dimension functions. The main result here is the relation between the Hölder behavior of a function and the fractal dimension of its graph.

Section 4 presents our main results: a generalization of local fractional derivatives by using generalized dimension functions as the order of differentiability instead of the more widely used powers. We develop the complete calculus for such derivatives.

Finally in section 5, we consider some possible generalizations, such as Cresson and Nottale's quantum difference operators, the local fractional calculus in several dimensions, and a generalization of Clarke's derivative to fractional order.

Notation 1. *We use the notation* $\mathcal{V}_\xi^+(\epsilon) = [\,\xi, \xi + \epsilon\,)$, $\mathcal{V}_\xi^-(\epsilon) = (\,\xi - \epsilon, \xi\,]$, *and* $\mathcal{V}_\xi(\epsilon) = \mathcal{V}_\xi^+(\epsilon) \cup \mathcal{V}_\xi^-(\epsilon) = (\xi - \epsilon, \xi + \epsilon)$, *for* $\epsilon > 0$.

The vector space of continuous real functions on an interval I *- denoted* $C^0(I)$ *- is a normed space with respect to the sup-norm*

$$||f||_0 := \sup_{x \in I} |f(x)|. \tag{1}$$

Let U *be any non-empty subset of a metric space* Ω *with metric* d. *The* **diameter** *of* U *is defined as* $|U| := sup\,\{d(x, y) : \ x, y \in U\}$. *The ball of radius* $r > 0$ *and center* $x \in U$ *is* $B(x, r) := \{y \in U : \ d(x, y) < r\}$.

We denote by $int(M)$ *the set of interior points of some subset* M *of a topological space* X *and by* $\overline{M}$ *its closure. The complement of* M *in* X *is* $M^c = X \backslash M$.

2. Non-differentiable and Non-analytic Functions

2.1. Baire Categories. Baire's Theorem

From the mathematical point of view there are various ways of assessing whether a certain set or property can be deemed as **negligible**. For instance, one may say that a finite subset of the integers is negligible in the set of integers, since the **cardinality** of $\mathbb{Z}$ is infinity. From the point of view of cardinality, $\mathbb{Z}$ is not a negligible subset of $\mathbb{Q}$ since both are countable. On the other hand, $\mathbb{Q}$ is a negligible subset of $\mathbb{R}$ both from the point of view of cardinality as well as from the **measure/probability** point of view. Indeed, the **measure** (see below) of $\mathbb{Q}$ is zero. One could therefore say that the probability of picking a rational number randomly in some interval is zero. This may not come as a surprise, given that $\mathbb{Q}$ is countable. Nevertheless, there are uncountable subsets of $\mathbb{R}$ which have equally a vanishing (Lebesgue) measure. The most famous example is no doubt Cantor's ternary set C. If we consider instead a Hausdorff measure of suitable dimension ($\log 2 / \log 3$), then Cantor's set no longer has vanishing measure. In spite of this, after a simple inspection, one realizes that Cantor's set is rarefied in $\mathbb{R}$. From the topological point of view it is nowhere dense. In this case, measure and cardinality do not adequately describe this sparseness, as C is *not* negligible from these points of view. If a metrizable topology is at our disposal in some larger set X, then one can characterize this latter notion of a subset $M \subset X$ being negligible, by saying that M is nowhere dense in X. In the case of Cantor's set, one could express this idea by saying that there is no interval, where all the elements of this interval could be approximated with arbitrary precision by elements of C.

These considerations lie at the heart of Baire's classification of sets into categories, describing whether a subset M of a metric space is a countable union of nowhere dense sets, or if the opposite holds. Within this framework we will reproduce here famous theorems which show that differentiable functions are *negligible* in the larger space of continuous functions and likewise that analytic functions are a *negligible* subset of the space of infinitely differentiable functions. From the point of view of mathematical modeling of natural phenomena, one could thus be tempted to say that over the years one has been resorting to a *negligible* sector of the tool kit at our disposal.

Let us then establish the relevant definitions.

Definition 2. *Let X be a metric space. A subset $M \subset X$ is said to be* **rare** *(or* **nowhere dense***) if its closure $\overline{M}$ has no interior points. It is said to be* **meager** *or of the* **first category** *if it is the union of countably many sets each of which is rare in X. If M is not meager it is said to be* **nonmeager** *in X or of the* **second category***. If the complement of M is meager, then M is called* **comeager** *or* **residual***. By abuse of language we will sometimes denote by* **negligible set** *a meager set.*

One way of assessing whether a set is of first category is through the following lemma.

Lemma 3. *A subset M of a metric space X is meager if and only if $M \subset \cup_{n\in\mathbb{N}} M_n$, with M_n closed and with empty interior for all $n \in \mathbb{N}$.*

Proof. Suppose that M is meager. Then $M = \cup_{n\in\mathbb{N}} F_n$, where $int(\overline{F_n}) = \emptyset$. Let $M_n := \overline{F_n}$ for all $n \in \mathbb{N}$. Then F_n is closed and $int(F_n) = \emptyset$. Since $F_n \subset \overline{F_n} = M_n$, we conclude $M \subset \cup_{n\in\mathbb{N}} M_n$.

Conversely, suppose $M \subset \cup_{n\in\mathbb{N}} M_n$, with $\overline{M_n} = M_n$ and $int(M_n) = \emptyset$. Define $F_n = M_n \cap M$ for $n \in \mathbb{N}$. We then have $int(\overline{F_n}) \subset int(M_n) = \emptyset$ and $M = \cup_{n\in\mathbb{N}} F_n$. Consequently, M is meager in X. ■

To prove Baire's Theorem (Theorem 5. below), we need the following preliminary result due to Cantor.

Lemma 4. (Cantor) *Let X be a complete metric space and $\{F_n\}_{n\in\mathbb{N}}$ a nested collection of nonempty closed subsets of X such that $X \supset F_1 \supset F_2 \supset \cdots$ and $|F_n| \to 0$ as $n \to \infty$. Then $\cap_{n\in\mathbb{N}} F_n \neq \emptyset$.*

Proof. For each $n \in \mathbb{N}$ let us choose $x_n \in F_n$. Since $F_1 \supset F_2 \supset \cdots$, we have $x_n \in F_k$, $\forall n \geq k$. By assumption, for each $\epsilon > 0$, there exists $k \in \mathbb{N}$ such that $|F_k| < \epsilon$. Since, for $n, m \geq k$, we have $x_n, x_m \in F_k$, we conclude that $d(x_n, x_m) \leq |F_k| < \epsilon$. In other words, $(x_n)_{n\in\mathbb{N}}$ is a Cauchy sequence in X. Given that X is complete there exists $x \in X$ such that $x_n \to x$ as $n \to \infty$. As all term of the sequence, after a certain order, belong to F_k and F_k is closed, we conclude that $x \in F_k$, that is $x \in \cap_{n\in\mathbb{N}} F_n$. ■

There are various equivalent ways of stating Baire's Theorem. A familiar formulation says that a complete metric space is of second category in itself. Here we shall consider the following convenient version of Baire's Theorem.

Theorem 5. (Baire) *Let X be a complete metric space and $B_1, B_2, B_3, \cdots$ a countable collection of closed subsets of X with no interior points. Then $B = \cup_{n\in\mathbb{N}} B_n$ has empty interior as well.*

Proof. To prove the theorem, we shall demonstrate an equivalent statement, namely that any countable intersection of open, dense subsets of X is dense in X.

Let then $A_1, A_2, A_3, \cdots$ be a countable collection of open and dense subsets of X. We want to prove that $A = \cap_{n\in\mathbb{N}} A_n$ is dense in X, that is, for any nonempty open subset U of X, $U \cap A \neq \emptyset$. Since A_1 is open and dense in X, $A_1 \cap U \neq \emptyset$. Moreover, $A_1 \cap U$ is open, as it is the intersection of two open sets. We can then find another open set $U_1 \subset A_1 \cap U$, such that $\overline{U_1} \subset A_1 \cap U$ and $|\overline{U_1}| \leq 1$. To justify this, let $a \in A_1 \cap U$. There exists $\epsilon > 0$ such that $B(a, \epsilon) \subset A_1 \cap U$. Obviously, we can then find another open ball centered at a of radius $\delta \leq 1$ contained in $B(a, \epsilon)$. Let $U_1 = B(a, \delta/2)$. We have: $B(a, \delta/2) \subset \overline{B(a, \delta/2)} \subset B(a, \delta) \subset B(a, \epsilon) \subset A_1 \cap U$. Moreover, $|\overline{U_1}| = |\overline{B(a, \delta/2)}| = 2 \cdot \delta/2 = \delta \leq 1$.

Similarly, since A_2 is open and dense in X, $A_2 \cap U_1 \neq \emptyset$ and $A_2 \cap U_1$ is an open set. We can thus find another open set $U_2 \subset A_2 \cap U_1$ such that $\overline{U_2} \subset A_2 \cap U_1$ and $|\overline{U_2}| \leq \frac{1}{2}$. By induction, we have a collection of closed sets $\overline{U_n} \neq \emptyset$, $n \in \mathbb{N}$ such that $U \supset \overline{U_1} \supset \overline{U_2} \supset \overline{U_3} \supset \cdots$ and $\overline{U_n} \to 0$ as $n \to \infty$. By Cantor's Lemma, there exists $x \in \cap_{n \in \mathbb{N}} \overline{U_n}$. In particular, $x \in \overline{U_1}$ and thus $x \in U$ and $x \in A_1$. Moreover, $\overline{U_{n+1}} \subset A_{n+1} \cap U_n$, $\forall n \in \mathbb{N}$ and thus $x \in A_n$, $\forall n \in \mathbb{N}$. Consequently, $x \in U \cap A$. ■

2.2. Continuous Nowhere Differentiable Functions

The following theorem due to Banach and Mazurkiewicz conveys the idea that the set of functions which are differentiable at some point in an interval are, from a topological point of view, a *negligible* subset of the space of continuous functions on that interval.

Theorem 6. (Banach, Mazurkiewicz) *Let M be the set of functions with a finite right-derivative at some point in $[\, a, b\,)$ regarded as a subset of the metric space $C^0\,[a, b]$. Then M is meager in $C^0\,[a, b]$.*

Proof. Let us denote by $\{M_n\}_{n \in \mathbb{N}}$ the collection of subsets of $C^0\,[a, b]$

$$M_n := \left\{ f \in C^0\,[a, b] : \exists t \in \left[a, b - \tfrac{(b-a)}{n}\right] \text{ such that } \left| \tfrac{f(t+h) - f(t)}{h} \right| \leq n, \text{ for all } h \in \left(\, 0, \tfrac{b-a}{n}\, \right] \right\}. \tag{2}$$

If $g \in M$, then $g \in M_n$ for some (sufficiently large) $n \in \mathbb{N}$, and thus $M \subset \cup_{n \in \mathbb{N}} M_n$. To justify this claim, suppose $\lim_{h \to 0^+} \frac{g(t+h) - g(t)}{h} = g'(t^+) \in \mathbb{R}$ for some $t \in [\, a, b\,)$. If we fix $K > |g'(t^+)|$, there exists $\delta > 0$ such that $t + \delta \leq b$ and

$$\left| \frac{g(t+h) - g(t)}{h} \right| \leq K, \tag{3}$$

for all $h \in (\, 0, \delta\,]$. If we choose $n \in \mathbb{N}$ such that $n > K$ and $\frac{b-a}{n} \leq \delta$, then $g \in M_n$.

Since $C^0\,[a, b]$ is complete, the result follows by Baire's Theorem, if we prove that F_n is closed with no interior points for all $n \in \mathbb{N}$.

1) We prove that F_n is closed. Let $f \in \overline{F_n}$. Then, there exists a sequence $(f_k)_{k \in \mathbb{N}}$ in F_n such that $f_k \to f$ uniformly as $k \to \infty$. Now let $g_k(t, h) = \frac{f_k(t+h) - f_k(t)}{h}$, and let $g(t, h) = \frac{f(t+h) - f(t)}{h}$ for $t \in \left[a, b - \frac{(b-a)}{n}\right]$ and $h \in (\, 0, \frac{b-a}{n}]$. Since $f_k \in F_n$, there exists $t_k \in \left[a, b - \frac{(b-a)}{n}\right]$, such that $|g_k(t_k, h)| \leq n$ for all $h \in (\, 0, \frac{b-a}{n}]$. Given the uniform convergence of $(f_k)_{k \in \mathbb{N}}$, $g_k(t, h)$ also converges uniformly to $g(t, h)$ in $\left[a, b - \frac{(b-a)}{n}\right]$ for fixed

h. On the other hand, the sequence $(t_k)_{k\in\mathbb{N}}$, being bounded, has a convergent subsequence $t_{k_m} \to c$ for some $c \in \left[a, b - \frac{(b-a)}{n}\right]$. We thus have

$$|g_{k_m}(t_{k_m}, h) - g(c,h)| \leq |g_{k_m}(t_{k_m}, h) - g_{k_m}(c,h)| + |g_{k_m}(c,h) - g(c,h)| \tag{4}$$

Since g_k and g are continuous functions and $g_k(t,h) \to g(t,h)$ uniformly, we conclude that $g_{k_m}(t_{k_m}) \to g(c,h)$ as $k_m \to \infty$. But, for all $h \in (\,0, \frac{b-a}{n}]$, we have $|g_k(t_k, h)| \leq n$, and thus $|g(c,h)| \leq n$, that is $\left|\frac{f(c+h)-f(c)}{h}\right| \leq n$ for all $h \in (\,0, \frac{b-a}{n}]$. In other words, $f \in F_n$ and F_n is closed.

2) Next we prove that $int(F_n) = \emptyset$. We want to show that for all $f \in F_n$ and any $\epsilon > 0$, $B(f,\epsilon) \cap F_n^c \neq \emptyset$.

By Weierstrass' approximation theorem [14], let $p(x)$ be a polynomial on $[a,b]$ such that $d(f,p) < \epsilon/2$. Let S denote the maximum slope of $p(x)$ on $[a,b]$, i.e.:

$$S := max\left\{\frac{p(x)-p(y)}{x-y} \;:\; x, y \in [a,b]\right\}. \tag{5}$$

We shall also consider a continuous function $q(x)$ on $[a,b]$ consisting of straight-line segments of slopes $\pm(S+n+1)$ constrained so that $|q(x)| < \epsilon/2$ for all $x \in [a,b]$. Define the function $g(x) = p(x) + q(x)$. It is straightforward to conclude that $g \in B(f,\epsilon)$. Indeed: $d(f,g) \leq d(f,p) + d(p,g) = d(f,p) + ||q|| < \frac{\epsilon}{2} + \frac{\epsilon}{2} = \epsilon$.

On the other hand, we have for any $t \in \left[a, b - \frac{(b-a)}{n}\right]$:

$$\begin{aligned} \left|\tfrac{g(t+h)-g(t)}{h}\right| &= \left|\tfrac{p(t+h)+q(t+h)-p(t)-q(t)}{h}\right| \geq \\ &\geq \left|\tfrac{q(t+h)-q(t)}{h}\right| - \left|\tfrac{p(t+h)-p(t)}{h}\right| \end{aligned} \tag{6}$$

One can then find a (sufficiently small) $h \in (\,0, \frac{b-a}{n}\,]$ for which the right-hand side is $\geq S+n+1-S = n+1 > n$ and thus $g \neq F_n$. ■

Remark 7. *Notice that if f admits a finite right-derivative at some $\xi \in [\,a, b\,)$, then $g(x) = f(a+b-x)$ has a finite left-derivative at $a+b-\xi$ and vice-versa. Consequently, the set of functions with a finite left-derivative at some $\zeta \in (\,a, b\,]$ is equally of first category in $C^0\,[a,b]$. Finally, the set of functions which are differentiable at some point in the interval is a subset of the former two sets, and hence must be of first category.*

Before we end this section, we shall present two celebrated functions which are continuous in an interval but nowhere differentiable in that interval: the Takagi and the Weierstrass functions.

Takagi's function was introduced in [39]. A recent survey of its properties and applications is [1].

Definition 8. *Let $\phi(x) = dist(x,\mathbb{Z})$ denote the distance from $x \in \mathbb{R}$ to the nearest integer. Then the* **Takagi function** *is defined on $[0,1]$ by:*

$$T(x) := \sum_{n=0}^{\infty} \frac{1}{2^n}\phi(2^n x). \tag{7}$$

The continuity of T is easily established.

Lemma 9. *Takagi's function is continuous in* $[0,1]$.

Proof. Notice that $0 \leq \phi(x) \leq \frac{1}{2}$ for any $x \in \mathbb{R}$. Consequently:

$$\left| \sum_{n=0}^{\infty} \frac{1}{2^n} \phi(2^n x) \right| \leq \sum_{n=0}^{\infty} \frac{1}{2^{n+1}} < \infty \tag{8}$$

for all $x \in [0,1]$, which establishes the uniform convergence of T in $[0,1]$. Since ϕ is continuous, the result follows. ■

Proposition 10. (Takagi) *The Takagi function does not possess a finite derivative at any point.*

Proof. We will reproduce here the simple proof due to Billingsley [5]. Let $\mathbb{Z}_+$ denote the set of non-negative integers. A **dyadic rational** of order m is a rational number of the form $x = \frac{k}{2^m}$ with $k \in \mathbb{Z}_+$ and $m \in \mathbb{N}$. Put $\phi_k(x) = 2^{-k}\phi(2^k x)$ for $k = 1, 2, \cdots$. Fix a point x, and, for each $n \in \mathbb{N}$, let u_n and v_n be dyadic rationals of order n with $v_n - u_n = 2^{-n}$ and $u_n \leq x < v_n$. Then:

$$\frac{T(v_n) - T(u_n)}{v_n - u_n} = \sum_{k=0}^{n-1} \frac{\phi_k(v_n) - \phi_k(u_n)}{v_n - u_n}, \tag{9}$$

since $\phi_k(v_n) = \phi_k(u_n) = 0$ for all $k \geq n$. But for $k < n$, ϕ_k is linear on $[u_k, v_k]$ with slope $\phi_k'(x^+)$, the right-derivative of ϕ_k at x. Thus

$$\frac{T(v_n) - T(u_n)}{v_n - u_n} = \sum_{k=0}^{n-1} \phi_k'(x^+). \tag{10}$$

Since $\phi_k'(x^+) = \pm 1$ for each k, this last sum cannot converge to finite limit as $n \to \infty$. However, if T were differentiable at x, then (10) would have to be a convergent sequence, and we have a contradiction. Hence T cannot have a finite derivative at x. Since x was chosen arbitrarily, the result follows. ■

It will also be relevant for the sequel to consider the Hölder behavior of the Takagi function. The following Proposition is due to Anderson and Pitt [3], and was proved in the more general context of functions belonging to the so-called Zygmund space Λ_d^*.

Proposition 11. (Anderson, Pitt) *There exists a constant* $C > 0$ *such that, for all* $x \in (0,1)$ *and sufficiently small* $h \neq 0$*:*

$$|T(x+h) - T(x)| \leq C|h| \log(1/|h|). \tag{11}$$

It is also interesting to enquire about the existence of the limit $\lim_{h \to 0} \frac{T(x+h)-T(x)}{-h\log_2(|h|)}$. The following theorem is due to Krüppel [23].

Theorem 12. (Krüppel) *Let $x \in [0,1]$ be a dyadic rational. Then:*

$$\lim_{h\to 0} \frac{T(x+h)-T(x)}{|h|\log_2\left(\frac{1}{|h|}\right)} = 1. \tag{12}$$

Another characterization proved by Allaart and Kawamura [1] requires the following definitions.

Definition 13. *The* **binary expansion** *of a point $x \in [\,0,1\,)$ is denoted by*

$$x = \sum_{n=1}^{\infty} \frac{\epsilon_n}{2^n} = 0.\epsilon_1\epsilon_2\cdots\epsilon_n\cdots, \qquad \epsilon_n \in \{0,1\} \tag{13}$$

where, obviously $\epsilon_n = \epsilon_n(x)$. If the limit

$$d_1(x) := \lim_{n\to\infty} \frac{1}{n}\sum_{k=1}^{n} \epsilon_k \tag{14}$$

exists, we call $d_1(x)$ the **density** *(or* **long-time frequency***) of the digit "1" in the binary expansion of x. In that case, the number*

$$d_0(x) := 1 - d_1(x) \tag{15}$$

is the density of the digit "0".

Definition 14. *Let $x \in [0,1]$ be non-dyadic, and let $(a_n)_{n\in\mathbb{N}}$ and $(b_n)_{n\in\mathbb{N}}$ be the (unique) strictly increasing sequences of positive integers such that $x = \sum_{n=1}^{\infty} 2^{-a_n}$, $1-x = \sum_{n=1}^{\infty} 2^{-b_n}$.*

We say that x is **density-regular** *if $d_1(x)$ exists and one of the following holds:*

(a) $0 < d_1(x) < 1$; or

(b) $d_1(x) = 0$ and $\frac{a_{n+1}}{a_n} \to 1$; or

(c) $d_1(x) = 1$ and $\frac{b_{n+1}}{b_n} \to 1$.

Theorem 15. (Allaart, Kawamura) *Let $\in [0,1]$ be non-dyadic. The limit in (12) exists if and only if x is density-regular, in which case the limit is equal to $d_0(x) - d_1(x)$.*

The Weierstrass function was the first function to be proved to be continuous and nowhere differentiable [41]. For a while it was regarded as a mathematical curiosity, but in his 1926 investigation of turbulence, Richardson [34] observed that the velocity field of the atmospheric wind is so erratic that it probably cannot be described by any analytic function or indeed a differentiable function. He suggested Weierstrass' function as a candidate to represent the velocity field, since its properties of continuity and non-differentiability were observed in the wind data. We have, since that time, come to realize that Richardson's intuition was superior to a half century of analysis regarding the nature of turbulence.

Definition 16. *Fix $\lambda > 1$ and $1 < s < 2$. Define the* **Weierstrass function** $W_{\lambda,s} : [0,1] \to \mathbb{R}$ *by:*

$$W_{\lambda,s}(x) := \sum_{k=1}^{\infty} \lambda^{(s-2)k} \sin(\lambda^k x). \tag{16}$$

Lemma 17. *The Weierstrass function is continuous in* $[0,1]$.

Proof. Again the continuity is an immediate consequence of the uniform convergence of the series in $[0,1]$:

$$\left|\sum_{k=1}^{\infty} \lambda^{(s-2)k} \sin(\lambda^k x)\right| \leq \sum_{k=1}^{\infty} \lambda^{(s-2)k} < \infty, \qquad \forall x \in [0,1]\,. \tag{17}$$

■

The following proposition expresses the Hölder and anti-Hölder behavior of the Weierstrass function.

Proposition 18. *Let $W_{\lambda,s}$ denote the Weierstrass function for $1 < s < 2$, $\lambda > 1$. Then:*
(i) there exists a constant $C_1 > 0$ such that

$$|W_{\lambda,s}(x+h) - W_{\lambda,s}(x)| \leq C_1 |h|^{2-s}, \quad \textit{for all } h \textit{ with } |h| < \frac{1}{\lambda}; \tag{18}$$

(ii) there exist positive constants C_2, δ_0 with the following property: for each $x \in [0,1]$ and $0 < \delta \leq \delta_0$ there exists y such that $|x - y| \leq \delta$ and

$$|W_{\lambda,s}(x) - W_{\lambda,s}(y)| \geq C_2 \delta^{2-s}, \tag{19}$$

for large enough λ.

Proof. The proof presented here is given in Falconer's book [12].

(i) Given $0 < |h| < \lambda^{-1}$, let $N \in \mathbb{N}$ such that

$$\lambda^{-(N+1)} \leq |h| < \lambda^{-N}. \tag{20}$$

Then

$$\begin{aligned} |W_{\lambda,s}(x+h) - W_{\lambda,s}(x)| &\leq \textstyle\sum_{k=1}^{N} \lambda^{(s-2)k} \left|\sin\left(\lambda^k(x+h)\right) - \sin(\lambda^k x)\right| + \\ &+ \textstyle\sum_{k=N+1}^{\infty} \lambda^{(s-2)k} \left|\sin\left(\lambda^k(x+h)\right) - \sin(\lambda^k x)\right| \leq \\ &\leq \textstyle\sum_{k=1}^{N} \lambda^{(s-2)k} \lambda^k |h| + \sum_{k=N+1}^{\infty} 2\lambda^{(s-2)k} \end{aligned} \tag{21}$$

using that $|\sin(u) - \sin(v)| \leq |u - v|$ (a consequence of the mean-value theorem) on the first N terms of the sum, and that $|\sin(u) - \sin(v)| \leq |\sin(u)| + \sin(v)| \leq 2$ on the remaining terms. Summing these geometric series, we obtain:

$$|W_{\lambda,s}(x+h) - W_{\lambda,s}(x)| \leq \frac{|h|\lambda^{(s-1)N}}{1-\lambda^{1-s}} + \frac{2\lambda^{(s-2)(N+1)}}{1-\lambda^{s-2}} \leq c|h|^{2-s} \tag{22}$$

where c is independent of h and we used (20).

(ii) In the same way, but splitting the sum into three parts - the first $(N-1)$ terms, the Nth term, and the rest - we get that:

$$\begin{aligned}&\left|W_{\lambda,s}(x+h)-W_{\lambda,s}(x)-\lambda^{(s-2)N}\left[\sin\left(\lambda^N(x+h)\right)-\sin(\lambda^k x)\right]\right|\leq\\&\qquad\leq\tfrac{\lambda^{(s-2)N-s+1}}{1-\lambda^{1-s}}+\tfrac{2\lambda^{(s-2)(N+1)}}{1-\lambda^{s-2}},\end{aligned}\tag{23}$$

where we used (20) again.

Suppose $\lambda > 2$ is large enough for the right-hand side of (23) to be less than $\frac{\lambda^{(s-2)N}}{20}$ for all N. For $\delta < \lambda^{-1}$, take N such that $\lambda^{-N} \leq \delta < \lambda^{-N+1}$. For each $x \in [0,1]$, we may choose h with $\lambda^{-N-1} \leq |h| < \lambda^{-N} < \delta$ such that $\left|\sin\left(\lambda^N(x+h)\right) - \sin(\lambda^k x)\right| > \frac{1}{10}$. So by (23)

$$\begin{aligned}&\left||W_{\lambda,s}(x+h)-W_{\lambda,s}(x)|-\left|\lambda^{(s-2)N}\left[\sin\left(\lambda^N(x+h)\right)-\sin(\lambda^k x)\right]\right|\right|\leq\\&\left|W_{\lambda,s}(x+h)-W_{\lambda,s}(x)-\lambda^{(s-2)N}\left[\sin\left(\lambda^N(x+h)\right)-\sin(\lambda^k x)\right]\right|\leq\tfrac{\lambda^{(s-2)N}}{20}\\&\Rightarrow|W_{\lambda,s}(x+h)-W_{\lambda,s}(x)|\geq\left|\lambda^{(s-2)N}\left[\sin\left(\lambda^N(x+h)\right)-\sin(\lambda^k x)\right]\right|-\tfrac{\lambda^{(s-2)N}}{20}\\&\geq\tfrac{\lambda^{(s-2)N}}{10}-\tfrac{\lambda^{(s-2)N}}{20}=\tfrac{\lambda^{(s-2)N}}{20}\geq\tfrac{\lambda^{s-2}\delta^{2-s}}{20},\end{aligned}\tag{24}$$

which concludes the proof. ■

The non-differentiability of $W_{\lambda,s}$ is then a trivial consequence of Proposition 18. (ii).

Corollary 19. *Let $\lambda > 1$ and $1 < s < 2$. Then the Weierstrass function $W_{\lambda,s}$ is nowhere differentiable in the interval $[0,1]$.*

Proof. From Proposition 18. (ii), for each $x \in [0,1]$ and sufficiently large $n \in \mathbb{N}$, we can find $y_n \in [0,1] \setminus \{x\}$ such that

$$|x - y_n| \leq \frac{1}{n}, \tag{25}$$

and

$$|W_{\lambda,s}(x) - W_{\lambda,s}(y_n)| \geq C_2 \left(\frac{1}{n}\right)^{2-s}. \tag{26}$$

From (25) and (26), we then have

$$\left|\frac{W_{\lambda,s}(x) - W_{\lambda,s}(y_n)}{x - y_n}\right| \geq C_2 n^{s-1}. \tag{27}$$

The right-hand side of the previous inequality diverges as $n \to \infty$ and thus $W_{\lambda,s}$ cannot be differentiable at x. ■

2.3. Infinitely Differentiable, Nowhere Analytic Functions

By the same token, one can prove that the set of analytic functions is a negligible subset of the set of infinitely differentiable functions. We will follow closely [6].

Let $C^\infty[a,b]$ denote the set of functions which admit continuous derivatives of all orders in $[a,b]$. The set $C^\infty[a,b]$ is a subspace of $C^0[a,b]$ which is incomplete with respect to the sup-norm of $C^0[a,b]$. Indeed, the set of polynomials on $[a,b]$ is a subset of $C^\infty[a,b]$ and, by Weierstrass' approximation theorem [14], dense in $C^0[a,b]$. Hence a sequence of polynomials can converge to, say, a continuous non-differentiable function.

We thus need an alternative metric in $C^\infty[a,b]$ to ensure completeness. For $f_1, f_2 \in C^\infty[a,b]$, we define

$$d_F(f_1,f_2) := \sum_{k=0}^{\infty} \frac{1}{2^k} \sup_{a\leq x\leq b} \frac{|f_1^{(k)}(x) - f_2^{(k)}(x)|}{1+|f_1^{(k)}(x) - f_2^{(k)}(x)|} \tag{28}$$

with the convention $f^{(0)} = f$.

Proposition 20. *The set $F = (C^\infty[a,b], d_F)$ is a complete, separable metric space.*

Proof. The proof proceeds in several steps.

1) We start by showing that d_F is a metric. First of all, since $0 \leq \frac{t}{1+t} \leq 1$ for all $t \geq 0$, we conclude that:

$$0 \leq d_F(f_1,f_2) \leq \sum_{k=0}^{\infty} \frac{1}{2^k} < \infty \tag{29}$$

for all $f_1, f_2 \in C^\infty[a,b]$. Clearly, $d_F(f_1,f_2) = 0$ if and only if $f_1 = f_2$ and $d_F(f_1,f_2) = d_F(f_2,f_1)$. It remains to prove the triangle inequality. Let $g(t) = \frac{t}{1+t}$ for $t \geq 0$. Since $g'(t) = \frac{1}{(1+t)^2} > 0$ for all $t \geq 0$, we conclude that g is strictly increasing. Since for any $u, v \in \mathbb{R}$, $|u+v| \leq |u| + |v|$, it follows that $g(|u+v|) \leq g(|u|+|v|)$ and thus:

$$\frac{|u+v|}{1+|u+v|} \leq \frac{|u|+|v|}{1+|u|+|v|} = \frac{|u|}{1+|u|+|v|} + \frac{|v|}{1+|u|+|v|} \leq \frac{|u|}{1+|u|} + \frac{|v|}{1+|v|}. \tag{30}$$

Consequently:

$$\begin{aligned} &\frac{\left|\left(f_1^{(k)}(x)-f_3^{(k)}(x)\right)+\left(f_3^{(k)}(x)-f_2^{(k)}(x)\right)\right|}{1+\left|\left(f_1^{(k)}(x)-f_3^{(k)}(x)\right)+\left(f_3^{(k)}(x)-f_2^{(k)}(x)\right)\right|} \leq \\ &\leq \frac{\left|f_1^{(k)}(x)-f_3^{(k)}(x)\right|}{1+\left|f_1^{(k)}(x)-f_3^{(k)}(x)\right|} + \frac{\left|f_3^{(k)}(x)-f_2^{(k)}(x)\right|}{1+\left|f_3^{(k)}(x)-f_2^{(k)}(x)\right|} \end{aligned} \tag{31}$$

and $d_F(f_1,f_2) \leq d_F(f_1,f_3) + d_F(f_3,f_2)$.

2) Let us now prove that F is complete. Let $(f_n)_{n\in\mathbb{N}}$ denote a Cauchy sequence in F, that is $\lim_{n,m\to\infty} d_F(f_n,f_m) = 0$. Then

$$\sup_{a\leq x\leq b} \frac{|f_n^{(k)}(x) - f_m^{(k)}(x)|}{1+|f_n^{(k)}(x) - f_m^{(k)}(x)|} \leq 2^k d_F(f_n,f_m) \tag{32}$$

Thus, for each fixed $k \in \mathbb{N}$, the sequence of continuous functions $(f_n^{(k)})_{n\in\mathbb{N}}$ converges uniformly on $[a,b]$, say to the continuous function g_k. Then:

$$\lim_{n\to\infty} \sup_{a\le x\le b} |f_n^{(k)}(x) - g_k(x)| = 0, \tag{33}$$

and

$$\lim_{n\to\infty} \sup_{a\le x\le b} \frac{|f_n^{(k)}(x) - g_k(x)|}{1 + |f_n^{(k)}(x) - g_k(x)|} = 0, \tag{34}$$

for each $k \geq 0$. By the uniform convergence, it is legitimate to interchange the limit with the summation and obtain:

$$\lim_{n\to\infty} \sum_{k=0}^{\infty} \frac{1}{2^k} \sup_{a\le x\le b} \frac{|f_n^{(k)}(x) - g_k(x)|}{1 + |f_n^{(k)}(x) - g_k(x)|} = 0. \tag{35}$$

It remains to prove that $g_0 \in F$ and $g_0^{(k)} = g_k$ for all $k = 0, 1, 2, 3, \cdots$ by induction on k. For $k = 0$, there is nothing to prove. We assume $g_0^{(N-1)} = g_{N-1}$ for some fixed $N \geq 1$. Clearly, the sequence $(f'_n)_{n\in\mathbb{N}}$ is also a Cauchy sequence in F, because $d_F(f'_n, f'_m) \leq 2d_F(f_n, f_m)$ by the definition of d_F. Moreover, $(f_n'^{(k)})_{n\in\mathbb{N}}$ converges uniformly to g_{k+1} for $k \geq 0$. By the induction hypothesis $g_0^{(N)} = g_0'^{(N-1)} = g_N$. This completes the induction. Thus $g_0 \in F$ and $\lim_{n\to\infty} d_F(f_n, g_0) = 0$ follows from (33).

3) We finally prove the separability of F. Our aim will be to show that the set of polynomials in x with rational coefficients is dense in F. For the sake of simplicity, we shall now assume that the functions of F are defined on the interval $[0, 1]$ (something which can be obtained by a straightforward translation and dilation).

Take any $\epsilon > 0$ and any $f \in F$. We prove by induction on n that for each $n \geq 0$ there is a polynomial p with rational coefficients such that

$$\sum_{i=0}^{n} \left| f^{(i)}(x) - p^{(i)}(x) \right| < \epsilon, \tag{36}$$

for $0 \leq x \leq 1$.

For $n = 0$, we use Weierstrass' approximation theorem [14] a find a polynomial $q(x)$ such that

$$|f(x) - q(x)| < \frac{\epsilon}{4}, \tag{37}$$

for $0 \leq x \leq 1$. We then approximate the coefficients of q with rational numbers to find the appropriate polynomial p such that

$$|p(x) - q(x)| < \frac{\epsilon}{4}, \tag{38}$$

for $0 \leq x \leq 1$. And hence:

$$|f(x) - p(x)| < \frac{\epsilon}{2}, \tag{39}$$

for $0 \leq x \leq 1$.

Now assume that p can be found for some $N \geq 0$. But, since $f' \in F$, there is a polynomial $p_0(x)$ with rational coefficients such that:

$$\sum_{i=0}^{N} \left| f^{(i+1)}(x) - p_0^{(i)}(x) \right| < \frac{\epsilon}{4}, \tag{40}$$

for $0 \leq x \leq 1$.

Let $\xi \in \mathbb{Q}$ such that $\left| f\left(\frac{1}{2}\right) - \xi \right| < \frac{\epsilon}{4}$. Put

$$p(x) = \xi + \int_{\frac{1}{2}}^{x} p_0(t)dt, \tag{41}$$

for $0 \leq x \leq 1$. Then

$$\begin{aligned} |p(x) - f(x)| &= \left| \xi + \textstyle\int_{\frac{1}{2}}^{x} p_0(t)dt - f\left(\frac{1}{2}\right) - \int_{\frac{1}{2}}^{x} f'(t)dt \right| \leq \\ &\leq \left|\xi - f\left(\tfrac{1}{2}\right)\right| + \left| \textstyle\int_{\frac{1}{2}}^{x} |p_0(t) - f'(t)|\, dt \right| < \tfrac{\epsilon}{4} + \tfrac{\epsilon}{4} = \tfrac{\epsilon}{2}, \end{aligned} \tag{42}$$

for $0 \leq x \leq 1$. So p' and $p^{(i+1)} = p_0^{(i)}$ and

$$\begin{aligned} \textstyle\sum_{i=0}^{N+1} \left| f^{(i)}(x) - p^{(i)}(x) \right| &= |f(x) - p(x)| + \\ + \textstyle\sum_{i=0}^{N} \left| f^{(i+1)}(x) - p^{(i+1)}(x) \right| &< \tfrac{\epsilon}{4} + \tfrac{\epsilon}{4} = \tfrac{\epsilon}{2}, \end{aligned} \tag{43}$$

for $0 \leq x \leq 1$, which completes the induction proof on n.

Let n be chosen so that $2^{-n} < \epsilon$. Then:

$$\begin{aligned} d(f,p) &= \textstyle\sum_{k=0}^{n} \frac{1}{2^k} \sup_{0\leq x\leq 1} \frac{\left|f^{(k)}(x)-p^{(k)}(x)\right|}{1+\left|f^{(k)}(x)-p^{(k)}(x)\right|} + \\ &+ \textstyle\sum_{k=n+1}^{\infty} \frac{1}{2^k} \sup_{0\leq x\leq 1} \frac{\left|f^{(k)}(x)-p^{(k)}(x)\right|}{1+\left|f^{(k)}(x)-p^{(k)}(x)\right|} \leq \\ &\leq \textstyle\sum_{k=0}^{n} \sup_{0\leq x\leq 1} \left| f^{(k)}(x) - p^{(k)}(x) \right| + \sum_{k=n+1}^{\infty} \frac{1}{2^k} < \\ &< \epsilon + \tfrac{1}{2^n} < \epsilon + \epsilon = 2\epsilon \end{aligned} \tag{44}$$

■

We next proceed to show that the set of functions which are analytic in some interval is of first category in F.

For future reference, let us first define the following element of $\mathcal{D}$, the Schwartz space of functions in C^∞ with compact support in some interval $[c,d]$. Let then

$$h(x) = \begin{cases} e^{-(x-c)^{-2}(x-d)^{-2}}, & \text{if } c \leq x \leq d \\ 0 & \text{otherwise} \end{cases} \tag{45}$$

This function will be dubbed a **spike** on (c, d). By a simple inspection, one realizes that h does indeed admit an infinite number of continuous derivatives in $\mathbb{R}$ with $h^{(k)}(x) = 0$ whenever $x \geq d$ or $x \leq c$. Consequently, h cannot be analytic on any interval centered around c or d.

Theorem 21. (Cater) *The set F_0 of nowhere analytic functions on $[a, b]$ forms a residual subset of F.*

Proof. If a function is analytic in some interval, then it is analytic in every other subinterval. It thus suffices to prove that the set of functions that are the limit of their Taylor series in $x - x_0$ on an interval I centered at x_0, with rational endpoints, is a first category set in F. Fix an interval $I \subset [a, b]$ with rational endpoints and center x_0.

For each $j \in \mathbb{N}$, let:

$$U_j := \left\{ f \in F : \left| \sum_{k=0}^{j} \frac{f^{(k)}(x_0)}{k!}(\xi - x_0)^k - f(\xi) \right| > 1 \text{ for some } \xi \in I \right\}. \tag{46}$$

1) The set U_j is open. Indeed, let $f \in U_j$ and define:

$$\epsilon = -1 + \left| \sum_{k=0}^{j} \frac{f^{(k)}(x_0)}{k!}(\xi - x_0)^k - f(\xi) \right| \tag{47}$$

where ξ is as in the definition of U_j. Now let $f_1 \in B\left(f, 2^{-j}\frac{e^{a-b}\epsilon/2}{1+e^{a-b}\epsilon/2}\right)$. Then, by the definition of d_F, we have:

$$\begin{aligned} &\frac{1}{2^k}\frac{\left|f^{(k)}(x_0)-f_1^{(k)}(x_0)\right|}{1+\left|f^{(k)}(x_0)-f_1^{(k)}(x_0)\right|} \leq \\ &\leq \sum_{m=0}^{\infty} \sup_{a\leq x\leq b} \frac{1}{2^m}\frac{\left|f^{(m)}(x)-f_1^{(m)}(x)\right|}{1+\left|f^{(m)}(x)-f_1^{(m)}(x)\right|} < \frac{1}{2^j}\frac{e^{a-b}\epsilon/2}{1+e^{a-b}\epsilon/2} \end{aligned} \tag{48}$$

for all $k \geq 0$. In particular, for $0 \leq k \leq j$, we have:

$$\frac{\left|f^{(k)}(x_0) - f_1^{(k)}(x_0)\right|}{1 + \left|f^{(k)}(x_0) - f_1^{(k)}(x_0)\right|} < \frac{e^{a-b}\epsilon/2}{1 + e^{a-b}\epsilon/2} \Leftrightarrow \left|f^{(k)}(x_0) - f_1^{(k)}(x_0)\right| < e^{a-b}\frac{\epsilon}{2}. \tag{49}$$

Consequently

$$\begin{aligned} &\left|\sum_{k=0}^{j} \frac{f^{(k)}(x_0)}{k!}(\xi - x_0)^k - \sum_{k=0}^{j} \frac{f_1^{(k)}(x_0)}{k!}(\xi - x_0)^k\right| \leq \\ &\leq \sum_{k=0}^{j} \left|f^{(k)}(x_0) - f_1^{(k)}(x_0)\right| \frac{|\xi-x_0|^k}{k!} < e^{a-b}\frac{\epsilon}{2}\sum_{k=0}^{j} \frac{(b-a)^k}{k!} < \frac{\epsilon}{2}. \end{aligned} \tag{50}$$

From (48), we also have:

$$|f(x) - f_1(x)| < e^{a-b}\frac{\epsilon}{2} < \frac{\epsilon}{2}, \tag{51}$$

for all $a \leq x \leq b$. Consequently, from (49) and (50), we have:

$$\left|\sum_{k=0}^{j} \frac{f_1^{(k)}(x_0)}{k!}(\xi - x_0)^k - f_1(\xi)\right| =$$

$$= \left|\sum_{k=0}^{j} \frac{f_1^{(k)}(x_0)}{k!}(\xi - x_0)^k - \sum_{k=0}^{j} \frac{f^{(k)}(x_0)}{k!}(\xi - x_0)^k + \right.$$

$$\left. + \sum_{k=0}^{j} \frac{f^{(k)}(x_0)}{k!}(\xi - x_0)^k - f_1(\xi)\right| \geq$$

$$\geq \left|\sum_{k=0}^{j} \frac{f^{(k)}(x_0)}{k!}(\xi - x_0)^k - f(\xi) + f(\xi) - f_1(\xi)\right| \tag{52}$$

$$- \left|\sum_{k=0}^{j} \frac{f_1^{(k)}(x_0)}{k!}(\xi - x_0)^k - \sum_{k=0}^{j} \frac{f^{(k)}(x_0)}{k!}(\xi - x_0)^k\right| >$$

$$> \left|\sum_{k=0}^{j} \frac{f^{(k)}(x_0)}{k!}(\xi - x_0)^k - f(\xi)\right| - |f(\xi) - f_1(\xi)| - \frac{\epsilon}{2} >$$

$$> \left|\sum_{k=0}^{j} \frac{f^{(k)}(x_0)}{k!}(\xi - x_0)^k - f(\xi)\right| - \epsilon = 1$$

and thus $f_1 \in U_j$. Since $f_1 \in B\left(f, 2^{-j}\frac{e^{a-b}\epsilon/2}{1+e^{a-b}\epsilon/2}\right)$ was chosen arbitrarily, we conclude that $B\left(f, 2^{-j}\frac{e^{a-b}\epsilon/2}{1+e^{a-b}\epsilon/2}\right) \subset U_j$, and U_j is open.

2) We now prove that for any fixed $J \in \mathbb{N}$, the set $\cup_{j=J}^{\infty} U_j$ is dense in F. Take any $f \in F$ and $\epsilon_1 > 0$. Let g be the spike on some interval $(x_0, x_0 + \delta) \subset I$. Let c be a number such that $cg\left(x_0 + \frac{\delta}{2}\right) > 2$. Let $p(x)$ be some polynomial in the countable and dense subset of F mentioned in the proof of Proposition 20. such that $d_F(p, cg) < \epsilon_1$. Put $h(x) = f(x) + cg(x) - p(x)$ for $a \leq x \leq b$. Then, $h \in B(f, \epsilon_1)$. Indeed:

$$d_F(f, h) = d_F(cg, p) < \epsilon_1. \tag{53}$$

Next, we notice that for large enough $j \geq J$, $p^{(j)} = 0$ and

$$\sum_{k=0}^{j} \frac{p^{(k)}(x_0)}{k!}(x - x_0)^k = p(x), \tag{54}$$

for all $a \leq x \leq b$. Moreover, since g is spike in $(x_0, x_0 + \delta)$, $g^{(j)}(x_0) = 0$ for all $j \in \mathbb{N}$ and

$$\sum_{k=0}^{j} \frac{g^{(k)}(x_0)}{k!}(x - x_0)^k = 0, \tag{55}$$

identically.

It follows that

$$\sum_{k=0}^{j} \frac{h^{(k)}(x_0)}{k!}\left(\frac{\delta}{2}\right)^k = \sum_{k=0}^{j} \frac{f^{(k)}(x_0)}{k!}\left(\frac{\delta}{2}\right)^k - p\left(x_0 + \frac{\delta}{2}\right). \tag{56}$$

Thus for such j:

$$\sum_{k=0}^{j} \frac{h^{(k)}(x_0)}{k!} \left(\frac{\delta}{2}\right)^k - h\left(x_0 + \frac{\delta}{2}\right) = \\ = \sum_{k=0}^{j} \frac{f^{(k)}(x_0)}{k!} \left(\frac{\delta}{2}\right)^k - f\left(x_0 + \frac{\delta}{2}\right) - cg\left(x_0 + \frac{\delta}{2}\right). \tag{57}$$

Given that, by assumption $cg\left(x_0 + \frac{\delta}{2}\right) > 2$, the previous equality can only hold provided either

$$\left|\sum_{k=0}^{j} \frac{f^{(k)}(x_0)}{k!} \left(\frac{\delta}{2}\right)^k - f\left(x_0 + \frac{\delta}{2}\right)\right| > 1 \tag{58}$$

or

$$\left|\sum_{k=0}^{j} \frac{h^{(k)}(x_0)}{k!} \left(\frac{\delta}{2}\right)^k - h\left(x_0 + \frac{\delta}{2}\right)\right| > 1. \tag{59}$$

In other words, either $f \in \cup_{j=J}^{\infty} U_j$ or $h \in \cup_{j=J}^{\infty} U_j$. Hence, $\cup_{j=J}^{\infty} U_j$ is an open and dense subset of F. By Baire's Theorem $\cap_{J=1}^{\infty} \left(\cup_{j=J}^{\infty} U_j\right)$ is a residual set. Finally, the set of all functions in F analytic on I is disjoint from $\cap_{J=1}^{\infty} \left(\cup_{j=J}^{\infty} U_j\right)$ and must be of first category. Since there are only countably many intervals $I \subset [a, b]$ with rational endpoints, it follows that $F \backslash F_0$ is a first category subset of F. ■

We shall now construct a family of nowhere analytic functions in $[a, b]$. But first we prove a lemma about subintervals of (a, b). Let $\{r_1, r_2, r_3, \cdots, r_n, \cdots\}$ be the set of rational numbers in (a, b) enumerated.

Lemma 22. *There exists a sequence of irrational positive numbers $(v_n)_{n\in\mathbb{N}}$, such that $v_n < \frac{b-a}{n}$ and $[r_n - v_n, r_n + v_n] \subset (a, b)$ for all $n \in \mathbb{N}$, and none of the points r_i, $r_i + \frac{v_i}{2}$, $r_i + v_i$ $(i = 1, 2, \cdots, j-1)$ lie in the closed interval $[r_j - v_j, r_j + v_j]$ for any $j \geq 2$.*

Proof. We proceed by induction. For $n = 1$, let v_1 be any irrational number such that $[r_1 - v_1, r_1 + v_1] \subset (a, b)$ and hence $v_1 < b - a$. Now suppose that $v_1, \cdots, v_{N-1}$ have been suitably selected for some $N > 1$. Then r_N is not any of the rational numbers $r_1, \cdots, r_{N-1}$ and is certainly not any of the irrational numbers $r_1 + \frac{v_1}{2}, \cdots, r_{N-1} + \frac{v_{N-1}}{2}$, $r_1 + v_1, \cdots, r_{N-1} + v_{N-1}$. It is hence possible to select a positive irrational number v_N such that $v_N < \frac{b-a}{N}$, $[r_N - v_N, r_N + v_N] \subset (a, b)$, and such that the closed interval $[r_N - v_N, r_N + v_N]$ excludes all the points $r_1, \cdots, r_{N-1}, r_1 + \frac{v_1}{2}, \cdots, r_{N-1} + \frac{v_{N-1}}{2}$ and $r_1 + v_1, \cdots, r_{N-1} + v_{N-1}$. This completes the induction. ■

Proposition 23. (Cater) *For each $n \in \mathbb{N}$, let c_n be a positive number so small that $d_F(0, c_n h_n) < \frac{1}{2^n}$, where h_n is the spike on the interval $(r_n, r_n + v_n)$ as in the previous lemma. Moreover, let $(a_n)_{n\in\mathbb{N}}$ be any bounded sequence of real numbers and $(w_n)_{n\in\mathbb{N}}$ such that $w_n = 1$ if $a_n \neq 0$ and $w_n = 0$ otherwise. Then the series $\sum_{n=1}^{\infty} a_n c_n h_n$ converges to a nowhere analytic function in F if the sequence $(w_n r_n)_{n\in\mathbb{N}}$ is dense in $[a, b]$.*

Proof. Let $(w_n r_n)_{n\in\mathbb{N}}$ be dense in $[a,b]$. Since $d_F(0, c_n h_n) < \frac{1}{2^n}$ and $|a_n| \leq M$ for some fixed M and all $n \in \mathbb{N}$, it follows that:

$$\begin{aligned} d_F(0, a_n c_n, h_n) &= \textstyle\sum_{k=0}^{\infty} \frac{1}{2^k} \sup_{a\leq x\leq b} \frac{|a_n c_n h_n(x)|}{1+|a_n c_n h_n(x)|} \leq \\ &\leq \textstyle\sum_{k=0}^{\infty} \frac{1}{2^k} \sup_{a\leq x\leq b} \frac{M|c_n h_n(x)|}{1+M|c_n h_n(x)|} \leq \\ &\leq \textstyle\sum_{k=0}^{\infty} \frac{1}{2^k} \sup_{a\leq x\leq b} (M+1) \frac{|c_n h_n(x)|}{1+|c_n h_n(x)|} = \\ &= (M+1) d_F(0, c_n h_n) < \frac{M+1}{2^n} \end{aligned} \tag{60}$$

Thus

$$d_F\left(0, \sum_{i=N}^{N+j} a_i c_i h_i\right) \leq \sum_{i=N}^{N+j} d_F(0, a_i c_i h_i) < \sum_{i=N}^{N+j} \frac{M+1}{2^i} < \frac{M+1}{2^{N-1}} \tag{61}$$

and $\sum_{n=1}^{\infty} a_n c_n h_n$ must converge in the complete metric space F, say to f.

Now fix any index $N > 1$ for which $w_N = 1$ and $a_N \neq 0$. By Lemma 22., for each $i = 1, \cdots, N-1$, the interval $[r_N - v_N, r_N + v_N]$ is a subinterval of $(-\infty, r_i)$ or $(r_i + v_i, \infty)$ or $(r_i, r_i + v_i)$. For $i = 1, \cdots, N-1$, each spike h_i is analytic in $(r_N - v_N, r_N + v_N)$ and so is $\sum_{i=1}^{N-1} a_i c_i h_i$. Note also that for $i > N$, $r_N \notin (r_i, r_i + v_i)$ and $r_N + \frac{v_N}{2} \notin (r_i, r_i + v_i)$. Thus:

$$h_i^{(k)}(r_N) = h_i^{(k)}\left(r_N + \frac{v_N}{2}\right) = 0, \quad \text{for } i > N \text{ and all } k \geq 0. \tag{62}$$

Of course $h_N^{(k)}(r_N) = 0$ for all $k \geq 0$. Notice that $\lim_{J\to\infty} \sum_{i=1}^{J} a_i c_i h_i^{(k)} = f^{(k)}$ uniformly for all $k \geq 0$, and

$$\sum_{i=1}^{N} a_i c_i h_i \left(r_n + \frac{v_N}{2}\right) = f\left(r_n + \frac{v_N}{2}\right), \tag{63}$$

so

$$\begin{aligned} &\lim_{j\to\infty} \textstyle\sum_{k=0}^{j} \frac{f^{(k)}(r_N)}{k!} \left(\frac{v_N}{2}\right)^k = \\ &= \lim_{j\to\infty} \textstyle\sum_{k=0}^{j} \frac{1}{k!} \left(\frac{v_N}{2}\right)^k \sum_{i=1}^{\infty} a_i c_i h_i^{(k)}(r_N) = \\ &= \lim_{j\to\infty} \textstyle\sum_{k=0}^{j} \frac{1}{k!} \left(\frac{v_N}{2}\right)^k \sum_{i=1}^{N-1} a_i c_i h_i^{(k)}(r_N) = \\ &= \textstyle\sum_{i=1}^{N-1} a_i c_i \sum_{k=0}^{\infty} \frac{1}{k!} \left(\frac{v_N}{2}\right)^k h_i^{(k)}(r_N) = \sum_{i=1}^{N-1} a_i c_i h_i \left(r_N + \frac{v_N}{2}\right). \end{aligned} \tag{64}$$

But

$$f\left(r_N + \frac{v_N}{2}\right) = \sum_{i=1}^{N-1} a_i c_i h_i \left(r_N + \frac{v_N}{2}\right) + a_N c_N h_N \left(r_N + \frac{v_N}{2}\right). \tag{65}$$

Since $h_N\left(r_N + \frac{v_N}{2}\right) > 0$, and $c_N \neq 0$, $a_N \neq 0$, we conclude that

$$\lim_{j\to\infty} \sum_{k=0}^{j} \frac{f^{(k)}(r_N)}{k!} \left(\frac{v_N}{2}\right)^k \neq f\left(r_N + \frac{v_N}{2}\right) \tag{66}$$

and f is not analytic in the interval $(r_N - v_N, r_N + v_N)$.

Finally, if I is any subinterval of $[a, b]$, then by hypothesis there is some index $N > 1$ for which $a_N \neq 0$ and $[r_N - v_N, r_N + v_N] \subset I$, so f is not analytic in I. Consequently, f is nowhere analytic. ■

3. Measure and Dimension

From the point of view of Baire's categories, we have come to the conclusion in the previous section, that non-differentiable and non-analytic functions should play an important role in modeling natural phenomena. In most complex phenomena where non-differentiability is a prevalent feature, fractals of various natures are ubiquitous. Typically non-differentiable functions are non-rectifiable and the length of an arc of its graph diverges [40]. This divergence is intimately related with the fact that the dimension of the graph of a such a function may be lower than 1. For each positive real number - the dimension of some fractal - we can associate a fractal measure.

It is well known [12, 27] that the fractal dimensions of certain phenomena are more suitably described by so-called *dimension functions* [12, 13] rather than real positive numbers (which are associated with the powers). For instance, with probability 1, Brownian trails in $\mathbb{R}^n$ $(n \geq 2)$ have Hausdorff dimension 2. However, more delicate arguments reveal that they have positive finite Hausdorff measure with respect to the functions [12]:

$$\begin{aligned} &x^2 \log\left(\tfrac{1}{x}\right) \log\log\log\left(\tfrac{1}{x}\right), \quad &&\text{if } n = 2 \\ &x^2 \log\log\left(\tfrac{1}{x}\right), &&\text{if } n \geq 3 \end{aligned} \tag{67}$$

With these ideas in mind we recapitulate various ideas on generalized dimensions and measure.

3.1. Dimension Functions

Definition 24. *Consider real valued functions ψ defined on some interval of the form $(0, \epsilon_\psi)$ with ϵ_ψ (which may be different for different functions) a positive constant or ∞. The set of* **dimension functions** *$\mathcal{D}$ is the set of such functions which in addition are continuous, strictly increasing and such that $\lim_{x\downarrow 0} \psi(x) = 0$.*

Definition 25. *In $\mathcal{D}$ we define the following binary relations. We say that $\psi, \phi \in \mathcal{D}$ are of the* **same order of growth** *and write $\phi \sim \psi$ if there exist real constants C_1, C_2 such that $0 < C_1 \leq \frac{\psi(x)}{\phi(x)} \leq C_2$ for all $0 < x <$ min $\{\epsilon_\psi, \epsilon_\phi\}$. In particular, this is valid if the limit $\lim_{x\downarrow 0} \frac{\psi(x)}{\phi(x)}$ exists and is a positive real number. Moreover, we say that the* **order of growth** *of ψ is* **greater** *than that of ϕ and write $\phi \prec \psi$, if $\lim_{x\downarrow 0} \frac{\psi(x)}{\phi(x)} = 0$. Finally, we have $\phi \preceq \psi$ if $\phi \sim \psi$ or $\phi \prec \psi$.*

The following proposition is straightforward to prove. We leave the proof to the reader.

Proposition 26. *The relations $\sim$ and $\preceq$ in $\mathcal{D}$ are an equivalence relation and a partial ordering, respectively.*

Remark 27. *We denote by $\mathcal{D}/\sim$ the set of equivalence classes in $\mathcal{D}$ and write $[\psi]$ for the equivalence class of $\psi \in \mathcal{D}$. We should however mention that $(\mathcal{D}, \preceq)$ is not a totally ordered set as there are pairs of elements in $\mathcal{D}$ which are not comparable. Indeed, let*

$$\psi(x) = \int_0^x \tau \left[1 + \sin\left(\frac{1}{\tau}\right)\right] d\tau + x^{\frac{3}{2}} \tag{68}$$

for $x > 0$. Clearly, ψ is continuous and $\lim_{x \downarrow 0} \psi(x) = 0$. Moreover

$$\psi'(x) = x \left[1 + \sin\left(\frac{1}{x}\right)\right] + \frac{3}{2}\sqrt{x} > 0, \qquad \forall x > 0 \tag{69}$$

and so ψ is strictly increasing. Altogether, $\psi \in \mathcal{D}$. Next, let $\phi(x) = x^{\frac{3}{2}} \in \mathcal{D}$. For $x > 0$:

$$\frac{\psi(x)}{\phi(x)} = x^{-\frac{3}{2}} \int_0^x \tau \left[1 + \sin\left(\frac{1}{\tau}\right)\right] d\tau + 1 \tag{70}$$

Consider the sequence $x_n = \left(\frac{\pi}{2}(2n+1)\right)^{-1}$. It follows that

$$\begin{aligned} \tfrac{\psi(x_n)}{\phi(x_n)} &= \left(\tfrac{\pi}{2}(2n+1)\right)^{\frac{3}{2}} \int_0^{\frac{2}{\pi(2n+1)}} \tau \left[1 + \sin\left(\tfrac{1}{\tau}\right)\right] d\tau + 1 = \\ &= \left(\tfrac{\pi}{2}(2n+1)\right)^{\frac{3}{2}} \left\{ \tfrac{2}{\pi(2n+1)} \left[1 + (-1)^n\right] + \mathcal{O}\left(\tfrac{1}{(2n+1)^2}\right)\right\} + 1 \end{aligned} \tag{71}$$

If n is odd, then $\frac{\psi(x_n)}{\phi(x_n)} \to 1$, whereas $\frac{\psi(x_n)}{\phi(x_n)} \to \infty$, when n is even. We conclude that ψ and ϕ are not comparable.

An important subset of $\mathcal{D}$ is the set of powers.

Example 28. *Let us consider the powers x^r, with $r > 0$. Clearly, these functions belong to $\mathcal{D}$. As a convention we shall use them as the representatives of the corresponding equivalence classes. We denote the set of functions belonging to the equivalence classes of powers by:*

$$\mathcal{P} := \{\phi \in \mathcal{D} : \exists r > 0 \text{ such that } \phi \sim x^r\}. \tag{72}$$

By abuse of language we shall call this the **set of powers**.

The following lemmata present useful methods for obtaining new elements with different orders of growth in $\mathcal{D}$.

Lemma 29. *Let $\phi_1, \phi_2 \in \mathcal{D}$ be defined on $(0, \epsilon_{\phi_1})$ and $(0, \epsilon_{\phi_2})$, respectively. Then their product $\phi_1 \cdot \phi_2 : (0, \epsilon) \longrightarrow \mathbb{R}$, with $\epsilon = min\{\epsilon_{\phi_1}, \epsilon_{\phi_2}\}$, belongs to $\mathcal{D}$. Moreover, $\phi_1 \prec \phi_1 \cdot \phi_2$ and $\phi_2 \prec \phi_1 \cdot \phi_2$.*

Proof. The product of two non-negative, continuous and strictly increasing functions is again continuous and strictly increasing. The product of two infinitesimals is again an infinitesimal. On the other hand, we have $\lim_{x \downarrow 0} \frac{\phi_1(x)\phi_2(x)}{\phi_1(x)} = \lim_{x \downarrow 0} \phi_2(x) = 0$ and $\lim_{x \downarrow 0} \frac{\phi_1(x)\phi_2(x)}{\phi_2(x)} = \lim_{x \downarrow 0} \phi_1(x) = 0$, that is $\phi_1 \prec \phi_1 \cdot \phi_2$ and $\phi_2 \prec \phi_1 \cdot \phi_2$. ■

Lemma 30. *Let $\psi_1 : (0, \epsilon_{\psi_1}) \longrightarrow \mathbb{R}$ and $\psi_2 : (0, \epsilon_{\psi_2}) \longrightarrow \mathbb{R}$ belong to $\mathcal{D}$ and be such that $\psi_2((0, \epsilon_{\psi_2})) \subset (0, \epsilon_{\psi_1})$. Then the composition $\psi_1 \circ \psi_2 : (0, \epsilon_{\psi_2}) \longrightarrow \mathbb{R}$ belongs to $\mathcal{D}$. Furthermore (i) $\psi_2 \preceq \psi_1 \circ \psi_2$ if and only if $x \preceq \psi_1$; (ii) $\psi_1 \preceq \psi_1 \circ \psi_2$ if $x \preceq \psi_1 \preceq \psi_2$; and (iii) $\psi_1 \circ \psi_2 \preceq \psi_1$ if $\psi_2 \preceq \psi_1 \preceq x$.*

Proof. The composition of two continuous and strictly increasing functions is a continuous and strictly increasing function. Moreover, we have that $\lim_{x\downarrow 0}(\psi_1 \circ \psi_2)(x) = \lim_{\tau\downarrow 0} \psi_1(\tau) = 0$, since $\lim_{x\downarrow 0} \psi_2(x) = 0^+$. Altogether, $\psi_1 \circ \psi_2 \in \mathcal{D}$. Let us now prove the remaining statements: (i) $\lim_{x\downarrow 0} \frac{(\psi_1\circ\psi_2)(x)}{\psi_2(x)} = \lim_{x\downarrow 0} \frac{\psi_1(\psi_2(x))}{\psi_2(x)} = \lim_{\tau\downarrow 0} \frac{\psi_1(\tau)}{\tau}$. We conclude that $\psi_2 \prec \psi_1 \circ \psi_2$ if and only if $x \prec \psi_1$. Now suppose that $\psi_1 \sim x$. Then there exist C_1, C_2 such that $0 < C_1 \leq \frac{\psi_1(x)}{x} \leq C_2$ for all $x \in (0, \epsilon_{\psi_1})$. Since, by assumption, $0 < \psi_2(x) \leq \epsilon_{\psi_1}$ for all $x \in (0, \epsilon_{\psi_2})$, we conclude that $0 < C_1 \leq \frac{(\psi_1\circ\psi_2)(x)}{\psi_2(x)} \leq C_2$ for all $x \in (0, \epsilon_{\psi_2})$. This means that $\psi_1 \circ \psi_2 \sim \psi_2$. Conversely, suppose that $\psi_1 \circ \psi_2 \sim \psi_2$, i.e. there exist K_1, K_2 such that $0 < K_1 \leq \frac{(\psi_1\circ\psi_2)(x)}{\psi_2(x)} \leq K_2$ for all $x \in (0, \epsilon_{\psi_2})$. This means that for all $\tau \in (0, \psi_2(\epsilon_{\psi_2}))$ we have $0 < K_1 \leq \frac{\psi_1(\tau)}{\tau} \leq K_2$. That is $\psi_1 \sim x$. (ii) Notice that $\frac{(\psi_1\circ\psi_2)(x)}{\psi_1(x)} = \frac{\psi_1(\psi_2(x))}{\psi_2(x)} \cdot \frac{\psi_2(x)}{\psi_1(x)}$. Then result is then proved in the same way as (i). (iii) The proof is analogous since $\frac{\psi_1(x)}{(\psi_1\circ\psi_2)(x)} = \frac{\psi_2(x)}{\psi_1(\psi_2(x))} \cdot \frac{\psi_1(x)}{\psi_2(x)}$. ■

Lemma 31. *Let $\psi \in \mathcal{D}$ with $\lim_{x\uparrow\epsilon_\psi} \psi(x) = \psi(\epsilon_\psi^-)$. Then its inverse function $\psi^{-1} : \left(0, \psi(\epsilon_\psi^-)\right) \longrightarrow (0, \epsilon_\psi)$ also belongs to $\mathcal{D}$. Moreover, if for a given $s > 0$, $\psi \prec x^s$, $\psi \sim x^s$ or $x^s \prec \psi$, then $x^{\frac{1}{s}} \prec \psi^{-1}$, $x^{\frac{1}{s}} \sim \psi^{-1}$ or $\psi^{-1} \prec x^{\frac{1}{s}}$, respectively.*

Proof. Since ψ is strictly increasing, it is injective, and therefore admits an inverse. As it is strictly increasing, continuous and $\lim_{x\downarrow 0} \psi(x) = 0$, $\lim_{x\uparrow\epsilon_\psi} = \psi(\epsilon_\psi^-)$, we have that $\psi((0, \epsilon_\psi)) = \left(0, \psi(\epsilon_\psi^-)\right)$. Consequently, ψ^{-1} is a function from $\left(0, \psi(\epsilon_\psi^-)\right)$ onto $(0, \epsilon_\psi)$. Moreover, as it is the inverse of a continuous and strictly increasing function, it is also continuous and strictly increasing. Finally, since $\lim_{x\downarrow 0} \psi(x) = 0^+$, we conclude that $\lim_{x\downarrow 0} \psi^{-1}(x) = 0$. Altogether, $\psi^{-1} \in \mathcal{D}$. Finally, notice that:

$$\left[\lim_{x\downarrow 0} \frac{\psi^{-1}(x)}{x^{\frac{1}{s}}}\right]^s = \left[\lim_{x\downarrow 0} \frac{\psi^{-1}(\psi(x))}{(\psi(x))^{\frac{1}{s}}}\right]^s = \left[\lim_{x\downarrow 0} \frac{x}{(\psi(x))^{\frac{1}{s}}}\right]^s = \lim_{x\downarrow 0} \frac{x^s}{\psi(x)} \tag{73}$$

This relation yields the desired result. ■

Let us consider some more examples of functions in $\mathcal{D}$.

Example 32. *It is easy to show that the functions*

$$\begin{cases} \sigma_1(x) = \left[\log\left(\log\left(1 + \frac{1}{x}\right)\right)\right]^{-1}, & x \in \left(0, \frac{1}{e-1}\right) \\ \sigma_2(x) = \left[\log\left(1 + \frac{1}{x}\right)\right]^{-1}, & x \in (0, +\infty) \\ \sigma_3(x) = \left[e^{\frac{1}{x}} - 1\right]^{-1}, & x \in (0, +\infty) \\ \sigma_4(x) = e^{-\frac{1}{x^2}}, & x \in (0, +\infty) \end{cases} \tag{74}$$

all belong to $\mathcal{D}$. Moreover, for any $s > 0$

$$\lim_{x\downarrow 0} \frac{x^s}{\sigma_2(x)} = \lim_{x\downarrow 0} \frac{\sigma_3(x)}{x^s} = \lim_{x\downarrow 0} \frac{\sigma_2(x)}{\sigma_1(x)} = \lim_{x\downarrow 0} \frac{\sigma_4(x)}{\sigma_3(x)} = 0. \tag{75}$$

This means that:

$$\sigma_1 \prec \sigma_2 \prec x^s \prec \sigma_3 \prec \sigma_4, \qquad \forall s > 0. \tag{76}$$

Let us briefly mention that $\sigma_3 = \sigma_2^{-1}$, which illustrates Lemma 31..

Remark 33. *At this stage it is important to remark that, while the product of two elements in $\mathcal{D}/\sim$ is a well defined operation, the composition and inversion are not. Indeed, let $\psi_1, \psi_2, \phi_1, \phi_2 \in \mathcal{D}$ be such that $\psi_1 \sim \psi_2$ and $\phi_1 \sim \phi_2$. That means that there exist positive constants C_1, C_2, K_1 and K_2 such that $0 < C_1 \leq \frac{\psi_1(x)}{\psi_2(x)} \leq C_2$ for all $0 < x < \epsilon_\psi = min\{\epsilon_{\psi_1}, \epsilon_{\psi_2}\}$, and $0 < K_1 \leq \frac{\phi_1(x)}{\phi_2(x)} \leq K_2$ for all $0 < x < \epsilon_\phi = min\{\epsilon_{\phi_1}, \epsilon_{\phi_2}\}$. Consequently, $0 < K_1C_1 \leq \frac{\psi_1(x)\cdot\phi_1(x)}{\psi_2(x)\cdot\phi_2(x)} \leq K_2C_2$ for all $0 < x < min\{\epsilon_\psi, \epsilon_\phi\}$, i.e. $\psi_1 \cdot \phi_1 \sim \psi_2 \cdot \phi_2$. The product is thus a well defined operation $(\mathcal{D}/\sim) \times (\mathcal{D}/\sim) \longrightarrow (\mathcal{D}/\sim)$. To prove that composition and inversion are not well defined in $\mathcal{D}/\sim$, we will give concrete examples. Let $\psi_1(x) = L\sigma_2(x)$, $\psi_2(x) = \sigma_2(x)$, $\phi_1(x) = M\sigma_3(x) = M\sigma_2^{-1}(x)$ and $\phi_2(x) = \sigma_3(x) = \sigma_2^{-1}(x)$, where σ_2 and σ_3 are as in Example 32.. We shall further assume that $0 < L < 1$ and $M > 0$. And thus $\psi_1 \sim \psi_2$, $\phi_1 \sim \phi_2$. Moreover, $\psi_1^{-1}(x) = \sigma_3\left(\frac{x}{L}\right)$ and $\psi_2^{-1}(x) = \sigma_3(x)$. We thus have $\lim_{x\downarrow 0} \frac{\psi_1^{-1}(x)}{\psi_2^{-1}(x)} = \lim_{x\downarrow 0} \frac{e^{\frac{1}{x}}-1}{e^{\frac{L}{x}}-1} = +\infty$. And thus ψ_1^{-1} is not of the same order of growth as ψ_2^{-1}. This means that inversion is not a well defined operation $(\mathcal{D}/\sim) \longrightarrow (\mathcal{D}/\sim)$. Likewise, we have $(\phi_1 \circ \psi_1)(x) = \frac{M}{(1+x^{-1})^{1/L}-1}$ and $(\phi_2 \circ \psi_2)(x) = x$. But $\lim_{x\downarrow 0} \frac{(\phi_2\circ\psi_2)(x)}{(\phi_1\circ\psi_1)(x)} = \lim_{x\downarrow 0} \frac{x}{M}\left[\left(1+\frac{1}{x}\right)^{\frac{1}{L}} - 1\right] = +\infty$. And thus $(\phi_1 \circ \psi_1)$ and $(\phi_2 \circ \psi_2)$ are not of the same order of growth, which shows that composition is not a well defined operation $(\mathcal{D}/\sim) \times (\mathcal{D}/\sim) \longrightarrow (\mathcal{D}/\sim)$.*

An immediate consequence of Example 32. is the following proposition.

Proposition 34. *The set of powers $\mathcal{P}$ is a bounded subset of $\mathcal{D}$.*

As for the set $\mathcal{D}$, the following proposition holds.

Proposition 35. *The set $\mathcal{D}$ has no maximal or minimal elements.*

Proof. Suppose that $\mathcal{D}$ had a maximal element ψ. Then this would mean that there exists no $\phi \in \mathcal{D}$ such that $\psi \prec \phi$. However, from Lemma 29., for $\phi \in \mathcal{D}$, we have that $\phi \cdot \psi \in \mathcal{D}$ and $\psi \prec \phi \cdot \psi$, which is a contradiction. And so, $\mathcal{D}$ cannot have a maximal element. Likewise, if $\mathcal{D}$ had a minimal element ρ, then there could be no $\phi \in \mathcal{D}$ such that $\phi \prec \rho$. Take $\phi \circ \rho$ for $\phi \in \mathcal{D}$ such that $\phi \prec x$. Then from Lemma 30., we conclude that $\phi \circ \rho \in \mathcal{D}$ and $\phi \circ \rho \prec \rho$, which is a contradiction. ■

Lemma 36. *Let $\phi_1, \phi_2 \in \mathcal{P}$ be such that $\phi_1 \prec \phi_2$. Then we can always find $\phi \in \mathcal{P}$ and $\psi \in \mathcal{D}\backslash\mathcal{P}$ such $\phi_1 \prec \phi \prec \phi_2$ and $\phi_1 \prec \psi \prec \phi_2$.*

Proof. The proof is obviously independent of the representative of each equivalence class, so we may safely assume $\phi_1(x) = x^s$ and $\phi_2(x) = x^r$, with $0 < s < r$. Let $\phi(x) = x^n$, with $s < n < r$. Then we have $\phi \in \mathcal{P}$ and $\phi_1 \prec \phi \prec \phi_2$. This proves the first claim. Let now $\psi(x) = \phi_1(x) \cdot \sigma_2(x)$, where $\sigma_2(x)$ is as in Example 32.. From Lemma 29. $\psi \in \mathcal{D}$ and $\phi_1 \prec \psi$. On the other hand, for $u > 0$:

$$\lim_{x\downarrow 0} \frac{\psi(x)}{x^u} = \lim_{x\downarrow 0} x^{s-u} \cdot \sigma_2(x) = \begin{cases} +\infty & \text{if } u > s \\ 0 & \text{if } u \leq s \end{cases} \tag{77}$$

This proves that $\psi \in \mathcal{D}\backslash\mathcal{P}$. Finally $\lim_{x\downarrow 0} \frac{\phi_2(x)}{\psi(x)} = \lim_{x\downarrow 0} \frac{x^{r-s}}{\sigma_2(x)} = 0$. Altogether $\phi_1 \prec \psi \prec \phi_2$. ■

Remark 37. *Lemma 36. states that between any two elements ϕ_1, ϕ_2 of $\mathcal{P}$, we can always find a third element of $\mathcal{P}$. However, if either ϕ_1 or ϕ_2 (or both) do not belong to $\mathcal{P}$, then there may not exist any element of $\mathcal{P}$ between them. For instance, from Example 32., we have that $\sigma_1 \prec \sigma_2 \prec x^s \prec \sigma_3 \prec \sigma_4$ for any $s > 0$. This means that there exist no elements of $\mathcal{P}$ between σ_1 and σ_2 and between σ_3 and σ_4. Another example is as follows. For any $0 < n \leq s < r$, we have: $x^n \prec x^s\sigma_1(x) \prec x^s\sigma_2(x) \prec x^r$, with $x^s\sigma_1(x), x^s\sigma_2(x) \in \mathcal{D}\backslash\mathcal{P}$. This means that there are no elements of $\mathcal{P}$ between $x^s\sigma_1(x)$ and $x^s\sigma_2(x)$.*

In summary, between two elements $\psi_1, \psi_2 \in \mathcal{D}\backslash\mathcal{P}$ there may or may not exist elements of $\mathcal{P}$. Similarly, the previous analysis also reveals that between an element $\phi \in \mathcal{P}$ and an element $\psi \in \mathcal{D}\backslash\mathcal{P}$ there may not exist an element of $\mathcal{P}$. Indeed, since $x^s \prec x^s\sigma_2(x) \prec x^r$ for any $r > s > 0$, then between $x^s \in \mathcal{P}$ and $x^s\sigma_2 \in \mathcal{D}\backslash\mathcal{P}$ there exists no element of $\mathcal{P}$.

On the contrary, between any two elements of $\mathcal{D}$ we can always find an element of $\mathcal{D}\backslash\mathcal{P}$. To prove this we need the following lemma.

Lemma 38. *Let $\phi : (0, \epsilon_\phi) \longrightarrow \mathbb{R}$, with $\epsilon_\phi > 0$ be continuous, strictly positive and such that $\lim_{x\downarrow 0} \phi(x) = 0$. Then there exists $\psi \in \mathcal{D}$ such that $\lim_{x\downarrow 0} \frac{\phi(x)}{\psi(x)} = 0$.*

Proof. Notice that we do not require ϕ to be strictly increasing, so that it may or may not belong to $\mathcal{D}$. Let us define $f(x) = \max_{0<u\leq x} \phi(u)$, for $x \in (0, \epsilon_\phi)$. Obviously, f is continuous, strictly positive, increasing (but not necessarily strictly) and $\lim_{x\downarrow 0} f(x) = 0$. Take any $\rho \in \mathcal{D}$ and define $\psi(x) = \rho(x)f(x) + \sqrt{f(x)}$, for $x \in (0, \epsilon_\psi)$ and $\epsilon_\psi = \min\{\epsilon_\phi, \epsilon_\rho\}$. By construction $\psi \in \mathcal{D}$. It remains to prove that $\lim_{x\downarrow 0} \frac{\phi(x)}{\psi(x)} = 0$. Indeed, for $0 < x < \epsilon_\psi$,

$$0 < \frac{\phi(x)}{\psi(x)} \leq \frac{f(x)}{\rho(x)f(x) + \sqrt{f(x)}} = \frac{\sqrt{f(x)}}{\sqrt{f(x)}\rho(x) + 1}. \tag{78}$$

As $x \downarrow 0$, the right-hand side vanishes. ■

Proposition 39. *Let $\psi_1, \psi_2 \in \mathcal{D}$ be such that $\psi_1 \prec \psi_2$. Then there exists $\eta \in \mathcal{D}\backslash\mathcal{P}$ such that $\psi_1 \prec \eta \prec \psi_2$.*

Proof. Consider the function $\phi(x) = \frac{\psi_2(x)}{\psi_1(x)}$ defined on $(0, \epsilon)$ with $\epsilon = \min\{\epsilon_{\psi_1}, \epsilon_{\psi_2}\}$. This function is continuous, strictly positive and moreover $\lim_{x\downarrow 0} \phi(x) = 0$. From Lemma 38., there exists $\psi \in \mathcal{D}$ such that $\lim_{x\downarrow 0} \frac{\phi(x)}{\psi(x)} = 0$. Consider the function $\eta_1(x) = \psi(x) \cdot \psi_1(x)$.

From Lemma 29., $\eta_1 \in \mathcal{D}$ and $\psi_1 \prec \eta_1$. On the other hand: $\lim_{x\downarrow 0} \frac{\psi_2(x)}{\eta_1(x)} = \lim_{x\downarrow 0} \frac{\phi(x)}{\psi(x)} = 0$, which means that $\eta_1 \prec \psi_2$. If $\eta_1 \in \mathcal{D}\backslash\mathcal{P}$, then the problem is settled. If $\eta_1 \in \mathcal{P}$, then we repeat the previous procedure and find $\eta_3 \in \mathcal{D}$ such that $\eta_1 \prec \eta_3 \prec \psi_2$. If $\eta_3 \in \mathcal{D}\backslash\mathcal{P}$, then we terminate the proof. If η_1, η_3 both belong to $\mathcal{P}$, then from Lemma 36., we can find $\eta_2 \in \mathcal{D}\backslash\mathcal{P}$ such that $\psi \prec \eta_1 \prec \eta_2 \prec \eta_3 \prec \psi_2$. ■

The fact that $(\mathcal{D}, \preceq)$ is not totally ordered is unsatisfactory. This leads us to the following definition.

Definition 40. *A* **scale** *of dimension functions is a subset $\mathcal{F}$ of $\mathcal{D}$ such that: (i) $\mathcal{P} \subset \mathcal{F}$, (ii) $(\mathcal{F}, \preceq)$ is a totally ordered bounded complete set, and (iii) for any $\psi, \phi \in \mathcal{F}$ such that $\psi \prec \phi$, there exists $\rho \in \mathcal{F}$ such that $\psi \prec \rho \prec \phi$. We refer to the latter property by saying that $\mathcal{F}$ is* **dense**. *Two scales $\mathcal{F}_1, \mathcal{F}_2$ are said to be compatible if $\mathcal{F}_1 \cup \mathcal{F}_2$ is another scale. A scale is* **maximal** *if there exists no other scale which contains it as a proper subset.*

Remark 41. *One may wonder about the existence of maximal scales. However, their existence is ensured by Hausdorff's Maximality Principle [29]. This principle is equivalent to Zorn's Lemma and to the Axiom of Choice, and it states that in any partially ordered set, every totally ordered subset is contained in a maximally totally ordered subset. A subset is maximally totally ordered if, when it is enlarged in any way, fails to remain totally ordered. One should nevertheless stress that the maximally totally ordered set containing the original totally ordered set may not be unique.*

Lemma 42. *Two scales $\mathcal{F}_1, \mathcal{F}_2$ are compatible if and only if every pair of elements $\phi_1 \in \mathcal{F}_1\backslash\mathcal{F}_2$ and $\phi_2 \in \mathcal{F}_2\backslash\mathcal{F}_1$ are comparable.*

Proof. This follows immediately from the definition and the fact that $\mathcal{F}_1 \cup \mathcal{F}_2 = (\mathcal{F}_1\backslash\mathcal{F}_2) \cup (\mathcal{F}_1 \cap \mathcal{F}_2) \cup (\mathcal{F}_2\backslash\mathcal{F}_1)$. ■

Example 43. *The set of powers $\mathcal{P}$ is a scale, the* **trivial scale**. *In fact there is an order preserving bijection from the set of positive reals onto $\mathcal{P}\backslash \sim$ defined by $r \mapsto [x^r]$. As the set $\mathbb{R}^+$ is totally ordered, bounded complete and dense, so is the set of powers.*

Example 44. *Another interesting example which finds applications in the context of Brownian motion (see (67)) is the scale $\mathcal{F}$ that consists of the equivalence classes of functions of the form [40]:*

$$\phi(x) = x^\alpha \left(\log_n \frac{1}{x}\right)^\beta, \qquad \alpha > 0,\ \beta \in \mathbb{R},\ n = 0, 1, 2, 3, \cdots \tag{79}$$

Here $\log_n$ indicates the iterated logarithm: $\log_0(x) = 1$, $\log_1(x) = \log(x)$, $\log_n(x) = \log\left(\log_{n-1}(x)\right)$, for $n \geq 2$.

3.2. Hölder-continuity

As an application, let us consider the so-called Hölder-continuity of a function in some set.

Definition 45. *Let Ω_1, Ω_2 be metric spaces with metrics d_1, d_2, respectively. A function $f : F \subset \Omega_1 \to \Omega_2$ is said to be* **locally ψ-Hölder continuous** *at $u \in F$ for $\psi \in \mathcal{D}$ if there exist constants $K > 0$ and $0 < \epsilon < \epsilon_\psi$ such that*

$$d_2(f(u), f(v)) \leq K\psi\left(d_1(u, v)\right) \tag{80}$$

for all $v \in F$ with $d_1(u, v) < \epsilon$. The function is said to be **ψ-Hölder continuous in** *F if there exists a constant $K > 0$ such that inequality (80) holds for all $u, v \in F$. Finally, if Ω_1 has a partial ordering $\trianglelefteq$, we say that f is* **locally right (resp. left) ψ-Hölder continuous** *at $u \in F$ if there exist constants $K > 0$ and $0 < \epsilon < \epsilon_\psi$ such that (80) is valid whenever $v \in F$ with $d_1(u, v) < \epsilon$ and $u \trianglelefteq v$ (resp. $v \trianglelefteq u$). By abuse of language if f is ψ-Hölder continuous for $\psi(x) = x^r$ for some $r > 0$, then we shall simply say that f is r-Hölder continuous.*

Proposition 46. *If $f : F \subset \Omega_1 \to \Omega_2$ is ψ-Hölder continuous in F, locally ψ-Hölder continuous at $u \in F$, locally right ψ-Hölder continuous at $u \in F$, or locally left ψ-Hölder continuous at $u \in F$ for $\psi \in \mathcal{D}$, then it is, respectively, ϕ-Hölder continuous in F, locally ϕ-Hölder continuous at $u \in F$, locally right ϕ-Hölder continuous at $u \in F$, or locally left ϕ-Hölder continuous at $u \in F$ for any $\phi \in \mathcal{D}$ with $\phi \preceq \psi$.*

Proof. We prove the result when f is locally ψ-Hölder continuous at $u \in F$. The rest is proved analogously. This means that there exist positive constants K, ϵ such that (80) holds for any $v \in F$ with $d_1(u, v) < \epsilon$. Now take an arbitrary $\phi \in \mathcal{D}$ such that $\phi \prec \psi$, i.e. $\lim_{x \downarrow 0} \frac{\psi(x)}{\phi(x)} = 0$. This means that there exists $0 < \rho < \min\{\epsilon_\psi, \epsilon_\phi, \epsilon\}$ such that $\psi(x) \leq \phi(x)$ for all $0 < x < \rho$. For any $v \in F$ such that $d_1(u, v) < \rho$, we have that:

$$d_2(f(u), f(v)) \leq K\psi\left(d_1(u, v)\right) \leq K\phi\left(d_1(u, v)\right) \tag{81}$$

which means that f is locally ϕ-Hölder continuous at $u \in F$. Finally, suppose that $\phi \in \mathcal{D}$ is such that $\phi \sim \psi$. Then there exist constants C_1, C_2 such that $0 < C_1 \leq \frac{\psi(x)}{\phi(x)} \leq C_2$, for all $0 < x < \delta = \min\{\epsilon_\phi, \epsilon_\psi\}$. And so, for any $v \in F$ such that $d_1(u, v) < \min\{\rho, \epsilon\}$, we have:

$$d_2(f(u), f(v)) \leq K\psi\left(d_1(u, v)\right) < C_2 K\phi\left(d_1(u, v)\right), \tag{82}$$

which proves that f is locally ϕ-Hölder continuous at $u \in F$. ■

Now remember that a scale of dimension functions $\mathcal{F}$ is a totally ordered bounded complete set. According to the previous proposition, if a function f is ψ-Hölder continuous for some $\psi \in \mathcal{F}$, then it is ϕ-Hölder continuous for all $\phi \in \mathcal{F}$ with $\phi \prec \psi$. This motivates the following definition.

Definition 47. *Let $\mathcal{F}$ be a scale of dimension functions and let $f : F \subset \Omega_1 \to \Omega_2$ be a continuous function. We define the* **critical Hölder order** *in F,* **critical local Hölder order** *at $u \in F$,* **critical local right Hölder order** *at $u \in F$, and* **critical local left Hölder order**

at $u \in F$ to be the suprema in $\mathcal{F}$ of each of the following sets, respectively.

$$\begin{aligned}
&\{\phi \in \mathcal{F} : \ f \text{ is } \phi\text{-Hölder continuous in } F\} \\
&\{\phi \in \mathcal{F} : \ f \text{ is locally } \phi\text{-Hölder continuous at } u \in F\} \\
&\{\phi \in \mathcal{F} : \ f \text{ is locally right } \phi\text{-Hölder continuous at } u \in F\} \\
&\{\phi \in \mathcal{F} : \ f \text{ is locally left } \phi\text{-Hölder continuous at } u \in F\}
\end{aligned} \tag{83}$$

Obviously, the suprema may or may not exist, depending on whether these sets have an upper bound or not, or on whether they are empty or not.

As an example, we consider the Weierstrass and the Takagi functions.

Lemma 48. *The Takagi function is locally ψ-Hölder continuous for $\psi(x) = x \log\left(\frac{1}{x}\right)$, with $x \in \left(0, e^{-1}\right)$.*

Proof. First of all notice that $\psi \in \mathcal{F}$, where $\mathcal{F}$ is the scale of Example 44. (with $\alpha = \beta = n = 1$ in (79)). Moreover, from (11), we conclude that T is indeed locally ψ-Hölder continuous at every $x \in [0, 1]$. ■

Likewise, for the Weierstrass function we have:

Lemma 49. *The Weierstrass function $W_{\lambda,s}$ has critical local Hölder order $2 - s$.*

Proof. From Proposition 18. (i) we know that $W_{\lambda,s}$ is indeed locally $(2 - s)$-Hölder continuous in $[0, 1]$.

On the other hand, suppose that $W_{\lambda,s}$ is locally r-Hölder continuous at some $x \in [0, 1]$ for $r > 2 - s$. Then there exist $\epsilon, K > 0$ such that

$$|W_{\lambda,s}(x) - W_{\lambda,s}(y)| \leq K|x - y|^r \tag{84}$$

for all $y \in [0, 1]$ with $|x - y| < \epsilon$.

Following the proof of Corollary 19., for each suffciently large $n \in \mathbb{N}$ we can find $y_n \in [0, 1] \setminus \{x\}$ such that:

$$|x - y_n| < \frac{1}{n}, \tag{85}$$

and

$$\frac{|W_{\lambda,s}(x) - W_{\lambda,s}(y_n)|}{|x - y_n|^r} \geq \frac{C_2}{|x - y_n|^r} \left(\frac{1}{n}\right)^{2-s} \geq C_2 n^{r-(s-2)} \tag{86}$$

As $n \to \infty$ the right-hand side of (86) diverges and this contradicts (84). Consequently $2 - s$ is the local critical Hölder order of $W_{\lambda,s}$ at every $x \in [0, 1]$. ■

3.3. Hausdorff Measure

Definition 50. *Let U be any non-empty subset of a metric space Ω with metric d. A countable (or finite) collection $\{U_i\}$ of sets of diameter at most $\delta > 0$ that cover $F \subset \Omega$ is called a δ-**cover** of F.*

Definition 51. *Let F be a subset of Ω and $\psi \in \mathcal{D}$. For any $\delta > 0$ we define*

$$\left(\mathcal{H}_\delta^\psi\right)^*(F) := inf\left\{\sum_{i=1}^{\infty} \psi\left(|U_i|\right) : \ \{U_i\} \text{ is a } \delta\text{-cover of } F\right\} \tag{87}$$

The set on the right-hand side is obviously a non-empty subset of $\mathbb{R}^+$ and thus the infimum exists. As δ decreases $\left(\mathcal{H}_\delta^\psi\right)^*(F)$ increases, and we thus define:

$$\left(\mathcal{H}^\psi\right)^*(F) := \lim_{\delta \downarrow 0}\left(\mathcal{H}_\delta^\psi\right)^*(F). \tag{88}$$

*We call $\left(\mathcal{H}^\psi\right)^*(F)$ the ψ-**Hausdorff measure** of F.*

Theorem 52. *The ψ-Hausdorff measure is an outer measure.*

Proof. Given $\delta > 0$ we may cover $\emptyset$ with a single set of diameter ϵ with $0 < \epsilon \leq \delta$. And thus $0 \leq \left(\mathcal{H}_\delta^\psi\right)^*(\emptyset) \leq \psi(\epsilon)$ for any $0 < \epsilon \leq \delta$, giving $\left(\mathcal{H}_\delta^\psi\right)^*(\emptyset) = 0$. Consequently, $\left(\mathcal{H}^\psi\right)^*(\emptyset) = \lim_{\delta\downarrow 0}\left(\mathcal{H}_\delta^\psi\right)^*(\emptyset) = 0$.
If $E \subset F \subset \Omega$, every δ-cover of F is also a δ-cover of E, so taking the infimum over all possible δ-covers gives $\left(\mathcal{H}_\delta^\psi\right)^*(E) \leq \left(\mathcal{H}_\delta^\psi\right)^*(F)$ for all $\delta > 0$. Letting $\delta \downarrow 0$ gives $\left(\mathcal{H}^\psi\right)^*(E) \leq \left(\mathcal{H}^\psi\right)^*(F)$.
Let $\{F_i\}$ be a countable collection of subsets of Ω. Without loss of generality, we may assume that $\sum_{i=1}^{\infty}\left(\mathcal{H}_\delta^\psi\right)^*(F_i) < \infty$. For $\epsilon > 0$, let $\{U_{i,j},\ j = 1, 2, \cdots\}$ be a δ-cover of F_i such that $\sum_{j=1}^{\infty}\psi\left(|U_{i,j}|\right) \leq \left(\mathcal{H}_\delta^\psi\right)^*(F_i) + \frac{\epsilon}{2^i}$. Then $\{U_{i,j} : \ i, j = 1, 2, \cdots\}$ is a δ-cover of $\cup_{i=1}^{\infty} F_i$ and

$$\begin{aligned}\left(\mathcal{H}_\delta^\psi\right)^*(\cup_{i=1}^{\infty} F_i) &\leq \textstyle\sum_{i=1}^{\infty}\sum_{i=j}^{\infty}\psi\left(|U_{i,j}|\right) \leq \sum_{i=1}^{\infty}\left(\left(\mathcal{H}_\delta^\psi\right)^*(F_i) + \frac{\epsilon}{2^i}\right) = \\ &= \epsilon + \textstyle\sum_{i=1}^{\infty}\left(\mathcal{H}_\delta^\psi\right)^*(F_i) \leq \epsilon + \sum_{i=1}^{\infty}\left(\mathcal{H}^\psi\right)^*(F_i).\end{aligned} \tag{89}$$

Since this is true for any $\epsilon > 0$, it follows that

$$\left(\mathcal{H}^\psi\right)^*(\cup_{i=1}^{\infty} F_i) = \lim_{\delta\downarrow 0}\left(\mathcal{H}_\delta^\psi\right)(\cup_{i=1}^{\infty} F_i) \leq \sum_{i=1}^{\infty}\left(\mathcal{H}^\psi\right)(F_i), \tag{90}$$

which concludes the proof. ■

Definition 53. *A set $F \subset \Omega$ is said to be $\mathcal{H}^\psi$-measurable for some $\psi \in \mathcal{D}$ if*

$$\left(\mathcal{H}^\psi\right)^* (A) = \left(\mathcal{H}^\psi\right)^* (A \cap F) + \left(\mathcal{H}^\psi\right)^* (A \cap F^c), \tag{91}$$

for all $A \subset \Omega$. Here F^c denotes the complement $\Omega \backslash F$.

The following theorem due to C. Carathéodory [22] then establishes our measure space.

Theorem 54. *Let $\mathcal{M}_\psi$ denote the set of measurable sets in Ω with respect to the outer measure $\left(\mathcal{H}^\psi\right)^*$. Then $\mathcal{M}_\psi$ is a σ-algebra and the restriction $\mathcal{H}^\psi$ of $\left(\mathcal{H}^\psi\right)^*$ to $\mathcal{M}_\psi$ is a measure.*

Remark 55. *The fact that $\mathcal{H}^\psi$ is a measure means that besides satisfying all the properties stated in the proof of Theorem 52., it also satisfies:*

$$\mathcal{H}^\psi \left(\cup_{i=1}^\infty F_i\right) = \sum_{i=1}^\infty \mathcal{H}^\psi(F_i), \tag{92}$$

for any countable family (F_i) of mutually disjoint measurable subsets of Ω. Altogether $\left(\Omega, \mathcal{M}_\psi, \mathcal{H}^\psi\right)$ is a measure space. Let us also mention that if $\Omega = \mathbb{R}^n$ and $\psi(x) = x^s$, $s > 0$, we recover the textbook Hausdorff measure of dimension s [12].

Remark 56. *For most applications one is usually content with the collection of Borel sets as a σ-algebra. This is the smallest σ-algebra containing all the open subsets of Ω. If, as assumed, Ω is a metric space, then the open sets are just the open sets in the metric topology. We denote the Borel set by $\mathcal{B}$. It is important to remark that every Borel set is measurable with respect to $\left(\mathcal{H}^\psi\right)^*$ for any $\psi \in \mathcal{D}$. In other words: $\mathcal{B} \subset \mathcal{M}_\psi$ for all $\psi \in \mathcal{D}$.*

Theorem 57. *Let $f : F \subset \Omega_1 \longrightarrow \Omega_2$ be of ψ-Hölder type in F, with $\psi \in \mathcal{D}$, i.e. $d_2\left(f(x), f(y)\right) \leq \psi\left(d_1(x,y)\right)$ for any $x, y \in F$. If F is a Borel subset of Ω_1, then $\mathcal{H}^{\phi \circ \psi^{-1}}\left(f(F)\right) \leq \mathcal{H}^\phi(F)$, for any $\phi \in \mathcal{D}$.*

Proof. First of all notice that since F is a Borel set in Ω_1 and f is a continuous function (as it is Hölderian), then $f(F)$ is also a Borel set in Ω_2. Let $\{U_i\}$ be an arbitrary δ-cover of F. Then we have

$$\begin{aligned} |f\left(F \cap U_i\right)| &= \sup\left\{d_2\left(f(x), f(y)\right),\ x, y \in F \cap U_i\right\} \leq \\ &\leq \sup\left\{\psi\left(d_1(x,y)\right),\ x, y \in F \cap U_i\right\} \leq \psi\left(|U_i|\right) \end{aligned} \tag{93}$$

where we used the fact that ψ is strictly increasing in the last step. We conclude that $\{f\left(F \cap U_i\right)\}$ constitutes a $\psi(\delta)$-cover of $f(F)$. Notice that as $\psi, \phi \in \mathcal{D}$, then also $\psi^{-1}, \phi \circ \psi^{-1} \in \mathcal{D}$. Next we have:

$$\sum_i \phi \circ \psi^{-1}\left(|f(F \cap U_i)|\right) \leq \sum_i \phi \circ \psi^{-1} \circ \psi(|U_i|) = \sum_i \phi(|U_i|). \tag{94}$$

Taking the infimum over all δ-covers of F, we obtain:

$$\inf\left\{\sum_i \phi\circ\psi^{-1}\left(|f(F\cap U_i)|\right),\ \{U_i\} \text{ is a } \delta\text{-cover of } F\right\} \leq \mathcal{H}_\delta^\phi(F) \tag{95}$$

Since there are more $\psi(\delta)$-covers of $f(F)$ than those of the form $\{f(F\cap U_i)\}$, we conclude that $\mathcal{H}_{\psi(\delta)}^{\phi\circ\psi^{-1}}(f(F))$ is lower or equal to the left-hand side of the previous equation. Altogether: $\mathcal{H}_{\psi(\delta)}^{\phi\circ\psi^{-1}}(f(F)) \leq \mathcal{H}_\delta^\phi(F)$. Taking the limit $\delta\downarrow 0$, we obtain the desired result. ■

The following proposition describes the behavior of the measure under a change of dimension function.

Proposition 58. *Let $\psi,\phi,\rho\in\mathcal{D}$ be such that $\phi\sim\rho$ and $\phi\prec\psi$. Then there exist constants $C_1, C_2 > 0$ such that for any $F\subset\mathcal{B}$:*

$$C_1\mathcal{H}^\rho(F) \leq \mathcal{H}^\phi(F) \leq C_2\mathcal{H}^\rho(F), \qquad \mathcal{H}^\psi(F) \leq \mathcal{H}^\phi(F). \tag{96}$$

Proof. Since $\phi\sim\rho$, there exist constants $C_1, C_2 > 0$ and $0<\delta\leq\min\{\epsilon_\phi,\epsilon_\rho\}$ such that $C_1 \leq \frac{\phi(x)}{\rho(x)} \leq C_2$ for all $x\in(0,\delta)$. It then follows that $C_1\sum_{i=1}^\infty \rho(|U_i|) \leq \sum_{i=1}^\infty \phi(|U_i|) \leq C_2\sum_{i=1}^\infty \rho(|U_i|)$ where $\{U_i\}$ is an arbitrary δ-cover of F. Taking infima over all δ-covers yields $C_1\mathcal{H}_\delta^\rho(F) \leq \mathcal{H}_\delta^\phi(F) \leq C_2\mathcal{H}_\delta^\rho(F)$. Letting $\delta\downarrow 0$ we obtain $C_1\mathcal{H}^\rho(F) \leq \mathcal{H}^\phi(F) \leq C_2\mathcal{H}^\rho(F)$, which proves the first part. The second part is proved similarly. If $\phi\prec\psi$, then $\lim_{x\downarrow 0}\frac{\psi(x)}{\phi(x)} = 0$. That means that there exists δ with $0<\delta\leq\min\{\epsilon_\psi,\epsilon_\phi\}$ such that $\psi(x)\leq\phi(x)$, $\forall x\in(0,\delta)$. For any δ-cover $\{U_i\}$, we thus have: $\sum_{i=1}^\infty \psi(|U_i|) \leq \sum_{i=1}^\infty \phi(|U_i|)$. Taking infima over all δ-covers yields $\mathcal{H}_\delta^\psi(F) \leq \mathcal{H}_\delta^\phi(F)$. Letting $\delta\downarrow 0$, we obtain $\mathcal{H}^\psi(F) \leq \mathcal{H}^\phi(F)$. ■

Remark 59. *If for $\phi,\rho\in\mathcal{D}$, $\phi\sim\rho$, then from the previous proposition, we may establish an equivalence relation for measures. We write $\mathcal{H}^\phi\sim\mathcal{H}^\rho$ if and only if there exist positive real constants C_1, C_2 such that $C_1\mathcal{H}^\rho(F) \leq \mathcal{H}^\phi(F) \leq C_2\mathcal{H}^\rho(F)$ for every $F\subset\mathcal{B}$. It is easy to prove that this is indeed an equivalence relation. As before, for each equivalence class in $\mathcal{D}/\sim$, we fix a representative $\phi\in[\phi]$, which in turn, will fix a representative of the equivalence class of $\mathcal{H}^\phi$. The relevant point here is that a set F is measurable with respect to $\mathcal{H}^\phi$ if and only if it is measurable with respect to any $\mathcal{H}^\rho$ with $\rho\sim\phi$.*

Proposition 60. *Let $\phi,\psi\in\mathcal{D}$ with $\phi\prec\psi$ and let $F\subset\mathcal{B}$ be such that $\mathcal{H}^\phi(F)<\infty$. Then $\mathcal{H}^\psi(F) = 0$.*

Proof. Let δ be such that $0<\delta\leq\min\{\epsilon_\phi,\epsilon_\psi\}$ and $\{U_i\}$ a δ-cover of F. Since $\phi\prec\psi$, we have $\lim_{x\downarrow 0}\frac{\psi(x)}{\phi(x)} = 0$, which means that $\frac{\psi(x)}{\phi(x)}$ is bounded in $(0,\delta)$. Let $\eta(\delta) = \sup_{0<x\leq\delta}\frac{\psi(x)}{\phi(x)}$. Obviously, $\lim_{\delta\downarrow 0}\eta(\delta) = 0$. We then have: $\sum_{i=1}^\infty \psi(|U_i|) = \sum_{i=1}^\infty \phi(|U_i|)\frac{\psi(|U_i|)}{\phi(|U_i|)} \leq \eta(\delta)\sum_{i=1}^\infty \phi(|U_i|)$. Taking infima over all δ-covers of F, we get: $0\leq\mathcal{H}_\delta^\psi(F)\leq\eta(\delta)\mathcal{H}_\delta^\phi(F)$. Since $\mathcal{H}^\phi(F)<\infty$, upon taking the limit $\delta\downarrow 0$, we obtain $\mathcal{H}^\psi(F) = 0$. ■

Propositions 58. and 60. reveal that, as we take different elements ϕ in some scale of functions $\mathcal{F}$ and evaluate the corresponding measures $\mathcal{H}^{\phi}(F)$ of a set $F \subset \mathcal{B}$, these will jump from 0 to ∞. This motivates the following definition.

Definition 61. *Let $F \subset \mathcal{B}$ and $\mathcal{F}$ be a scale of dimension functions. We define the* **Hausdorff dimension function** *of F to be the element of $\mathcal{F}$:*

$$\begin{aligned} \psi_H(F) &:= sup_{\mathcal{F}} \left\{ \psi \in \mathcal{F}/\sim \colon \ \mathcal{H}^{\psi}(F) = \infty \right\} = \\ &= inf_{\mathcal{F}} \left\{ \psi \in \mathcal{F}/\sim \colon \ \mathcal{H}^{\psi}(F) = 0 \right\} \end{aligned} \tag{97}$$

Notice that $\mathcal{H}^{\psi_H}(F)$ may be 0, ∞ or any real positive number.

Definition 62. *The* **Hausdorff-Besicovitch dimension** *of a set $F \subset \mathcal{B}$ with Hausdorff dimension function $\psi_H(F) := \psi_H$ is defined by:*

$$dim_H(F) := \lim_{x \downarrow 0} \frac{\log\left(\psi_H(x)\right)}{\log(x)}. \tag{98}$$

Let us briefly verify that this definition makes sense. But before that let us define the following quantities:

$$s_H := \sup\left\{s > 0 : \ x^s \prec \psi_H\right\}, \qquad r_H := \inf\left\{r > 0 : \ \psi_H \prec x^r\right\}. \tag{99}$$

Notice that s_H may be ∞, or it may not exist, if $\psi_H \prec x^s$ for all $s > 0$. Likewise, r_H may be equal to zero, or it may not exist if $x^s \prec \psi_H$ for all $s > 0$.

Proposition 63. *The Hausdorff-Besicovitch dimension has the following properties. (i) For $\psi_H(x) = x^s$ $(s > 0)$, we have: $dim_H(F) = s$.*
(ii) The Hausdorff-Besicovitch dimension is independent of the representative of the equivalence class of ψ_H.
(iii) The Hausdorff-Besicovitch dimension is determined by the element of $\mathcal{P}$ "closest" to ψ_H. To be more specific: $dim_H(F) = s_H$ or $dim_H(F) = r_H$.

Proof. (i) By definition $\dim_H(F) = \lim_{x \downarrow 0} \frac{\log(x^s)}{\log(x)} = s$, as expected. (ii) Next, suppose that $\phi \in \mathcal{D}$ is such that $\phi \sim \psi_H$. Then there exist $C_1, C_2 > 0$ such that $C_1 \leq \frac{\phi(x)}{\psi_H(x)} \leq C_2$, for $x \in (0, \delta)$ and $\delta = \min\left\{\epsilon_{\phi}, \epsilon_{\psi_H}\right\}$. It then follows that:

$$\begin{aligned} \lim_{x \downarrow 0} \frac{\log(\phi(x))}{\log(x)} &= \lim_{x \downarrow 0} \frac{\log\left(\frac{\phi(x)}{\psi_H(x)} \psi_H(x)\right)}{\log(x)} = \\ &= \lim_{x \downarrow 0} \frac{\log(\psi_H(x))}{\log(x)} + \lim_{x \downarrow 0} \frac{\log\left(\frac{\phi(x)}{\psi_H(x)}\right)}{\log(x)} = \dim_H(F). \end{aligned} \tag{100}$$

This means that the Hausdorff dimension does not depend on the representative of the equivalence class to which ψ_H belongs. (iii) Finally, if $\psi_H(x) = x^u$ for some $u > 0$, then $0 < \dim_H(F) = s_H = r_H = u < \infty$. Alternatively, suppose that $\psi_H \notin \mathcal{P}$. Since the scale $\mathcal{F}$ is a well-ordered set, then at least one of the sets $\{s > 0 : \ x^s \prec \psi_H\}$,

$\{r > 0 : \psi_H \prec x^r\}$ is non-empty. Suppose that $\{s > 0 : x^s \prec \psi_H\}$ is non-empty. This means that $0 < s_H \leq \infty$. We first assume that $s_H < \infty$. Let us check that $\dim_H(F) = s_H$. Since $\psi_H \notin \mathcal{P}$, we have $\lim_{x\downarrow 0} \frac{\psi_H(x)}{x^{s_H}} = 0$.

$$\dim_H(F) = \lim_{x\downarrow 0} \frac{\log(\psi_H(x))}{\log(x)} = \lim_{x\downarrow 0} \frac{\log\left(x^{s_H}\frac{\psi_H(x)}{x^{s_H}}\right)}{\log(x)} = s_H + \lim_{x\downarrow 0} \frac{\log\left(\frac{\psi_H(x)}{x^{s_H}}\right)}{\log(x)}. \quad (101)$$

It remains to prove that $\lim_{x\downarrow 0} \frac{\log\left(\frac{\psi_H(x)}{x^{s_H}}\right)}{\log(x)} = 0$. Let us write $\phi(x) = \frac{\psi_H(x)}{x^{s_H}}$. By definition of s_H (and as $\psi_H \notin \mathcal{P}$), we have $\lim_{x\downarrow 0} \frac{\phi(x)}{x^\delta} = \infty$ for any $\delta > 0$. That is, for any $\delta > 0$, we can find $0 < \epsilon < 1$ such that $x^\delta < \phi(x) < 1 \Leftrightarrow \left|\log\left(x^\delta\right)\right| > |\log(\phi(x))|$, $\forall x \in (0, \epsilon)$. And thus, we have:

$$\left|\frac{\log(\phi(x))}{\log(x)}\right| < \delta \left|\frac{\log(x)}{\log(x)}\right| = \delta. \quad (102)$$

In summary $\dim_H(F) = s_H$. If however, $s_H = \infty$, then $\lim_{x\downarrow 0} \frac{\psi_H(x)}{x^s} = 0$ for all $s > 0$. For sufficiently small $|x| < 1$, we have $(\log(x) < 0)$ and $\psi_H(x) \leq x^s$. Consequently, $\frac{\log(\psi_H(x))}{\log(x)} \geq \frac{\log(x^s)}{\log(x)}$. Taking the limit $x \downarrow 0$, we obtain $\dim_H(F) \geq s$ for any $s > 0$. And thus $\dim_H(F) = s_H = \infty$.
If we assume instead that $\{r > 0 : \psi_H \prec x^r\}$ is non-empty, then $0 \leq r_H < \infty$. Following the same steps, we can prove that $0 < \dim_H(F) = r_H < \infty$ or $0 = \dim_H(F) = r_H$. ∎

Let us now explore some properties of the Hausdorff dimension and of Hausdorff dimension functions.

The following proposition is an immediate consequence of Theorem 57..

Proposition 64. *Let $f : F \subset \Omega_1 \longrightarrow \Omega_2$ be ψ-Hölder continuous in $F \in \mathcal{B}$ with $\psi \in \mathcal{F}$ and suppose that F has Hausdorff dimension function $\phi_H \in \mathcal{F}$. Then the Hausdorff dimension function of $f(F)$ is some $\rho_H \in \mathcal{F}$ such that $\rho_H \preceq \phi_H \circ \psi^{-1}$. Likewise: $dim_H(f(F)) \leq \lim_{x\downarrow 0} \frac{\log(\phi_H \circ \psi^{-1}(x))}{\log(x)}$.*

Proof. Let $\phi \in \mathcal{F}$ be such that $\phi_H \prec \phi$. Then from Theorem 57. and Proposition 60., we have that $\mathcal{H}^{\phi\circ\psi^{-1}}(f(F)) \leq \mathcal{H}^\phi(F) = 0$, which implies that the Hausdorff dimension function ρ_H of $f(F)$ satisfies $\rho_H \preceq \phi \circ \psi^{-1}$ for all $\phi \in \mathcal{F}$ with $\phi_H \prec \phi$. Consequently $\rho_H \preceq \phi_H \circ \psi^{-1}$. Finally, we prove the last statement. By definition $\dim_H(f(F)) = \lim_{x\downarrow 0} \frac{\log(\rho_H(x))}{\log(x)}$. If $\rho_H \sim \phi_H \circ \psi^{-1}$, then from Proposition 63. (ii), the result follows immediately. Let us assume that $\rho_H \prec \phi_H \circ \psi^{-1}$. For sufficiently small x, we may assume that $|x| < 1$ $(\log(x) < 0)$ and $\rho_H(x) \geq \phi_H \circ \psi^{-1}(x)$. It then follows that $\frac{\log(\rho_H(x))}{\log(x)} \leq \frac{\log(\phi_H \circ \psi^{-1}(x))}{\log(x)}$. Letting $x \downarrow 0$, we obtain the desired result. ∎

Theorem 65. *Let Ω be a metric space. The Hausdorff dimension and the Hausdorff dimension functions with respect to a scale $\mathcal{F}$ of Borel subsets of Ω satisfy the following properties.*

(monotonicity) *If $E \subset F \subset \mathcal{B}$, then $\psi_H(E) \preceq \psi_H(F)$ and $dim_H(E) \leq dim_H(F)$.*

(countable stability) *If $F_1, F_2, \cdots$ is a (countable) sequence of Borel subsets of Ω, then $\psi_H\left(\cup_{i=1}^{\infty} F_i\right) = sup_{1\le i<\infty}\{\psi_H(F_i)\}$ and $dim_H\left(\cup_{i=1}^{\infty} F_i\right) = sup_{1\le i<\infty}\{dim_H(F_i)\}$.*

Proof. Let $E \subset F$. As the ψ-Hausdorff measure is a measure, $\mathcal{H}^{\psi}(E) \le \mathcal{H}^{\psi}(F)$ for any $\psi \in \mathcal{F}$. It then follows that $\psi_H(E) \preceq \psi_H(F)$ and $\dim_H(E) \le \dim_H(F)$. This proves monotonicity. Next consider a (countable) sequence $F_1, F_2, \cdots$ of Borel subsets of Ω . From the monotonicity property, we immediately have that

$$\sup_{1\le i<\infty}\{\psi_H(F_i)\} \preceq \psi_H\left(\cup_{i=1}^{\infty} F_i\right) \tag{103}$$

and

$$\sup_{1\le i<\infty}\{\dim_H(F_i)\} \le \dim_H\left(\cup_{i=1}^{\infty} F_i\right) \tag{104}$$

for each i. Here $\sup_{1\le i<\infty}\{\psi_H(F_i)\}$ is the supremum in the scale of dimension functions $\mathcal{F}$. On the other hand, given $\psi \in \mathcal{F}$, if $\psi_H(F_i) \prec \psi$ for all i, then $\mathcal{H}^{\psi}(F_i) = 0$ for all i and thus $\mathcal{H}^{\psi}\left(\cup_{i=1}^{\infty} F_i\right) = 0$. This gives the opposite inequality $\psi_H\left(\cup_{i=1}^{\infty} F_i\right) \preceq \sup_{1\le i<\infty}\{\psi_H(F_i)\}$. The same reasoning leads to $\dim_H\left(\cup_{i=1}^{\infty} F_i\right) \le \sup_{1\le i<\infty}\{\dim_H(F_i)\}$. ■

Regarding countable sets, we make the following conjecture.

Conjecture 66. *Let $F \subset \mathcal{B}$ and $\mathcal{F}$ a maximal scale of dimension functions. Then $\mathcal{H}^{\psi}(F) = 0$ for all $\psi \in \mathcal{F}$ if and only if F is countable.*

Proof. We prove only sufficiency. Let F be a countable set. Then we may write it as a sequence of points $F = \{x_1, x_2, \cdots\}$. Since the ψ-Hausdorff measure is a measure, we have for any $\psi \in \mathcal{F}$: $\mathcal{H}^{\psi}(F) = \mathcal{H}^{\psi}\left(\cup_{i=1}^{\infty}\{x_i\}\right) \le \sum_{i=1}^{\infty} \mathcal{H}^{\psi}(\{x_i\})$. As the ψ-Hausdorff measure of a set with a single point is zero for any $\psi \in \mathcal{F}$, we obtain the result. ■

Proposition 67. *Suppose that F can be covered by n_k sets of diameter at most δ_k with $\delta_k \downarrow 0$ as $k \to \infty$. Moreover assume that $n_k\psi(\delta_k)$ stays bounded as $k \to \infty$ for some $\psi \in \mathcal{D}$. Then $\mathcal{H}^{\psi}(F) < \infty$.*

Proof. If F is covered by n_k sets of diameter at most δ_k, then by definition, $0 \le \mathcal{H}^{\psi}_{\delta_k}(F) \le n_k\psi(\delta_k)$. If the right-hand side of this inequality stays bounded as $k \to \infty$, then we conclude that $\mathcal{H}^{\psi}(F) < \infty$. ■

Remark 68. *Under the conditions of the previous proposition, we conclude that the Hausdorff dimension function ϕ_H of F must be such that $\phi_H \preceq \psi$.*

Usually Hausdorff dimensions are difficult to determine. One sometimes resorts to alternative definitions of fractal dimension. For subsets of $\mathbb{R}^n$, the following definition is useful.

Definition 69. *Let $F \in \mathbb{R}^n$ and for each $\delta > 0$ let N_δ denote the number of δ-mesh cubes that intersect F. The* **lower and upper box-counting dimensions** *of F are given by:*

$$\underline{dim}_B(F) = \underline{\lim}_{\delta\downarrow 0}\frac{\log N_\delta(F)}{-\log\delta}, \qquad \overline{dim}_B(F) = \overline{\lim}_{\delta\downarrow 0}\frac{\log N_\delta(F)}{-\log\delta}, \tag{105}$$

respectively. The **box-counting dimension** *of F is:*

$$dim_B(F) = \lim_{\delta \downarrow 0} \frac{\log N_\delta(F)}{-\log \delta}, \tag{106}$$

whenever the limit exists.

From the definition it follows immediately that:

$$\dim_H(F) \leq \underline{\dim}_B(F) \leq \overline{\dim}_B(F). \tag{107}$$

Let us now state some results concerning the dimension of graph of a function.

Proposition 70. *Let $f : I \to \mathbb{R}$ be a continuous function. Let $0 < \delta \leq 1$, and m is the least integer greater than or equal to $\frac{1}{\delta}$. Then if N_δ is the number of squares of the δ-mesh that intersect the graph of f,*

$$\delta^{-1} \sum_{i=0}^{m-1} R_f\left[i\delta, (i+1)\delta\right] \leq N_f \leq 2m + \delta^{-1} \sum_{i=0}^{m-1} R_f\left[i\delta, (i+1)\delta\right]. \tag{108}$$

Here $R_f[x_1, x_2] = sup_{x_1 \leq x, y \leq x_2} |f(x) - f(y)|$ is the range of the function f in the interval $[x_1, x_2]$.

Proof. See [12]. ■

Theorem 71. *Let $f : I \to \mathbb{R}$ be a continuous function, $\mathcal{F}$ a scale of functions and $\psi \in \mathcal{F}$ such that $x \prec \psi \prec x^2$ and $\frac{x^2}{\psi(x)} \in \mathcal{F}$. Moreover, suppose that there exist constants $0 < \delta \leq 1$ and $K > 0$ such that:*

$$|f(x) - f(y)| \leq K \frac{|x-y|^2}{\psi(|x-y|)}, \tag{109}$$

for all $x, y \in I$ such that $|x - y| \leq \delta$. Then the graph Γ of f satisfies $\mathcal{H}^\psi(\Gamma) < \infty$ and has a Hausdorff dimension function $\phi_H(\Gamma) \preceq \psi$.

Proof. If $x_1, x_2 \in I$ are such that $|x_1 - x_2| \leq \delta$, then we have $R_f[x_1, x_2] \leq K\frac{|x_1-x_2|^2}{\psi(|x_1-x_2|)} \leq K\frac{\delta^2}{\psi(\delta)}$, where we used the fact that $\frac{x^2}{\psi(x)}$ is an increasing function as it belongs to $\mathcal{F}$. If m is the least integer greater than or equal to $\frac{1}{\delta}$ then $m < 1 + \frac{1}{\delta}$. From Proposition 70. it then follows that:

$$\begin{aligned} N_\delta &\leq 2m + m\delta^{-1} K \frac{\delta^2}{\psi(\delta)} \leq \\ &\leq (1+\delta^{-1})\left[2 + K\frac{\delta}{\psi(\delta)}\right] = \frac{1}{\psi(\delta)}\left[2\psi(\delta) + \delta K + 2\frac{\psi(\delta)}{\delta} + K\right]. \end{aligned} \tag{110}$$

Under the assumptions of the theorem, the terms inside the bracket stay bounded as $\delta \to 0$. We conclude that there exists some constant $C > 0$ such that $N_\delta \psi(\delta) \leq C$. From Proposition 67. the result of the theorem follows. ■

Theorem 72. *Suppose that there exist real numbers* $C > 0$, $\delta_0 > 0$, *and* $\psi \in \mathcal{F}$ *with* $x \preceq \psi \prec x^2$ *and* $\frac{x^2}{\psi} \in \mathcal{F}$. *Moreover, let* $f : I \to \mathbb{R}$ *be a continuous function for which the following property holds: for each* $x \in I$ *and* $0 < \delta \leq \delta_0$ *there exists* u *such that* $|x - u| \leq \delta$ *and* $|f(x) - f(u)| \geq C\frac{\delta^2}{\psi(\delta)}$. *Then* $\underline{dim}_B(\Gamma) \geq \lim_{\delta\downarrow 0} \frac{\log(\psi(\delta))}{\log(\delta)}$.

Proof. We shall consider again a δ-mesh and use the same notation as in Proposition 70. and Theorem 71.. Under the assumptions of the theorem, we have $R_f\,[x_1, x_2] \geq C\frac{\delta^2}{\psi(\delta)}$. Since $m \geq \delta^{-1}$ and $\delta < \delta_0$, we conclude from Proposition 70. that $N_\delta \geq m\delta^{-1}C\frac{\delta^2}{\psi(\delta)} \geq \frac{C}{\psi(\delta)}$. By definition we conclude that $\underline{\dim}_B \geq \lim_{\delta\downarrow 0} \frac{\log(\psi(\delta))}{\log(\delta)}$. ∎

Using the previous two theorems, we can easily obtain the box dimension of the graph of the Weierstrass function.

Lemma 73. *Let* $\lambda > 1$ *and* $1 < s < 2$. *The graph* $\Gamma_{\lambda,s}$ *of the Weierstrass function* $W_{\lambda,s}$ *has box dimension* $dim_B(\Gamma_{\lambda,s}) = s$.

Proof. From Proposition 18. we realize that the conditions of Theorems 71. and 72. are verified for $\psi(x) = x^s$. ∎

4. Fractional Calculus

In the previous sections we came across functions such as Takagi's or Weierstrass' functions which are continuous everywhere in an interval, but nowhere differentiable in that interval. If such functions are to represent some physical dynamical system or some other process, then we would like to be able to write down equations for this sort of motion. We are nevertheless faced with a complicated problem. Since these functions are nowhere differentiable, we cannot write down differential equations to which they would be solutions. There is an old answer to this problem. It is called the **fractional calculus** [16, 28, 31, 33, 36].

Definition 74. *Let* f *be a continuous function on* $I = [a, b]$. *The* **right and left Riemann-Liouville fractional integrals** *of order* $\alpha \geq 0$ *of* f *at* $x \in I$ *are defined by:*

$$\begin{aligned} I^\alpha_{a,-}(f)(x) &:= \frac{1}{\Gamma(\alpha)} \int_a^x (x-y)^{\alpha-1} f(y)dy \\ I^\alpha_{b,+}(f)(x) &:= \frac{1}{\Gamma(\alpha)} \int_x^b (y-x)^{\alpha-1} f(y)dy \end{aligned} \tag{111}$$

where $\Gamma(\cdot)$ *denotes the Gamma function.*

These integrals regarded as linear operators from $C^0(I)$ to $C^0(I)$ are bounded.

Lemma 75. *Let* $f \in C^0(I)$. *Then*

$$||I^\alpha_{a,-}(f)||_0 \leq \frac{(b-a)^\alpha}{|\Gamma(\alpha+1)|}||f||_0, \quad ||I^\alpha_{b,+}(f)||_0 \leq \frac{(b-a)^\alpha}{|\Gamma(\alpha+1)|}||f||_0. \tag{112}$$

Proof. For any $x \in [a,b]$, we have:

$$\begin{aligned} \left|I^{\alpha}_{a,-}(f)(x)\right| &\leq \tfrac{1}{|\Gamma(\alpha)|}\int_a^x (x-y)^{\alpha-1}|f(y)|dy \leq \\ &\leq \tfrac{||f||_0}{|\Gamma(\alpha)|}\tfrac{(x-a)^{\alpha}}{\alpha} \leq \tfrac{||f||_0}{|\Gamma(\alpha+1)|}(b-a)^{\alpha}, \end{aligned} \tag{113}$$

where we used $\alpha\Gamma(\alpha) = \Gamma(\alpha+1)$. The first inequality in (112) follows immediately. The second inequality is proved in a similar fashion. ■

Definition 76. *The* **left and right Riemann-Liouville fractional derivatives** *of order $\alpha \geq 0$ at $x \in I$ are defined by:*

$$\begin{aligned} D^{\alpha}_{a,-}(f)(x) &:= \tfrac{d^n}{dx^n} I^{n-\alpha}_{a,-}(f)(x) \\ D^{\alpha}_{b,+}(f)(x) &:= (-1)^n \tfrac{d^n}{dx^n} I^{n-\alpha}_{b,+}(f)(x) \end{aligned} \tag{114}$$

Here $n-1$ denotes the integral part $[\alpha]$ of α.

Fractional derivatives partially solve our problem: it may be that a function is nowhere differentiable in an interval, but that it admits nevertheless a fractional derivative of some order $0 < \alpha < 1$. Indeed this what happens with the Weierstrass function. It can be proved that it admits fractional derivatives (left and right) of order $0 < \alpha \leq 2 - r < 1$ [26, 35]. Likewise Takagi's function admits fractional derivatives (left and right) of any order $0 < \alpha < 1$ [37].

However, fractional derivatives pose equally some problems. Let us consider a simple example.

Example 77. *Let $f(x) = A(x-a)^{\beta}$ and $g(x) = B(b-x)^{\beta}$, with $\beta > -1$, $x \in I$ and A, B arbitrary real constants. For $\alpha \geq 0$ and $n = [\alpha]+1$, we obtain [33]:*

$$\begin{aligned} D^{\alpha}_{a,-}(f)(x) &= \tfrac{A\Gamma(\beta+1)}{\Gamma(\beta+1-\alpha)}(x-a)^{\beta-\alpha} \\ D^{\alpha}_{b,+}(g)(x) &= \tfrac{B\Gamma(\beta+1)}{\Gamma(\beta+1-\alpha)}(b-x)^{\beta-\alpha}. \end{aligned} \tag{115}$$

This example illustrates the kind of problem that one faces. For $\beta = 0$, we conclude that the fractional derivatives of non-integer order of constant functions do not vanish. Moreover, from (115), we realize that the fractional derivatives depend crucially on the choice of terminals a and b. This means that they have a non-local nature. Non-locality may play to our advantage when trying to model certain memory and hereditary properties of various materials and processes[1]. However, it is very difficult to devise a connection between fractional derivatives and the local geometry of the graph of a function as in standard differential calculus. Moreover, there are some more practical difficulties: the fractional derivatives of products obey a complicated rule (not the Leibniz rule) and those of the composition of functions follow a generalization of the intricate Faà di Bruno formula.

These facts led Kolwankar and Gangal to introduce the so-called local fractional derivatives [4, 7, 17, 18, 19].

[1] See chapter 10 of [33] for a survey of applications of fractional calculus and [42, 44].

4.1. Local Fractional Derivatives

Definition 78. *Let $f \in C^0(I)$ and $\alpha \in (0,1]$. The right ($\sigma = +$) or left ($\sigma = -$)* **Kolwankar-Gangal local fractional derivative** *of order α of f at $\xi \in I$ is given by:*

$$\frac{d^\alpha f}{dx^\alpha}(\xi^\sigma) := \frac{1}{\Gamma(1+\alpha)} \lim_{x \to \xi^\sigma} D^\alpha_{\xi,-\sigma}\left[\sigma\left(f - f(\xi)\right)\right](x). \tag{116}$$

If the limits exist, we say that f is **right or left α-differentiable in the sense of Kolwankar and Gangal** *(KG) at ξ.*

Perhaps the most remarkable aspect of this definition is the fact that, under certain circumstances, one may write a generalized Taylor formula. In the following theorem we denote:

$$F_\sigma(\xi, x) := \frac{1}{\Gamma(1+\alpha)} D^\alpha_{\xi,-\sigma}\left[\sigma\left(f - f(\xi)\right)\right](x+\xi). \tag{117}$$

Theorem 79. *Let $f \in C^0(I)$ be right ($\sigma = +$) or left ($\sigma = -$) α-differentiable at $\xi \in I$ in the KG sense. Then:*

$$f(x) = f(\xi) + \sigma \frac{d^\alpha f}{dx^\alpha}(\xi^\sigma)|x-\xi|^\alpha + R_\sigma(x,\xi). \tag{118}$$

The remainder satisfies

$$\lim_{x \to \xi^\sigma} \frac{R_\sigma(x,\xi)}{|x-\xi|^\alpha} = 0, \tag{119}$$

and is given by

$$R_\sigma(x,\xi) = \sigma \int_0^{x-\xi} \frac{dF_\sigma(\xi,\tau)}{d\tau} |x-\xi-\tau|^\alpha d\tau, \tag{120}$$

whenever this integral is well defined.

Corollary 80. *Under the conditions of Theorem 79., we may write [9]:*

$$\frac{d^\alpha f}{dx^\alpha}(\xi^\sigma) = (D^\alpha_x f)(\xi^\sigma) := \lim_{x \to \xi^\sigma} \frac{\sigma\left(f(x) - f(\xi)\right)}{|x-\xi|^\alpha}. \tag{121}$$

Definition 81. *If the limit in (121) exists or, equivalently, the Taylor formula (118,119) holds, then we say that f is* **right** *($\sigma = +$)* **or left** *($\sigma = -$)* **α-differentiable at ξ in the sense of Ben Adda and Cresson** *(BAC). We call $(D^\alpha_x f)(\xi^\sigma)$ the right ($\sigma = +$) or left ($\sigma = -$) α-derivative of f at ξ in the sense of BAC.*

There is a crucial fact that may have been overlooked in [9, 17, 18, 19]. The condition that the integral expression in (120) is well defined is in fact hard to fulfill. Thus, in general, α-differentiability in the sense of KG and BAC are not equivalent. Concomitantly, the Taylor formula (118,119) may not hold for α-differentiable functions in the sense of KG. Indeed in (120), we need the quantity:

$$\frac{dF_\sigma}{d\tau}(\xi,\tau) = \frac{1}{\Gamma(1+\alpha)} \frac{d}{d\tau} D^\alpha_{\xi,-\sigma}\left[\sigma\left(f - f(\xi)\right)\right](\tau+\xi). \tag{122}$$

Since $f(x) - f(\xi)$ vanishes for $x = \xi$, it is well known that the derivatives $D^{\alpha}_{\xi,-\sigma}$ and $\frac{d}{d\tau}$ commute [33]. We thus get:

$$\frac{dF_{\sigma}}{d\tau}(\xi, \tau) = \frac{1}{\Gamma(1+\alpha)} D^{\alpha}_{\xi,-\sigma} \frac{d}{d\tau} \left[\sigma \left(f - f(\xi)\right)\right](\tau + \xi).$$

However, the function $f(x) - f(\xi)$ is, by assumption, non-differentiable. And so, in general, the quantity (122) is not well defined.

Alternatively, we may require a (weaker) integrability condition for KG-differentiability to imply BAC-differentiability [7].

Proposition 82. (Chen, Yan, Zhang) *Let $f \in C^0[a,b]$ be such that the KG derivative $\frac{d^{\alpha} f}{dx^{\alpha}}(\xi^{\sigma})$ exists at $\xi \in (a,b)$ and $D^{\alpha}_{\xi,-\sigma}\left[\sigma\left(f - f(\xi)\right)\right](x)$ belongs to $L^{\infty}(\xi, \xi+\delta)$ if $\sigma = +$ or to $L^{\infty}(\xi - \delta, \xi)$ if $\sigma = -$ for some $\delta > 0$. Then:*

$$\frac{d^{\alpha} f}{dx^{\alpha}}(\xi^{\sigma}) = \left(D^{\alpha}_{x} f\right)(\xi^{\sigma}). \tag{123}$$

Proof. We will prove the result for $\sigma = +$. The case $\sigma = -$ is proved in a similar fashion. Since $D^{\alpha}_{\xi,-}\left[\left(f - f(\xi)\right)\right](x)$ belongs to $L^{\infty}(\xi, \xi+\delta)$, then it is integrable and the condition stated in (2.113), p.71 of [33] is satisfied and we have, for $x \in (\xi, \xi+\delta)$ that:

$$\begin{aligned} D^{-\alpha}_{\xi,-}\left(D^{\alpha}_{\xi,-}\left[\left(f - f(\xi)\right)\right](x)\right) = \\ = f(x) - f(\xi) - \left[D^{\alpha-1}_{\xi,-}\left[\left(f - f(\xi)\right)\right](x)\right]\Big|_{x=y} \frac{(x-y)^{\alpha-1}}{\Gamma(\alpha)} \end{aligned} \tag{124}$$

where

$$D^{-\alpha}_{\xi,-}\left(g(x)\right) = \frac{1}{\Gamma(\alpha)} \int_{\xi}^{x} (x-y)^{\alpha-1} g(y)\,dy \tag{125}$$

and the evaluation here $|_{x=y}$ is understood as the limit $x \to \xi^{+}$. Moreover, since f is continuous in $[a,b]$, we have

$$D^{\alpha-1}_{\xi,-}\left[\left(f - f(\xi)\right)\right](x) = \frac{1}{\Gamma(1-\alpha)} \int_{\xi}^{x} \frac{f(y) - f(\xi)}{(x-y)^{\alpha}}\,dy \to 0 \tag{126}$$

as $x \to \xi^{+}$. Consequently:

$$f(x) - f(\xi) = D^{-\alpha}_{\xi,-}\left(D^{\alpha}_{\xi,-}\left[\left(f - f(\xi)\right)\right](x)\right), \qquad x \in (\xi, \xi+\delta). \tag{127}$$

Now the proof follows from the fact that $\frac{d^{\alpha} f}{dx^{\alpha}}(\xi^{+}) = \frac{1}{\Gamma(1+\alpha)} \lim_{x\to\xi^{+}} D^{\alpha}_{\xi,-}\left[\left(f - f(\xi)\right)\right](x)$. We have

$$\begin{aligned} \tfrac{f(x)-f(\xi)}{(x-\xi)^{\alpha}} = \tfrac{1}{(x-\xi)^{\alpha}} D^{-\alpha}_{\xi,-}\left(D^{\alpha}_{\xi,-}\left[\left(f - f(\xi)\right)\right](x)\right) = \\ = \tfrac{1}{(x-\xi)^{\alpha}\Gamma(\alpha)} \int_{\xi}^{x} (x-y)^{\alpha-1} D^{\alpha}_{\xi,-}\left[\left(f - f(\xi)\right)(y)\right] dy = \\ = \tfrac{1}{(x-\xi)^{\alpha}\Gamma(\alpha)} \int_{\xi}^{x} (x-y)^{\alpha-1} \left(D^{\alpha}_{\xi,-}\left[\left(f - f(\xi)\right)(y)\right] - \Gamma(1+\alpha)\tfrac{d^{\alpha} f}{dx^{\alpha}}(\xi^{+})\right) dy + \\ + \tfrac{\Gamma(1+\alpha)}{(x-\xi)^{\alpha}\Gamma(\alpha)} \int_{\xi}^{x} (x-y)^{\alpha-1} \tfrac{d^{\alpha} f}{dx^{\alpha}}(\xi^{+})\,dy := I_1 + I_2. \end{aligned} \tag{128}$$

By a straightforward calculation, we have:

$$\begin{aligned} I_2 &= \tfrac{\Gamma(1+\alpha)}{(x-\xi)^\alpha\Gamma(\alpha)}\int_\xi^x (x-y)^{\alpha-1}\tfrac{d^\alpha f}{dx^\alpha}(\xi^+)dy = \\ &= \tfrac{\alpha\Gamma(\alpha)}{(x-\xi)^\alpha\Gamma(\alpha)}\tfrac{d^\alpha f}{dx^\alpha}(\xi^+)\tfrac{(x-\xi)^\alpha}{\alpha} = \tfrac{d^\alpha f}{dx^\alpha}(\xi^+). \end{aligned} \tag{129}$$

Thus we only need to show that $I_1 \to 0$ as $x \to \xi^+$. By definition we have that, for any $\epsilon > 0$, there exists $0 < \eta < \epsilon$ such that

$$\left| \frac{1}{\Gamma(1+\alpha)} D^\alpha_{\xi,-}\left[\left(f - f(\xi)\right)(y)\right] - \frac{d^\alpha f}{dx^\alpha}(\xi^+) \right| < \epsilon \tag{130}$$

whenever $y \in (\xi, \xi+\eta)$.

Therefore for any $x \in (\xi, \xi+\eta)$:

$$\begin{aligned} |I_1| &= \left| \tfrac{\Gamma(1+\alpha)}{(x-\xi)^\alpha\Gamma(\alpha)}\int_\xi^x (x-y)^{\alpha-1}\left(\tfrac{1}{\Gamma(1+\alpha)}D^\alpha_{\xi,-}\left[\left(f-f(\xi)\right)(y)\right] - \tfrac{d^\alpha f}{dx^\alpha}(\xi^+)\right)dy \right| \le \\ &\le \tfrac{\alpha}{(x-\xi)^\alpha}\int_\xi^x (x-y)^{\alpha-1}\left|\tfrac{1}{\Gamma(1+\alpha)}D^\alpha_{\xi,-}\left[\left(f-f(\xi)\right)(y)\right] - \tfrac{d^\alpha f}{dx^\alpha}(\xi^+)\right| dy \le \\ &\le \tfrac{\alpha}{(x-\xi)^\alpha}\int_\xi^x (x-y)^{\alpha-1}\epsilon dy = \epsilon \end{aligned} \tag{131}$$

which completes the proof. ■

There is an important structure theorem of Chen, Yan and Zhang that states that if the local KG fractional derivatives of a function of order $0 < \alpha < 1$ exist for a Hölder continuous function a.e. in an interval, then they vanish a.e. in that interval.

To prove that theorem we need some preliminary results and definitions [7].

Lemma 83. (Chen, Yan, Zhang) *Let $f : [a,b] \to \mathbb{R}$ be continuous and such that $\frac{d^\alpha f}{dx^\alpha}(\xi^+)$ exists for some $\xi \in (a,b)$. Then*

$$\lim_{h \downarrow 0} \int_0^1 (1-t)^{-\alpha} \frac{f(ht+\xi) - f(\xi)}{h^\alpha} dt \tag{132}$$

exists and

$$\frac{d^\alpha f}{dx^\alpha}(\xi^+) = \frac{1}{\Gamma(1+\alpha)\Gamma(1-\alpha)} \lim_{h \downarrow 0} \int_0^1 (1-t)^{-\alpha} \frac{f(ht+\xi) - f(\xi)}{h^\alpha} dt \tag{133}$$

Proof. By definition, we have:

$$\frac{d^\alpha f}{dx^\alpha}(\xi^+) = \frac{1}{\Gamma(1+\alpha)} \lim_{x \downarrow \xi} \frac{1}{\Gamma(1-\alpha)} \frac{d}{dx} \int_\xi^x \frac{f(t) - f(\xi)}{(x-t)^\alpha} dt. \tag{134}$$

Let

$$F_\xi(x) := \frac{1}{\Gamma(1-\alpha)} \int_\xi^x \frac{f(t) - f(\xi)}{(x-t)^\alpha} dt, \tag{135}$$

which can also be written as

$$F_\xi(x) = \frac{(x-\xi)^{1-\alpha}}{\Gamma(1-\alpha)} \int_0^1 \frac{(\xi + (x-\xi)s) - f(\xi)}{(1-s)^\alpha} ds, \tag{136}$$

if we change the variable to $s = \frac{(t-\xi)}{(x-\xi)}$. We observe from the definition of $\frac{d^\alpha f}{dx^\alpha}(\xi^+)$ that $F'_\xi(x)$ must exist in a small interval $(\ \xi, \xi+\delta\ ]$. Clearly, $F_\xi(x)$ is also continuous in $[\xi, \xi+\delta]$ if we define $F_\xi(\xi) = 0$. Thus by the mean value theorem, we have, on one hand that for each fixed $h \in (0, \delta)$, there is some $\rho_h \in (0, h)$, such that

$$F'_\xi(\xi + \rho_h) = \frac{F_\xi(\xi + h) - F_\xi(\xi)}{h}. \tag{137}$$

On the other hand, we have, by the equivalent definition of $F_\xi(x)$ above, that

$$\begin{aligned} &\tfrac{F_\xi(\xi+h) - F_\xi(\xi)}{h} = \tfrac{F_\xi(\xi+h)}{h} = \\ &= \tfrac{h^{1-\alpha}}{\Gamma(1-\alpha)} \int_0^1 \tfrac{f(\xi+(x-\xi)s) - f(\xi)}{h(1-s)^\alpha} ds = \\ &= \tfrac{1}{\Gamma(1-\alpha)} \int_0^1 (1-s)^{-\alpha} \tfrac{f(sh+\xi) - f(\xi)}{h^\alpha} ds \end{aligned} \tag{138}$$

Since by definition, the existence of $\frac{d^\alpha f}{dx^\alpha}(\xi^+)$ is equivalent to $\frac{1}{\Gamma(1+\alpha)} \lim_{x \downarrow \xi} F'_\xi(x)$, we have $\frac{1}{\Gamma(1+\alpha)} \lim_{h \downarrow 0} F'_\xi(\xi + \rho_h) = \frac{d^\alpha f}{dx^\alpha}(\xi^+)$. Therefore

$$\lim_{h \downarrow 0} \int_0^1 (1-t)^{-\alpha} \frac{f(ht+\xi) - f(\xi)}{h^\alpha} dt \tag{139}$$

exists and

$$\frac{d^\alpha f}{dx^\alpha}(\xi^+) = \frac{1}{\Gamma(1+\alpha)\Gamma(1-\alpha)} \lim_{h \downarrow 0} \int_0^1 (1-t)^{-\alpha} \frac{f(ht+\xi) - f(\xi)}{h^\alpha} dt \tag{140}$$

■

Stein and Zygmund [38] considered the α-fractional derivative in the sense of M. Riesz for functions defined on $\mathbb{R}$ and its variations including the Weyl fractional derivative.

Definition 84. *Let $0 < \alpha < 1$ and $\beta = 1 - \alpha$. Given a measurable function $f : \mathbb{R} \to \mathbb{R}$, the β-***th integral** *f_β for f is defined by*

$$f_\beta(x) := \int_{\mathbb{R}} \frac{f(y)}{|x-y|^{1-\beta}} dy = (f \star K_{1-\beta})(x), \tag{141}$$

*which is the convolution between f and $K_\gamma(x) = |x|^{-\gamma}$. The α-***fractional derivative** *of f at x, denoted by $f^{(\alpha)}(x)$, is defined by*

$$f^{(\alpha)}(x) := \frac{d}{dx} f_\beta(x). \tag{142}$$

To characterize the existence of such α-fractional derivatives requires the following definition [38].

Definition 85. *The function $f : \mathbb{R} \to \mathbb{R}$ is said to satisfy Λ_α $(0 < \alpha < 1)$ at x if*

$$R_x(t) := f(x+t) - f(x) = \mathcal{O}\left(|t|^\alpha\right) \qquad \text{as } t \to \infty. \tag{143}$$

On the other hand it satisfies the condition N_α^2 at x if

$$\int_{-\delta}^{\delta} \frac{[R_x(t)]^2}{|t|^{1+2\alpha}} dt < +\infty \qquad \text{for some } \delta > 0. \tag{144}$$

Stein and Zygmund proved the following result [38].

Proposition 86. (Stein, Zygmund) *Suppose $f \in L^1(\mathbb{R})$ and satisfies the condition Λ_α for each point x of a set $E \subset \mathbb{R}$ of positive measure. Then $f^{(\alpha)}(x)$ exists almost everywhere in E if and only if f satisfies N_α^2 almost everywhere in E.*

Remark 87. *All the results of [38] remain valid if one replaces f_β by* **Weyl's α-fractional derivative***:*

$$\frac{d}{dx} I_\beta(x), \qquad \text{where } I_\beta(x) := \int_\infty^x \frac{f(y)}{|x-y|^{1-\beta}} dy. \tag{145}$$

It is easy to see [20] that the KG local fractional derivative can be related to Weyl's fractional derivative as follows.

Remark 88. *Let f be locally α-Hölder continuous in $[a,b]$. For any fixed $\xi \in [a,b]$ we define*

$$f_\xi^+(x) := \begin{cases} f(x) - f(\xi), & \xi < x < b, \\ 0, & x \le \xi \text{ or } x \ge b, \end{cases} \tag{146}$$

and

$$f_\xi^-(x) := \begin{cases} f(x) - f(\xi), & a < x < \xi, \\ 0, & x \ge \xi \text{ or } x \le a. \end{cases} \tag{147}$$

Let

$$I_\beta^+ f_\xi^+(x) = \int_{-\infty}^x \frac{f_\xi^+(t)}{(x-t)^{1-\beta}} dt, \qquad I_\beta^- f_\xi^-(x) = \int_x^\infty \frac{f_\xi^-(t)}{(t-x)^{1-\beta}} dt, \tag{148}$$

and as usual $\beta = 1 - \alpha$. Then the right $(\sigma = +)$ and left $(\sigma = -)$ KG local fractional derivatives at $\xi \in [a,b]$ are given by

$$\frac{d^\alpha f}{dx^\alpha} f(\xi^\sigma) = \frac{1}{\Gamma(1+\alpha)\Gamma(1-\alpha)} \lim_{x \to \xi^\sigma} \frac{d}{dx} I_\beta^\sigma f_\xi^\sigma(x). \tag{149}$$

We then have the following remarkable theorem [7].

Theorem 89. (Chen, Yan, Zhang) *Let $f : [a,b] \to \mathbb{R}$ be locally α-Hölder continuous in $[a,b]$ for $0 < \alpha < 1$. Moreover, suppose that the KG local fractional derivatives $\frac{d^\alpha f}{dx^\alpha} f(\xi^\pm)$ exist for almost all $\xi \in (a,b)$. Then $\frac{d^\alpha f}{dx^\alpha} f(\xi^+) = \frac{d^\alpha f}{dx^\alpha} f(\xi^-) = 0$ for almost all $\xi \in (a,b)$.*

Proof. From Remark 88. we realize that if $\frac{d^\alpha f}{dx^\alpha} f(\xi^\pm)$ exist at some $\xi \in (a,b)$, there is a neighborhood $\mathcal{V}_\xi(\tau) = (\xi - \tau, \xi + \tau) \subset (a,b)$ with $\tau > 0$, such that

$$\frac{d}{dx} I_\beta^+ f_\xi^+(x) \text{ exists for } x \in \mathcal{V}_\xi^+(\tau), \qquad \frac{d}{dx} I_\beta^- f_\xi^-(x) \text{ exists for } x \in \mathcal{V}_\xi^-(\tau), \tag{150}$$

and $\beta = 1 - \alpha$.

Given that, by assumption, f is locally α-Hölder continuous, we then conclude from Proposition 86. that f_ξ^+ and f_ξ^- satisfy condition N_α^2 a.e. in $\mathcal{V}_\xi^+(\tau)$ and $\mathcal{V}_\xi^-(\tau)$, respectively. Note that the function $R_x(t)$ (143) when applied to f_ξ^+ reads: $R_x(t) = f_\xi^+(t+x) - f_\xi^+(x) = f(t+x) - f(\xi) - f(x) + f(\xi) = f(t+x) - f(x)$ if $x \in \mathcal{V}_\xi^+(\tau)$ and sufficiently small t. A similar result holds for f_ξ^-. This means that $R_x(t)$ is independent of $f(\xi)$ for $x \in \mathcal{V}_\xi(\tau)$. Thus $R_x(t) = f(t+x) - f(x)$, and hence N_α^2 holds for f at almost everywhere in $\mathcal{V}_\xi(\tau)$.

For those $x \in [a,b]$ for which N_α^2 holds, there exists a sufficiently small $\delta > 0$ such that $(x - \delta, x + \delta) \subset (x - \tau, x + \tau)$ and $\int_{-\delta}^{\delta} \frac{[R_x(t)]^2}{|t|^{2\alpha+1}} dt < +\infty$. By Vitali's equi-integrability theorem for Lebesgue integrals [24], we have

$$\lim_{h \downarrow 0} \int_0^h \frac{[R_x(t)]^2}{|t|^{2\alpha+1}} dt = 0. \tag{151}$$

Since $\frac{d^\alpha f}{dx^\alpha}(x^+)$ exists, we have by Lemma 83. that

$$\frac{d^\alpha f}{dx^\alpha}(x^+) = \frac{1}{\Gamma(1+\alpha)\Gamma(1-\alpha)} \lim_{h \downarrow 0} \int_0^1 (1-t)^{-\alpha} \frac{f(ht+x) - f(x)}{h^\alpha} dt. \tag{152}$$

We will now prove that

$$\lim_{h \downarrow 0} \int_0^1 (1-t)^{-\alpha} \left| \frac{f(ht+x) - f(x)}{h^\alpha} \right| dt = 0, \tag{153}$$

which completes the proof.

Since f is locally α-Hölder continuous in $[a,b]$, there is constant $M > 0$ such that $|f(x) - f(y)| \le M|x-y|^\alpha$ for y sufficiently close to x. For $0 < \epsilon < 1$, we have:

$$\begin{aligned} &\int_0^1 (1-t)^{-\alpha} \left| \tfrac{f(ht+x)-f(x)}{h^\alpha} \right| dt = \\ &= \int_{1-\epsilon}^1 (1-t)^{-\alpha} \left| \tfrac{f(ht+x)-f(x)}{h^\alpha} \right| dt + \int_0^{1-\epsilon} (1-t)^{-\alpha} \left| \tfrac{f(ht+x)-f(x)}{h^\alpha} \right| dt \\ &:= I_1 + I_2. \end{aligned} \tag{154}$$

We then have for sufficiently small ϵ

$$\begin{aligned} I_1 = \int_{1-\epsilon}^1 (1-t)^{-\alpha} \left| \tfrac{f(ht+x)-f(x)}{h^\alpha} \right| dt \le M \int_{1-\epsilon}^1 (1-t)^{-\alpha} t^\alpha dt \le \\ \le M \int_{1-\epsilon}^1 (1-t)^{-\alpha} dt = \tfrac{M}{1-\alpha} \epsilon^{1-\alpha}. \end{aligned} \tag{155}$$

On the other hand, by applying the Cauchy-Schwarz inequality we obtain:

$$I_2 = \int_0^{1-\epsilon}(1-t)^{-\alpha}\left|\frac{f(ht+x)-f(x)}{h^\alpha}\right| dt \le$$

$$\le \frac{1}{\epsilon^\alpha}\int_0^{1-\epsilon}\left|\frac{f(ht+x)-f(x)}{h^\alpha}\right| dt = \frac{1}{\epsilon^\alpha}\int_0^{(1-\epsilon)h}\left|\frac{f(s+x)-f(x)}{h^\alpha}\right| \frac{ds}{h} \le$$

$$\le \frac{1}{\epsilon^\alpha h^{1+\alpha}}\int_0^h s^{\alpha+1/2}\frac{|f(s+x)-f(x)|}{s^{\alpha+1/2}}ds \le \tag{156}$$

$$\le \frac{1}{\epsilon^\alpha h^{1+\alpha}}\left(\int_0^h s^{2\alpha+1}ds\right)^{\frac{1}{2}}\left(\int_0^h \frac{|f(s+x)-f(x)|^2}{s^{2\alpha+1}}ds\right)^{\frac{1}{2}} =$$

$$= \frac{1}{\epsilon^\alpha\sqrt{2\alpha+2}}\left(\int_0^h \frac{|f(s+x)-f(x)|^2}{s^{2\alpha+1}}ds\right)^{\frac{1}{2}}.$$

Since N_α^2 holds for f at x, we have from (155) and (156)

$$0 \le \lim_{h\downarrow 0}\int_0^1 (1-t)^{-\alpha}\left|\frac{f(ht+x)-f(x)}{h^\alpha}\right| dt \le \frac{M}{1-\alpha}\epsilon^{1-\alpha} \tag{157}$$

for any sufficiently small $\epsilon > 0$, which proves that (153) holds. ■

From Lemma 83. and Theorem 89., we realize that dividing the difference $f(x+h) - f(x)$ by the powers h^α may not be sufficient to "resolve" the details of certain types of Hölder regularity. The fact that local fractional derivatives vanish almost everywhere, may be an indication that a refinement may be required. Just as in the definition of Hausdorff dimension, dimension functions other than the powers may play an important role for certain sets, it may be that difference-quotients of the form $\frac{f(x+h)-f(x)}{\psi(h)}$ may be more adequate for certain types of regularity.

Let $\mathcal{F}$ be a scale of dimension functions and define $\mathcal{Q}$ to be the subset of functions ϕ of $\mathcal{F}/\sim$ such that:

$$\phi \preceq x \tag{158}$$

Definition 90. *A function $f : I \to \mathbb{R}$ is* **right** $(\sigma = +)$ **or left** $(\sigma = -)$ **differentiable** *to order $\psi \in \mathcal{Q}$ at $\xi \in I$, if there exists a real number λ_σ such that for $x \in \mathcal{V}_\xi^\sigma(\epsilon_\psi) \cap I$:*

$$f(x) = f(\xi) + \sigma\lambda_\sigma\psi(|x-\xi|) + R_\sigma(x,\xi), \tag{159}$$

with

$$\lim_{x\to\xi^\sigma}\frac{R_\sigma(x,\xi)}{\psi(|x-\xi|)} = 0 \tag{160}$$

We denote by $\mathcal{C}_+^\psi(I)$(resp. $\mathcal{C}_-^\psi(I)$) the set of right (resp. left) differentiable functions to order ψ in $[\,a,b\,)$ (resp. $(\,a,b\,]$). Finally, we say that f is **ψ-differentiable** *if it is both right and left ψ-differentiable. In this case, we write $\mathcal{C}^\psi(I) = \mathcal{C}_+^\psi(I)\cap\mathcal{C}_-^\psi(I)$.*

Definition 91. *From the Taylor formula (159,160) it follows immediately that*

$$\lambda_\sigma = \lim_{x\to\xi^\sigma}\frac{\sigma(f(x)-f(\xi))}{\psi(|x-\xi|)} = \lim_{h\downarrow 0}\frac{\sigma(f(\xi+\sigma h)-f(\xi))}{\psi(h)}, \qquad \sigma = \pm \tag{161}$$

We denote these limits by $\left(D_t^{\psi}f\right)(\xi^{\sigma})$ *and call them the* **right** $(\sigma=+)$ **and left** $(\sigma=-)$ ψ**-derivatives** *or* ψ**-velocities** *of* f *at* ξ*.*

Notice that for $\psi(x)=x$ we recover the ordinary right-sided and left-sided derivatives and for $\psi(x)=x^r$ $(0<r<1)$ we obtain the local fractional derivatives of Ben Adda and Cresson.

Remark 92. *A comment is now in order concerning the dependence of* ψ*-differentiability on the representative of the equivalence class* $[\psi]\in\mathcal{Q}$*. Take* $\psi_1,\psi_2\in[\psi]$*. Then there exist constants* C_1,C_2 *such that* $0<C_1\le\frac{\psi_2(x)}{\psi_1(x)}\le C_2$ *for* $0<x\le\epsilon=\min\{\epsilon_{\psi_1},\epsilon_{\psi_2}\}$*. Then we have for* $0<|x-\xi|<\epsilon$*:*

$$\frac{\sigma\left(f(x)-f(\xi)\right)}{\psi_1\left(|x-\xi|\right)}=\frac{\sigma\left(f(x)-f(\xi)\right)}{\psi_2\left(|x-\xi|\right)}\frac{\psi_2\left(|x-\xi|\right)}{\psi_1\left(|x-\xi|\right)} \tag{162}$$

We conclude that:

(i) if $\left(D_x^{\psi_2}f\right)(\xi^{\sigma})\ge 0$*, then*
$0\le C_1\left(D_x^{\psi_2}f\right)(\xi^{\sigma})\le\left(D_x^{\psi_1}f\right)(\xi^{\sigma})\le C_2\left(D_x^{\psi_2}f\right)(\xi^{\sigma})$*;*

(ii) if $\left(D_x^{\psi_2}f\right)(\xi^{\sigma})<0$*,*
then $0\le C_2\left(D_x^{\psi_2}f\right)(\xi^{\sigma})\ge\left(D_x^{\psi_1}f\right)(\xi^{\sigma})\ge C_1\left(D_x^{\psi_2}f\right)(\xi^{\sigma})$*.*

For all practical purposes, we can extract exactly the same information from the two derivatives $\left(D_x^{\psi_1}f\right)(\xi^{\sigma})$ *and* $\left(D_x^{\psi_2}f\right)(\xi^{\sigma})$*. Indeed,* f *is* ψ_1*-differentiable if and only if it is* ψ_2*-differentiable for all* $\psi_2\sim\psi_1$*. Moreover, the sign of the derivatives is the same (which means that we will be able to draw the same conclusions concerning the monotonicity of the function* f *and the existence of extrema). We thus fix once and for all a representative of each equivalence class in* $\mathcal{Q}$*.*

In this work we shall focus on ψ-diferentiability for $\psi\prec x$. We may nevertheless envisage the possibility of higher order differentiability:

Definition 93. *Let* $f:I\to\mathbb{R}$ *be of class* C^n $(n\ge 1)$ *at* $\xi\in I$*, but not* $(n+1)$*-times differentiable at* ξ*. Moreover, let* $\psi\in\mathcal{F}$ *be such that* $x^n\prec\psi\prec x^{n+1}$*. We say that* f *is* **right** $(\sigma=+)$ **or left** $(\sigma=-)$ ψ**-differentiable** *at* ξ *if there exists a real number* λ_{σ} *such that for any* $x\in\mathcal{V}_{\xi}^{\sigma}(\epsilon_{\psi})\cap I$*:*

$$f(x)=\sum_{k=0}^{n}\frac{f^{(k)}(\xi)}{k!}(x-\xi)^k+\sigma\lambda_{\sigma}\psi\left(|x-\xi|\right)+R_{\sigma}(x,\xi), \tag{163}$$

with

$$\lim_{x\to\xi^{\sigma}}\frac{R_{\sigma}(x,\xi)}{\psi\left(|x-\xi|\right)}=0. \tag{164}$$

The numbers λ_{σ} *are the* ψ*-derivatives of* f *at* ξ*:*

$$\lambda_{\sigma}=\left(D_x^{\psi}f\right)(\xi^{\sigma}):=\lim_{x\to\xi^{\sigma}}\frac{\sigma\left(f(x)-\sum_{k=0}^{n}\frac{f^{(k)}(\xi)}{k!}(x-\xi)^k\right)}{\psi\left(|x-\xi|\right)},\qquad\sigma=\pm. \tag{165}$$

4.2. ψ-differential Calculus

In this section we explore some of the properties of local fractional derivatives and the main theorems of ψ-differential calculus. Some of them were derived in various papers [4, 9, 10, 11] for differentiable functions to order $\alpha \in (0,1)$. We include them here for completeness and with the additional generalization to functions which are differentiable to order $\psi \notin \mathcal{P}$.

As in ordinary calculus, right (resp. left) ψ-differentiability entails right (resp. left) continuity.

Proposition 94. *Let $f \in C^0(I)$, $\psi \in \mathcal{Q}$ and $\xi \in I$.*
(i) If f is right (resp. left) ψ-differentiable at ξ, then f is right (resp. left) continuous at ξ.
(ii) If f is ψ-differentiable at ξ, then f is continuous at ξ.

Proof. Let $\rho_\sigma(x) := \frac{\sigma(f(x)-f(\xi))}{\psi(|x-\xi|)}$, for $x \in \left(\mathcal{V}_\xi^\sigma(\epsilon_\psi)\backslash\{\xi\}\right) \cap I$. If f is right ($\sigma = +$) or left ($\sigma = -$) ψ-differentiable at ξ, then the limit $\lim_{x\to\xi^\sigma} \rho_\sigma(x) = \left(D_x^\psi f\right)(\xi^\sigma)$ exists. We thus have:

$$\lim_{x\to\xi^\sigma} \sigma\left(f(x) - f(\xi)\right) = \lim_{x\to\xi^\sigma} \rho_\sigma(x) \cdot \psi\left(|x-\xi|\right) = \left(D_x^\psi f\right)(\xi^\sigma) \cdot 0 = 0,$$

which means that f is right ($\sigma = +$) or left ($\sigma = -$) continuous at ξ. This proves statement (i). If f is ψ-differentiable at ξ, then from (i) f is right and left continuous at ξ, which proves (ii). ■

The following result reveals that ψ-derivatives are true derivatives, in the sense that they are linear operators which obey the Leibniz rule. Moreover, they satisfy the standard quotient rule. We leave the simple proofs to the reader.

Proposition 95. *Let $f, g \in C^0(I)$ be right ($\sigma = +$) (resp. left ($\sigma = -$)) ψ-differentiable at $\xi \in I$, with $\psi \in \mathcal{Q}$, and $K \in \mathbb{R}$. Then, $f+g$, Kf and $f \cdot g$ are right ($\sigma = +$) (resp. left ($\sigma = -$)) ψ-differentiable at $\xi \in I$ and we have:*

$$\begin{aligned} \left(D_x^\psi(f+g)\right)(\xi^\sigma) &= \left(D_x^\psi f\right)(\xi^\sigma) + \left(D_x^\psi g\right)(\xi^\sigma) \\ \left(D_x^\psi(Kf)\right)(\xi^\sigma) &= K\left(D_x^\psi f\right)(\xi^\sigma) \\ \left(D_x^\psi(f\cdot g)\right)(\xi^\sigma) &= \left(D_x^\psi f\right)(\xi^\sigma)\cdot g(\xi) + f(\xi)\cdot\left(D_x^\psi g\right)(\xi^\sigma) \end{aligned} \tag{166}$$

Proposition 96. *Let $f, g \in C^0(I)$ be right ($\sigma = +$) (resp. left ($\sigma = -$)) ψ-differentiable at $\xi \in I$, with $\psi \in \mathcal{Q}$ and $g(\xi) \neq 0$. Then $\frac{f}{g}$ is right (resp. left) ψ-differentiable at ξ, and we have:*

$$\left(D_x^\psi\left(\frac{f}{g}\right)\right)(\xi^\sigma) = \frac{\left(D_x^\psi f\right)(\xi^\sigma)\cdot g(\xi) - f(\xi)\cdot\left(D_x^\psi f\right)(\xi^\sigma)}{[g(\xi)]^2}. \tag{167}$$

In the general case a chain rule is very difficult to derive. There is a particular case where this is possible.

Proposition 97. *Let g and f be continuous functions on $[a,b]$ and $[g(a),g(b)]$, respectively. Suppose that $\left(D_x^{\phi} g\right)(\xi^\sigma)$ and $(D_x^r f)\left(g(\xi)^{s^\sigma}\right)$ exist for $\xi \in (a,b)$, $\phi \in \mathcal{Q}$, $0 < r \leq 1$ and $s^\sigma = sign\left(\left(D_x^{\phi} g\right)(\xi^\sigma)\right)$. Then, $f \circ g$ is σ differentiable to order $t^r \circ \phi = \phi^r$ at $g(\xi)$, and we have:*

$$\left(D_x^{\phi^r}(f \circ g)\right)(\xi^\sigma) = s^\sigma \left[(D_x^r f)\left(g(\xi)^{\sigma s^\sigma}\right)\right] \left|\left(D_x^{\phi} g\right)(\xi^\sigma)\right|^r \tag{168}$$

Proof. Notice that

$$\begin{aligned}\frac{\sigma(f(g(x))-f(g(\xi)))}{\phi^r(|x-\xi|)} &= \\ &= s^\sigma \frac{\sigma s^\sigma(f(g(x))-f(g(\xi)))}{|g(x)-g(\xi)|^r} \left|\frac{g(x)-g(\xi)}{\phi(|x-\xi|)}\right|^r.\end{aligned} \tag{169}$$

Since $\left(D_x^{\phi} g\right)(\xi^\sigma)$ exists, we have from from (159,160):

$$g(x) = g(\xi) + \sigma s^\sigma \left|\left(D_x^{\phi} g\right)(\xi^\sigma)\right| \phi(|x-\xi|) + R_\sigma(x,\xi),$$

with $\lim_{x\to\xi^\sigma} \frac{R_\sigma(x,\xi)}{\phi(|x-\xi|)} = 0$. We conclude that $g(x) \to (g(\xi))^{\sigma s^\sigma}$ as $x \to \xi^\sigma$. Taking the limit $x \to \xi^\sigma$ in (169), we recover (168). ■

There are some particular cases of interest:

Corollary 98. *Let g and f be continuous in $[a,b]$ and $[g(a),g(b)]$, respectively. Let $\xi \in (a,b)$.*

(i) If g is σ-differentiable to order $\phi \in \mathcal{Q}$ at ξ and f is differentiable at $g(\xi)$, then $f \circ g$ is σ-differentiable to order ϕ at ξ and:

$$\left(D_x^{\phi}(f \circ g)\right)(\xi^\sigma) = f'(g(\xi))\left(D_x^{\phi} g\right)(\xi^\sigma). \tag{170}$$

(ii) If g is differentiable at ξ with $s = sign(g'(\xi))$, and $(D_x^r f)(g(\xi)^{\sigma s})$ exists, for $0 < r < 1$, then $f \circ g$ is σ-differentiable to order x^r at ξ and:

$$\left(D_x^r(f \circ g)\right)(\xi^\sigma) = s\left[(D_x^r f)(g(\xi)^{\sigma s})\right]|g'(\xi)|^r. \tag{171}$$

(iii) It g is the scale transformation $g(x) = \lambda x$ $(\lambda > 0)$ and $(D_x^r f)(g(\xi)^\sigma)$ exists, then:

$$\left(D_x^r(f \circ g)\right)(\xi^\sigma) = \lambda^r (D_x^r f)(\lambda\xi^\sigma). \tag{172}$$

Proof. The proof follows immediately from Proposition 97. ■

Let us now analyze the structure of the ψ-derivatives of a function as ψ runs through $\mathcal{Q}$.

Theorem 99. *Let $f \in C^0(I)$ and $\psi \in \mathcal{Q}$. If f is right $(\sigma = +)$ or left $(\sigma = -)$ ψ-differentiable at $\xi \in I$, then*

$$\left(D_x^{\phi} f\right)(\xi^\sigma) = 0, \qquad \text{for all } \phi \in \mathcal{Q} \text{ such that } \phi \prec \psi. \tag{173}$$

Proof. Let f be right or left ψ-differentiable at ξ and let $\phi \in \mathcal{Q}$ be such that $\phi \prec \psi$. We then have:

$$\left(D_x^{\phi} f\right)(\xi^{\sigma}) = \lim_{x \to \xi^{\sigma}} \frac{\sigma(f(x)-f(\xi))}{\phi(|x-\xi|)} = \lim_{x \to \xi^{\sigma}} \frac{\sigma(f(x)-f(\xi))}{\psi(|x-\xi|)} \times$$

$$\times \lim_{x \to \xi^{\sigma}} \frac{\psi(|x-\xi|)}{\phi(|x-\xi|)} = \left(D_x^{\psi} f\right)(\xi^{\sigma}) \times \lim_{x \to \xi^{\sigma}} \frac{\psi(|x-\xi|)}{\phi(|x-\xi|)} = 0.$$

■

In view of this theorem, as $\mathcal{Q} \subset \mathcal{F}$ with $\mathcal{F}$ a scale (which is bounded complete), the set $\left\{\phi \in \mathcal{Q} : \left(D_x^{\phi} f\right)(\xi^{\sigma}) = 0\right\}$ regarded as a subset of the chain $\mathcal{Q}$ has a supremum.

Definition 100. *The* **right** $(\sigma = +)$ **or left** $(\sigma = -)$ **critical order of differentiability** *of a function $f \in C^0(I)$ at a point $\xi \in I$ is given by:*

$$\psi_{\xi,c}^{\sigma} := sup_{\mathcal{F}} \left\{\phi \in \mathcal{Q} : \left(D_x^{\phi} f\right)(\xi^{\sigma}) = 0\right\} \tag{174}$$

If the critical order is the same throughout the interval I, then we may simply write ψ_c^{σ}.

A simple consequence of this definition and of Theorem 99. is:

Proposition 101. *Let $f \in C^0(I)$ have critical order of differentiability $\psi_{\xi,c}^{\sigma}$ at $\xi \in I$ with $\psi_{\xi,c}^{\sigma} \prec x$. Then $\left(D_x^{\phi} f\right)(\xi^{\sigma})$ does not exist for $\phi \in \mathcal{Q}$ with $\psi_{\xi,c}^{\sigma} \prec \phi$.*

Proof. Suppose that $\left(D_x^{\phi} f\right)(\xi^{\sigma})$ exists for $\psi_{\xi,c}^{\sigma} \prec \phi$. Then, from Theorem 99., we would have $(D_x^{\rho} f)(\xi^{\sigma}) = 0$ for any $\rho \in \mathcal{Q}$ such that $\psi_{\xi,c}^{\sigma} \prec \rho \prec \phi$. Note that such a ρ exists because $\mathcal{F}$ is a scale (and hence dense). However, this is contradictory with the assumption of $\psi_{\xi,c}^{\sigma}$ being the critical order of differentiability. ■

This behavior is easily illustrated with the following example.

Example 102. *Consider the function $f(x) = |x|^r$, $0 < r < 1$. From (161), we easily obtain for $\psi \in \mathcal{Q}$:*

$$\left(D_x^{\psi} f\right)(0^{\sigma}) = \begin{cases} 0, & \text{if } \psi \prec x^r \\ \sigma, & \text{if } \psi = x^r \\ \infty, & \text{if } x^r \prec \psi \preceq x \end{cases}, \quad \text{for } \sigma = \pm$$

Also for the derivatives of order r, we have:

$$\left(D_x^{r} f\right)(x^{\sigma}) = \begin{cases} \sigma, & \text{if } x = 0 \\ 0, & \text{if } x \neq 0 \end{cases}, \quad \text{for } \sigma = \pm$$

The last equation reveals that, contrary to ordinary derivatives, local fractional derivatives do not obey Darboux's theorem.

Proposition 103. *Let $f \in C^0(I)$ be differentiable at $\xi \in (a,b)$ to order $\psi \in \mathcal{Q}$. Then*

(i) $r\left(D_x^{\psi} f\right)(\xi^-) < 0 \Rightarrow \exists \epsilon > 0$, *such that* $\forall x \in (\xi - \epsilon, \xi)$, *we have* $r(f(x) - f(\xi)) > 0$, $r = \pm$.

(ii) $r\left(D_x^{\psi} f\right)(\xi^+) < 0 \Rightarrow \exists \epsilon > 0$, *such that* $\forall x \in (\xi, \xi + \epsilon)$, *we have* $r(f(x) - f(\xi)) < 0$, $r = \pm$.

Proof. From (159,160), we have, for sufficiently small $\epsilon > 0$, and for $x \in \mathcal{V}_\xi^\sigma(\epsilon) \backslash \{\xi\}$:

$$f(x) - f(\xi) = \sigma s^\sigma \left| \left(D_x^{\psi} f\right)(\xi^\sigma) \right| \psi(|x - \xi|) + R_\sigma(x, \xi),$$

with $s^\sigma = \text{sign}\left(\left(D_x^{\psi} f\right)(\xi^\sigma)\right)$, and $\frac{R_\sigma(x,\xi)}{\psi(x-\xi)} \to 0$ as $x \to \xi^\sigma$.

If $\left(D_x^{\psi} f\right)(\xi^\sigma) \neq 0$, then the sign of $\sigma(f(x) - f(\xi))$ is s^σ (for sufficiently small ϵ). ■

The converse result of this theorem is as follows:

Proposition 104. *Let $f \in C^0(I)$ be differentiable at $\xi \in (a,b)$ to order $\psi \in \mathcal{Q}$. Then*

(i) $\exists \epsilon > 0$, *such that* $\forall x \in (\xi - \epsilon, \xi)$, *we have* $r(f(x) - f(\xi)) > 0$, $r = \pm \Rightarrow r\left(D_x^{\psi} f\right)(\xi^-) \leq 0$.

(ii) $\exists \epsilon > 0$, *such that* $\forall x \in (\xi, \xi + \epsilon)$, *we have* $r(f(x) - f(\xi)) < 0$, $r = \pm \Rightarrow r\left(D_x^{\psi} f\right)(\xi^+) \leq 0$.

Proof. The proof follows the same *rationale* as the previous one. However, we cannot discard the possibility that $\left(D_x^{\psi} f\right)(\xi^\sigma)$ may vanish. This is certainly what happens if ψ is below the critical order of differentiability. And even for critical orders a vanishing derivative is not precluded. ■

An immediate consequence of this concerns the local extrema of functions.

Proposition 105. *Let $f \in C^0(I)$ be differentiable at $\xi \in (a,b)$ to order $\psi \in \mathcal{Q}$.*

(i) If ξ is a local maximum, then $\sigma\left(D_x^{\psi} f\right)(\xi^\sigma) \leq 0$, $\sigma = \pm$.

(ii) If ξ is a local minimum, then $\sigma\left(D_x^{\psi} f\right)(\xi^\sigma) \geq 0$, $\sigma = \pm$.

Proof. (i) Suppose that ξ is a local maximum. Then there exists $\epsilon > 0$ such that $f(x) - f(\xi) \leq 0$ for all $x \in \mathcal{V}_\xi(\epsilon)$. For $x \in (\xi - \epsilon, \xi)$, we have $\frac{(f(\xi)-f(x))}{\psi(|x-\xi|)} \geq 0$. Taking the limit $x \to \xi^-$, we obtain $\left(D_x^{\psi} f\right)(\xi^-) \geq 0$. Likewise, for $x \in (\xi, \xi + \epsilon)$, we have $\frac{(f(x)-f(\xi))}{\psi(|x-\xi|)} \leq 0$. If we take the limit $x \to \xi^+$, we conclude that $\left(D_x^{\psi} f\right)(\xi^+) \leq 0$.
(ii) This is proved in a similar way. ■

We leave the proof of the converse result to the reader:

Proposition 106. *Let $f \in C^0(I)$ be differentiable at $\xi \in (a,b)$ to order $\psi \in \mathcal{Q}$.*

(i) If $\sigma \left(D_x^{\psi} f\right)(\xi^{\sigma}) < 0,\ \sigma = \pm$, *then ξ is a local maximum.*

(ii) If $\sigma \left(D_x^{\psi} f\right)(\xi^{\sigma}) > 0,\ \sigma = \pm$, *then ξ is a local minimum.*

We are now in a position to generalize some of the fundamental theorems of differential calculus.

Theorem 107. (Rolle) *Let $f \in C^0(I)$ be differentiable in (a,b) to order $\psi \in \mathcal{Q}$. Moreover suppose that $f(a) = f(b)$. Then, there exists $\xi \in (a,b)$ such that:*

$$\left(D_x^{\psi} f\right)(\xi^+) \cdot \left(D_x^{\psi} f\right)(\xi^-) \leq 0. \tag{175}$$

Proof. Since f is continuous in I, by Weierstrass' Theorem, there exist, at least, two points $u, v \in I$, such that $f(u) \leq f(x) \leq f(v),\ \forall x \in I$. That is, $f(u),\ f(v)$ are a global minimum and maximum of f in I, respectively. If u and v coincide with the a and b, then from $f(a) = f(b)$, we conclude that f is constant in I, and thus $\left(D_x^{\psi} f\right)(\xi^{\sigma}) = 0$, for any $\xi \in (a,b)$. If, u or $v \in (a,b)$, then from Proposition 105., (175) follows immediately. ■

There is an alternative form of Rolle's theorem, which will be useful in the sequel. Its main advantage is the fact that we need to require only left or only right ψ-differentiability.

Theorem 108. *Let $f \in C^0(I)$ with $f(a) = f(b)$ and let $\psi \in \mathcal{Q}$.*
(i) If f is right ψ-differentiable at a and $\left(D_x^{\psi} f\right)(a^+) > 0$ (resp. $\left(D_x^{\psi} f\right)(a^+) < 0$), then there exists a local maximum (resp. minimum) of f in (a,b).
(ii) If f is left ψ-differentiable at b and $\left(D_x^{\psi} f\right)(b^-) > 0$ (resp. $\left(D_x^{\psi} f\right)(b^-) < 0$), then there exists a local minimum (resp. maximum) of f in (a,b).

Proof. Since f is continuous in $I = [a,b]$, there exist $u, v \in I$, such that $f(u) \leq f(x) \leq f(v),\ \forall x \in I$.
(i) Suppose that $\left(D_x^{\psi} f\right)(a^+) > 0$. If $v = a$ or $v = b$, then from Proposition 105. (i), we would have a contradiction. And thus $v \in (a,b)$. Conversely, let us assume that $\left(D_t^{\psi} f\right)(a^+) < 0$. If $u = a$ or $u = b$, then from Proposition 105. (ii), we would again have a contradiction. And thus $u \in (a,b)$.
(ii) This is proved in a similar way. ■

An immediate consequence of this is the following.

Theorem 109. *Let $\psi \in \mathcal{Q}$ with $\psi \prec x$, and let $f \in C_+^{\psi}(I)$ (resp. $f \in C_-^{\psi}(I)$) be such that $\left(D_x^{\psi} f\right)(x^+) \neq 0,\ \forall x \in [\,a,b\,)$ (resp. $\left(D_x^{\psi} f\right)(x^-) \neq 0,\ \forall x \in (\,a,b\,]$).*

(i) The sets of points $F_r^{(+)} = \left\{x \in [\, a, b\,) : sign\left(\left(D_x^{\psi} f\right)(x^+)\right) = r\right\}$, *with* $r = \pm$ *are both dense in* I.
(ii) The sets of points $F_r^{(-)} = \left\{x \in (\, a, b\,] \; : \; sign\left(\left(D_x^{\psi} f\right)(x^-)\right) = r\right\}$, *with* $r = \pm$ *are both dense in* I.

Proof. We start by proving that if the sign of the ψ-derivative remains constant throughout a non-degenerate interval, then the function is strictly monotone in that interval. Suppose that $\left(D_x^{\psi} f\right)(x^+) > 0, \forall x \in [\, a, b\,)$, but f is not a strictly increasing function in I. Then, by continuity, it is always possible to find two points x_1, x_2, such that $a \leq x_1 < x_2 \leq b$ and $f(x_1) = f(x_2)$. Consequently, in the interval $[x_1, x_2]$, f satisfies the conditions of Theorem 108. with $\left(D_x^{\psi} f\right)(x_1^+) > 0$. It then follows that there exists a local maximum at some $v \in (x_1, x_2)$. However, from Proposition 105. (i), we must have $\left(D_x^{\psi} f\right)(v^+) \leq 0$, which is a contradiction. And so, f is strictly increasing. Conversely, suppose that $\left(D_x^{\psi} f\right)(x^+) < 0, \forall x \in [\, a, b\,)$, but that f is not strictly decreasing in I. Again, there exist $a \leq x_3 < x_4 \leq b$ such that $f(x_3) = f(x_4)$ and $\left(D_x^{\psi} f\right)(x_3^+) < 0$. By Theorem 108., there exists a local minimum at some $u \in (x_3, x_4)$. From Proposition 105. (ii), $\left(D_x^{\psi} f\right)(u^+) \geq 0$, which is a contradiction. And thus, f is strictly decreasing. The proof for $\left(D_x^{\psi} f\right)(x^-)$ is analogous.

Now, suppose that there exists some non-degenerate interval $J \subset I$, such that sign$\left(\left(D_x^{\psi} f\right)(x^{\sigma})\right)$ is constant and non-zero in J. From the previous discussion, we conclude that f is strictly monotone in J. But by Lebesgue's Theorem, f must differentiable almost everywhere in J, which contradicts the hypothesis $\psi \prec x$. This means that, given $x \in I$ and an arbitrary neighborhood of x, we can find points where $\left(D_x^{\psi} f\right)(x^{\sigma})$ is positive and points where it is negative. ■

This leads to the fact that non-vanishing derivatives of order $\psi \prec x$ cannot be continuous.

Corollary 110. *Let* $\psi \in \mathcal{Q}$ *with* $\psi \prec x$, *and let* $f \in C_+^{\psi}(I)$ *(resp.* $f \in C_-^{\psi}(I)$*). If the local fractional derivative* $\left(D_x^{\psi} f\right)(x^{\sigma})$ *is continuous at* $\xi \in I$, *then* $\left(D_x^{\psi} f\right)(\xi^{\sigma}) = 0$.

Proof. Suppose that $\left(D_x^{\psi} f\right)(x^{\sigma})$ is continuous and non-zero at ξ. Then, by continuity, there exists $\epsilon > 0$, such that $sign\left(\left(D_x^{\psi} f\right)(x^{\sigma})\right)$ remains constant and non-zero in $\mathcal{V}_{\xi}(\epsilon) \cap I$. However, this contradicts Theorem 109.. And thus $\left(D_x^{\psi} f\right)(\xi^{\sigma}) = 0$. ■

From these results, we can derive the Theorem 113. below concerning the fractal (Hausdorff) dimension of positive and negative parts of the graph of a local ψ-derivative function. But first we need the following proposition and theorem. Here $\dim_H(F)$ will denote the Hausdorff dimension of set F and L_{θ} is the line through the origin of $\mathbb{R}^2$ that makes an angle θ with the horizontal axis. Finally, we denote the orthogonal projection onto L_{θ} by proj_{θ}.

Proposition 111. *A set $F \subset \mathbb{R}^n$ $(n \geq 1)$ with $dim_H(F) < 1$ is totally disconnected.*

Proof. The proof of this proposition can be found in [12]. ■

Theorem 112. (Projection Theorem) *Let $F \subset \mathbb{R}^2$ be a Borel set.*
(i) If $dim_H(F) \leq 1$ then $dim_H(proj_\theta F) = dim_H(F)$ for almost all $\theta \in [\,0,\pi\,)$.
(ii) If $dim_H(F) > 1$ then $proj_\theta F$ has positive length (as a subset of L_θ) and so has dimension 1 for almost all $\theta \in [\,0,\pi\,)$.

Proof. The proof of this theorem can be found in [12]. ■

Theorem 113. *Let $\psi \in \mathcal{Q}$ with $\psi \prec x$, and let $f \in C_+^{\psi}(I)$ (resp. $f \in C_-^{\psi}(I)$) be such that $\left(D_x^{\psi} f\right)(x^+) \neq 0$, $\forall x \in [\,a,b\,)$ (resp. $\left(D_x^{\psi} f\right)(x^-) \neq 0$, $\forall x \in (\,a,b\,]$). Then the positive $\Gamma_+^{\sigma} := \left\{\left(x, \left(D_x^{\psi} f\right)(x^{\sigma})\right) : x \in F_+^{(\sigma)}\right\}$ part of the graph of $\left(D_x^{\psi} f\right)(x^{\sigma})$ and the negative part*
$\Gamma_-^{\sigma} := \left\{\left(x, \left(D_x^{\psi} f\right)(x^{\sigma})\right) : x \in F_-^{(\sigma)}\right\}$ both have Hausdorff dimension greater or equal to 1.

Proof. Let $\left(D_x^{\psi} f\right)(x^{\sigma}) \neq 0$, for $x \in [\,a,b\,)$ if $\sigma = +$ or $x \in (\,a,b\,]$ if $\sigma = -$. Since, from Theorem 109., $F_r^{(\sigma)}$ is dense in I, it cannot be totally disconnected. From Proposition 111., we conclude that $\dim_H F_r^{(\sigma)} = 1$.

Let us now define

$$\Gamma_r^{\sigma} := \left\{\left(x, \left(D_x^{\psi} f\right)(x^{\sigma})\right) : x \in F_r^{(\sigma)}\right\}, \quad r,\sigma = \pm \tag{176}$$

Obviously, the graph of $\left(D_x^{\psi} f\right)(x^{\sigma})$ is $\Gamma^{\sigma} = \Gamma_+^{\sigma} \cup \Gamma_-^{\sigma}$. Consequently, from the Projection Theorem 112., we conclude that:

$$\dim_H \Gamma_r^{\sigma} \geq \dim_H F_r^{\sigma} = 1.$$

■

Likewise, we easily derive a generalized version of Cauchy's mean value theorem.

Theorem 114. (Cauchy) *Let $f,g \in C^0(I)$ be differentiable in (a,b) to order $\psi \in \mathcal{Q}$. Then, there exists $\xi \in (a,b)$, such that:*

$$\sigma\,[f(b)-f(a)]\left(D_x^{\psi} g\right)(\xi^{\sigma}) \geq \sigma\,[g(b)-g(a)]\left(D_x^{\psi} f\right)(\xi^{\sigma}), \tag{177}$$

or

$$\sigma\,[f(b)-f(a)]\left(D_x^{\psi} g\right)(\xi^{\sigma}) \leq \sigma\,[g(b)-g(a)]\left(D_x^{\psi} f\right)(\xi^{\sigma}), \tag{178}$$

Proof. Let us consider the function $h(x) = f(x)(g(b)-g(a)) - g(x)(f(b)-f(a))$. By linearity, this function is continuous in I and ψ-differentiable in (a,b). Moreover, $h(a) = h(b)$. From the generalized Rolle Theorem 107., there exists $\xi \in (a,b)$, such that

$$\sigma\left(D_x^{\psi} h\right)(\xi^{\sigma}) \leq 0, \qquad \text{or} \qquad \sigma\left(D_x^{\psi} h\right)(\xi^{\sigma}) \geq 0, \qquad \text{for } \sigma = \pm.$$

The first inequality is equivalent to (177) and the second one to (178). ■

The following result generalizes L'Hopital's rule.

Proposition 115. (L'Hôpital) *Let $f, g \in C^0(I)$ be differentiable functions at $\xi \in (a, b)$ to order $\psi \in \mathcal{Q}$. Suppose that there exists $\epsilon > 0$ such that $g(x) \neq 0$, $\forall x \in ((\xi - \epsilon, \xi + \epsilon) \setminus \{\xi\}) \cap I$. Moreover, suppose that $f(\xi) = g(\xi) = 0$ and that $\left(D_x^{\psi} g\right)(\xi^{\sigma}) \neq 0$. Then, the limit $\lim_{x \to \xi^{\sigma}} \frac{f(x)}{g(x)}$ exists and we have:*

$$\lim_{x \to \xi^{\sigma}} \frac{f(x)}{g(x)} = \frac{\left(D_x^{\psi} f\right)(\xi^{\sigma})}{\left(D_x^{\psi} g\right)(\xi^{\sigma})}. \tag{179}$$

Proof. For $x \in ((\xi - \epsilon, \xi + \epsilon) \setminus \{\xi\}) \cap I$, we have:

$$\frac{f(x)}{g(x)} = \frac{f(x) - f(\xi)}{g(x) - g(\xi)} = \frac{\sigma\left(f(x) - f(\xi)\right)}{\psi\left(|x - \xi|\right)} \frac{\psi\left(|x - \xi|\right)}{\sigma\left(g(x) - g(\xi)\right)}.$$

If we take the limit $x \to \xi^{\sigma}$, we obtain (179). ■

Example 116. *Let us consider the limit*

$$\lim_{x \downarrow 0} \frac{T(x)}{\psi(x)},$$

where $T(\cdot)$ denotes Takagi's function and

$$\psi(x) = \begin{cases} x \log_2\left(\frac{1}{x}\right) & \text{if } x > 0 \\ 0 & \text{if } x = 0 \end{cases}.$$

Clearly $T(0) = \psi(0) = 0$ and, from Krüppel's Theorem 12., both are right ψ-differentiable at $x = 0$. Since $D_x^{\psi} T(0^+) = D_x^{\psi} \psi(0^+) = 1$, from Proposition 115., we have:

$$lim_{x \downarrow 0} \frac{T(x)}{\psi(x)} = 1.$$

Let us now relate the critical order of differentiability with the critical order of Hölder continuity. We start by proving the following lemma.

Lemma 117. *If $f \in C^0(I)$ is locally right (resp. left) ψ-Hölder continuous at $\xi \in I$ for $\psi \in \mathcal{Q}$, then it is right (resp. left) differentiable at ξ to all orders $\phi \in \mathcal{Q}$ such that $\phi \prec \psi$. Moreover, we have $\left(D_x^{\phi} f\right)(\xi^{\sigma}) = 0$, for $\sigma = +$ (resp. $\sigma = -$).*

Proof. If f is locally right ($\sigma = +$) or left ($\sigma = -$) ψ-Hölder continuous at $\xi \in I$ with $\psi \in \mathcal{Q}$, then there exist positive constants K and ϵ such that:

$$|f(x) - f(\xi)| \leq K\psi\left(|x - \xi|\right), \qquad \forall x \in \mathcal{V}_{\xi}^{\sigma}(\epsilon). \tag{180}$$

For $x \in \mathcal{V}_{\xi}^{\sigma}(\epsilon) \setminus \{\xi\}$ we thus have:

$$0 \leq \frac{|f(x) - f(\xi)|}{\phi\left(|x - \xi|\right)} \leq K \frac{\psi\left(|x - \xi|\right)}{\phi\left(|x - \xi|\right)}. \tag{181}$$

Since $\phi \prec \psi$, upon taking the limit $x \to \xi^{\sigma}$, we obtain $\left|\left(D_x^{\phi} f\right)(\xi^{\sigma})\right| = 0$. ■

Conversely, local fractional derivatives describe the local Hölder continuity of the function.

Proposition 118. *Let $f \in C^0(I)$ be such that $\left(D_x^\psi f\right)(\xi^\sigma)$ exists for $\xi \in I$, $\sigma = \pm$ and $\psi \in \mathcal{Q}$. Then f is right ($\sigma = +$) or left ($\sigma = -$) locally ψ-Hölder-continuous at ξ.*

Proof. Since $\left(D_x^\psi f\right)(\xi^\sigma)$ exists, we have from (159,160) for $x \in I$:

$$f(x) - f(\xi) = \sigma \left(D_x^\psi f\right)(\xi^\sigma)\psi\left(|x-\xi|\right) + R_\sigma(x,\xi), \tag{182}$$

where $\frac{R_\sigma(x,\xi)}{\psi(|x-\xi|)} \to 0$, as $x \to \xi^\sigma$. The latter means that there exists $\epsilon > 0$, such that $\frac{R_\sigma(x,\xi)}{\psi(|x-\xi|)}$ is bounded in $\mathcal{V}_\xi^\sigma(\epsilon)$. In other words, there exist positive constants ϵ, C, such that:

$$|R_\sigma(x,\xi)| \leq C\psi\left(|x-\xi|\right), \qquad \forall x \in \mathcal{V}_\xi^\sigma(\epsilon). \tag{183}$$

We then have from (182,183) for $x \in \mathcal{V}_\xi^\sigma(\epsilon)$:

$$|f(x) - f(\xi)| \leq \left[\left|\left(D_x^\psi f\right)(\xi^\sigma)\right| + C\right]\psi\left(|x-\xi|\right). \tag{184}$$

This means that f is right ($\sigma = +$) or left ($\sigma = -$) locally ψ-Hölder continuous at ξ. ■

We may now establish the relation between critical order of differentiability and critical Hölder order.

Theorem 119. *A function $f \in C^0(I)$ has right (resp. left) critical order of differentiability $\psi \in \mathcal{Q}$ at $\xi \in I$ if and only if it has right (resp. left) critical Hölder order ψ at ξ.*

Proof. Let f have right ($\sigma = +$) or left ($\sigma = -$) critical Hölder order $\psi \in \mathcal{Q}$ at $\xi \in I$. Then f is right (resp. left) ϕ-Hölder continuous at ξ to all orders $\phi \in \mathcal{Q}$ with $\phi \prec \psi$, and we have $\left(D_x^\phi f\right)(\xi^\sigma) = 0$. We conclude that the critical order of differentiability $\psi_{\xi,c}^\sigma$ is such that $\psi \preceq \psi_{\xi,c}^\sigma$. Suppose that $\psi \prec \psi_{\xi,c}^\sigma$. Then $\left(D_x^\phi f\right)(\xi^\sigma)$ exists for all $\phi \in \mathcal{Q}$ with $\psi \prec \phi \prec \psi_{\xi,c}^\sigma$. Notice that such ϕ always exist given that scales of functions are dense. From Proposition 46., f is locally right (resp. left) ϕ-Hölder continuous at ξ. But this contradicts the assumption that ψ is the right (resp. left) critical Hölder order. We conclude that $\psi \sim \psi_{\xi,c}^\sigma$.

Conversely, suppose that f has right ($\sigma = +$) or left ($\sigma = -$) critical critical order of differentiability $\psi_{\xi,c}^\sigma$ at ξ. Then f is locally right (resp. left) ϕ-Hölder continuous at ξ for all $\phi \in \mathcal{Q}$ such that $\phi \prec \psi_{\xi,c}^\sigma$. We conclude that the critical local right (resp. left) Hölder order ψ at ξ, satisfies $\psi_{\xi,c}^\sigma \preceq \psi$. Suppose that $\psi_{\xi,c}^\sigma \prec \psi$. Then for all $\phi \in \mathcal{Q}$ with $\psi_{\xi,c}^\sigma \prec \phi \prec \psi$, f is locally right (resp. left) ϕ-Hölder continuous at ξ. This means that f is right (resp. left) differentiable at ξ to all orders $\rho \in \mathcal{Q}$ such that $\psi_{\xi,c}^\sigma \prec \rho \prec \phi$. But then $\psi_{\xi,c}^\sigma$ cannot be the critical order of differentibility. ■

Proposition 120. *Let $f \in C^0(I)$ be such that:*

$$0 < \left|\lim_{n\to\infty} \frac{\sigma\left(f(\xi + \sigma h_n) - f(\xi)\right)}{\psi(h_n)}\right| < +\infty, \qquad \sigma = \pm \tag{185}$$

for some $\psi \in \mathcal{D}$ with $\psi \prec x$ and some sequence $(h_n)_{n\in\mathbb{N}}$ of positive numbers such that $h_n \downarrow 0$. Moreover suppose that f is right ($\sigma = +$) or left ($\sigma = -$) ψ-Hölder continuous. Then the right ($\sigma = +$) or left ($\sigma = -$) critical order of differentiability is ψ for all scales $\mathcal{F}$ containing ψ. In particular, if $\psi \in \mathcal{P}$ this is true for all scales.

Proof. Let $\mathcal{F}$ be a scale containing ψ. Suppose that $\phi \in \mathcal{F}$ is the critical order of differentiability. Since f is ψ-Hölder continuous at ξ, from Lemma 48., the critical order of differentiability must satisfy $\psi \preceq \phi$. If $\psi \prec \phi$, then by Theorem 99., we should have $(D_x^{\psi} f)(\xi^{\sigma}) = 0$, which contradicts (185).

Finally, suppose that $\psi \in \mathcal{P}$. Since, by definition, every scale contains $\mathcal{P}$ (and hence ψ) the result follows. ∎

As an example, we consider the Weierstrass function.

Lemma 121. *Let* $W_{\lambda,s} : [0,1] \to \mathbb{R}$ *denote the Weierstrass function for* $\lambda > 1$ *and* $1 < s < 2$*, and* $h_n = \lambda^{-n}$*. Then we have:*

$$\lim_{n\to\infty} \frac{W_{\lambda,s}(h_n)}{h_n^{2-s}} = \sum_{k=-\infty}^{+\infty} \lambda^{k(s-2)} \sin\left(\lambda^k\right) = \\ = W_{\lambda,s}(1) + \sin(1) + W_{\frac{1}{\lambda},s}(1) \tag{186}$$

Moreover, $s - 2$ *is the right critical order of differentiability at* $x = 0$ *with respect to all scales.*

Proof.

$$\lim_{n\to\infty} \frac{W_{\lambda,s}(h_n)}{h_n^{2-s}} = \lim_{n\to\infty} h_n^{s-2} \left\{ \sum_{k=1}^{n-1} \lambda^{k(s-2)} \sin(\lambda^k h_n) + \right. \\ \left. + \lambda^{n(s-2)} \sin(\lambda^n h_n) + \sum_{k=n+1}^{\infty} \lambda^{k(s-2)} \sin(\lambda^k h_n) \right\} = \\ = \lim_{n\to\infty} \left\{ \sum_{k=1}^{n-1} \lambda^{(k-n)(s-2)} \sin(\lambda^{k-n}) + \sin(1) + \sum_{k=n+1}^{\infty} \lambda^{(k-n)(s-2)} \sin(\lambda^{k-n}) \right\} \tag{187}$$

In the first sum we perform the substitution $l = n - k$, whereas in the second sum we set $l = k - n$. We thus obtain;:

$$\lim_{n\to\infty} \frac{W_{\lambda,s}(h_n)}{h_n^{2-s}} = \sum_{l=1}^{\infty} \left(\frac{1}{\lambda}\right)^{l(s-2)} \sin\left(\frac{1}{\lambda^l}\right) + \sin(1) + \sum_{l=1}^{\infty} \lambda^{l(s-2)} \sin(\lambda^l) = \\ = W_{\frac{1}{\lambda},s}(1) + \sin(1) + W_{\lambda,s}(1) \tag{188}$$

which concludes the proof of (186).

Since $W_{\lambda,s}$ is $(2-s)$-Hölder continuous, we have from Proposition 120. that $2 - s$ is the right critical order of differentiability of $W_{\lambda,s}$ at 0 with respect to all scales. ∎

5. Generalizations

There are various generalizations of the previous definitions worth investigating.

5.1. Quantum Difference Operators

A basic property of differentiable functions is that the quantities

$$\nabla_{\sigma}^{\epsilon} f(x) := \frac{\sigma\left(f(x + \sigma\epsilon) - f(x)\right)}{\epsilon} \tag{189}$$

with $\epsilon > 0$ and $\sigma = \pm 1$ converge and are equal when $\epsilon \downarrow 0$.

As a consequence the quantity

$$a_\epsilon f(x) = \left| \frac{f(x+\epsilon) + f(x-\epsilon) - 2f(x)}{\epsilon} \right| \tag{190}$$

converges to zero when $\epsilon \downarrow 0$.

Definition 122. *Let $h > 0$ be a given real number. A function $f \in C^0$ is said to be $\epsilon - h$-* **differentiable** *at a point ξ if*

$$a_\epsilon f(\xi) < h. \tag{191}$$

We can detect the non-differentiable character of a function by investigating its $\epsilon - h$-differentiability.

Definition 123. *Let $h > 0$ be a given real number and $f \in C^0(I)$. The* h**-minimal resolution of** f **at a point** x*, denoted $\epsilon(f,h)(x)$ is defined as*

$$\epsilon(f,h)(x) := \inf_\epsilon \{a_\epsilon f(x) < h\}. \tag{192}$$

Similarly, the h**-minimal resolution of** f *is*

$$\epsilon(f,h) := \sup_{x \in I} \epsilon(f,h)(x). \tag{193}$$

Of course, if at some $\xi \in I$ and for a given h, the h-minimal resolution is non-zero, then f is non-differentiable at ξ.

In (193) we choose the sup, because if f is differentiable on a small set of points, the h-minimal resolution is however non-zero.

Following Nottale [30] and Cresson [11], we then define

Definition 124. *Let $h > 0$ and $f \in C^0(I)$. If $\epsilon(f,h) > 0$, then for all $0 < \epsilon < \epsilon(f,h)$, we define the* **right** $(\sigma = +)$ **and left** $(\sigma = -)$ **quantum difference operators** *of f at $x \in I$ as in (189). If $\epsilon(f,h) = 0$, then $\nabla_+^0 f(x) = \nabla_-^0 f(x) = f'(x)$.*

We then come to the following interesting result.

Proposition 125. *Let $Y : I \to \mathbb{R}$ be right $(\sigma = +)$ or left $(\sigma = -)$ locally ψ-Hölder continuous in I with $\psi \in \mathcal{Q}$ such that $x^s \prec \psi \preceq x^{\frac{1}{n}}$ for some $n \in \mathbb{N}$ with $n \geq 2$ and all $0 < s < \frac{1}{n}$. Moreover, let $f : \mathbb{R}^2 \to \mathbb{R}$ be of class C^{n+1} in $\mathcal{R} \times I$, where $\mathcal{R}$ denotes the range of Y on I. Then we have for sufficiently small ϵ:*

$$\begin{aligned} \nabla_\sigma^\epsilon f(Y(x),x) &= \tfrac{\partial f}{\partial x}(Y(x),x) + \\ &+ \sigma \textstyle\sum_{j=1}^n \frac{1}{j!} \frac{\partial^j f}{\partial y^j}(Y(x),x)\, \epsilon^{-1} \left(\sigma \epsilon \nabla_\sigma^\epsilon Y(x)\right)^j + \mathcal{O}(\psi(\epsilon)) \end{aligned} \tag{194}$$

for $\sigma = \pm$ and where $\mathcal{O}(\psi(\epsilon))$ include all the terms of order ψ and higher.

Proof. We prove the result for $\sigma = +$. The proof for $\sigma = -$ follows *mutatis mutandis*. We have:

$$f\left(Y(x+\epsilon), x+\epsilon\right) = f\left(Y(x) + \epsilon \nabla_+^\epsilon Y(x), x+\epsilon\right) \tag{195}$$

Since Y is locally right ψ-Hölder continuous, we have for sufficiently small ϵ:

$$\left|\epsilon \nabla_+^\epsilon Y(x)\right| = |Y(x+\epsilon) - Y(x)| = \mathcal{O}\left(\psi(\epsilon)\right). \tag{196}$$

Given that f is of class C^{n+1}, we conclude from (195,196) that we have the controlled Taylor expansion up to order n:

$$\begin{gathered} f\left(Y(x+\epsilon), x+\epsilon\right) = f\left(Y(x), x\right) + \\ + \sum_{k=1}^{n} \frac{1}{k!} \sum_{i+j=k} \left(\epsilon \nabla_+^\epsilon Y(x)\right)^i \epsilon^j \frac{\partial^k f}{\partial y^i \partial x^j}\left(Y(x), x\right) + \mathcal{O}\left(\left(\epsilon \nabla_+^\epsilon Y(x)\right)^{n+1}\right). \end{gathered} \tag{197}$$

As a consequence, it follows that

$$\begin{gathered} \epsilon \nabla_+^\epsilon f\left(Y(x+\epsilon), x+\epsilon\right) = \\ = \sum_{k=1}^{n} \frac{1}{k!} \sum_{i+j=k} \left(\epsilon \nabla_+^\epsilon Y(x)\right)^i \epsilon^j \frac{\partial^k f}{\partial y^i \partial x^j}\left(Y(x), x\right) + \mathcal{O}\left(\left(\epsilon \nabla_+^\epsilon Y(x)\right)^{n+1}\right). \end{gathered} \tag{198}$$

By selecting terms of order less than or equal to one in ϵ on the right-hand side of the previous equation and by taking into account (196), we obtain:

$$\begin{gathered} \epsilon \nabla_+^\epsilon f\left(Y(x+\epsilon), x+\epsilon\right) = \\ = \epsilon \left[\frac{\partial f}{\partial x}\left(Y(x), x\right) + \sum_{i=1}^{n} \frac{1}{i!} \frac{\partial^i f}{\partial y^i}\left(Y(x), x\right) \epsilon^{-1} \left(\epsilon \nabla_+^\epsilon Y(x)\right)^i\right] + \mathcal{O}\left(\epsilon^2 \nabla_+^\epsilon Y(x)\right) \end{gathered} \tag{199}$$

Dividing by ϵ, the result follows. ■

Remark 126. *It is interesting to remark that the case* $n = 2$

$$\begin{gathered} \nabla_\sigma^\epsilon f\left(Y(x), x\right) = \frac{\partial f}{\partial x}\left(Y(x), x\right) + \\ + \frac{\partial f}{\partial y}\left(Y(x), x\right)\left(\nabla_\sigma^\epsilon Y(x)\right) + \frac{\sigma\epsilon}{2!} \frac{\partial^2 f}{\partial y^2}\left(Y(x), x\right)\left(\nabla_\sigma^\epsilon Y(x)\right)^2 + \mathcal{O}\left(\psi(\epsilon)\right) \end{gathered} \tag{200}$$

is reminiscent of Itô's formula.

We also note that Proposition 125. is the starting point for deriving the Schrödinger equation in the context of scale relativity [11].

5.2. Local Fractional Directional Derivatives

Definition 127. *Let* X *be a vector space,* $(Y, ||\cdot||)$ *a normed vector space,* S *a nonempty subset of* X*,* $\alpha \in (\,0, 1\,]$ *and* $f : S \to Y$ *a given mapping. If for two elements* $\bar{x} \in S$ *and* $h \in X$ *the limit*

$$f^{(\alpha)}(\bar{x})(h) := \lim_{\lambda \downarrow 0} \frac{f(\bar{x} + \lambda h) - f(\bar{x})}{\lambda^\alpha} \tag{201}$$

exists, then $f^{(\alpha)}(\bar{x})(h)$ *is called the* **directional derivative of order** α *of* f *at* $\bar{x}$ *in the direction* h*. If this limit exists for all* $h \in X$*, then* f *is called* **directionally differentiable to order** α *at* $\bar{x}$*.*

Definition 128. *A nonempty subset S of a vector space is said to be* **starshaped** *with respect to some $\bar{x} \in S$, if for all $x \in S$ and all $\lambda \in [0,1]$:*

$$\lambda x + (1-\lambda)\bar{x} \in S \tag{202}$$

Definition 129. *Let X be a vector space, $S \subset X$ starshaped with respect to some $\bar{x} \in S$, $\alpha \in (\,0,1\,]$ and $f : S \to \mathbb{R}$ a functional. The functional f is said to be* **α-convex** *at $\bar{x}$ if*

$$f\left(\lambda x + (1-\lambda)\bar{x}\right) \leq \lambda^\alpha f(x) + (1-\lambda^\alpha) f(\bar{x}) \tag{203}$$

for all $x \in S$ and all $\lambda \in [0,1]$.

Lemma 130. *Let X be a vector space, $S \subset X$ starshaped with respect to some $\bar{x} \in S$, $\alpha \in (\,0,1\,]$ and $f : S \to \mathbb{R}$ a α-convex functional at $\bar{x}$. Then for arbitrary $x \in S$, the function $\varphi : (\,0,1\,] \to \mathbb{R}$ with*

$$\varphi(\lambda) = \frac{f(\bar{x} + \lambda(x-\bar{x})) - f(\bar{x})}{\lambda^\alpha} \tag{204}$$

is monotonically increasing.

Proof. Let $0 < s \leq t \leq 1$. From the α-convexity at $\bar{x}$, we have:

$$\begin{aligned} f(\bar{x} + s(x-\bar{x})) - f(\bar{x}) &= f\left(\tfrac{s}{t}(\bar{x} + t(x-\bar{x})) + \tfrac{t-s}{t}\bar{x}\right) - f(\bar{x}) \leq \\ &\leq \left(\tfrac{s}{t}\right)^\alpha f(\bar{x} + t(x-\bar{x})) + \left(1 - \left(\tfrac{s}{t}\right)^\alpha\right) f(\bar{x}) - f(\bar{x}) = \\ &= \left(\tfrac{s}{t}\right)^\alpha \left(f(\bar{x} + t(x-\bar{x})) - f(\bar{x})\right) \end{aligned} \tag{205}$$

Consequently

$$\varphi(s) \leq \varphi(t) \tag{206}$$

and the result follows. ■

Theorem 131. *Let X be a vector space, $\alpha \in (\,0,1\,]$ and $f : X \to \mathbb{R}$ a α-convex functional at $\bar{x}$ and at $\bar{x} - h$, for some $h \in X$. Then the derivative $f^{(\alpha)}(\bar{x})(h)$ exists.*

Proof. As before, we define

$$\varphi(\lambda) = \frac{f(\bar{x} + \lambda h) - f(\bar{x})}{\lambda^\alpha}, \qquad \text{for all } \lambda > 0. \tag{207}$$

Because of the α-convexity of f at $\bar{x} - h$, we have for all $\lambda > 0$

$$\begin{aligned} f(\bar{x}) &= f\left(\tfrac{1}{1+\lambda}(\bar{x} + \lambda h) + \tfrac{\lambda}{1+\lambda}(\bar{x} - h)\right) \leq \\ &\leq \left(\tfrac{1}{1+\lambda}\right)^\alpha f(\bar{x} + \lambda h) + \left(1 - \left(\tfrac{1}{1+\lambda}\right)^\alpha\right) f(\bar{x} - h) \end{aligned} \tag{208}$$

Therefore, we have:

$$\varphi(\lambda) = \frac{f(\bar{x} + \lambda h) - f(\bar{x})}{\lambda^\alpha} \geq \left(\frac{(1+\lambda)^\alpha - 1}{\lambda^\alpha}\right)\left(f(\bar{x}) - f(\bar{x} - h)\right) \tag{209}$$

A simple investigation reveals that

$$0 < \frac{(1+\lambda)^\alpha - 1}{\lambda^\alpha} < 1, \qquad \text{for all } \lambda > 0. \tag{210}$$

Consequently:

$$\left(\frac{(1+\lambda)^\alpha - 1}{\lambda^\alpha}\right)(f(\bar{x}) - f(\bar{x}-h)) \geq \min\{0, f(\bar{x}) - f(\bar{x}-h)\} \tag{211}$$

This means that the function φ is bounded from below. On the other hand, since f is α-convex at $\bar{x}$, we have from the previous Lemma that φ is monotonically increasing. Consequently the limit

$$\lim_{\lambda \downarrow 0} \varphi(\lambda) = f^{(\alpha)}(\bar{x})(h) \tag{212}$$

exists. ∎

Theorem 132. *Let S be a nonempty subset of a vector space, $\alpha \in (\,0,1\,]$ and $f : S \to \mathbb{R}$ a given functional.*

(a) Let $\bar{x} \in S$ be a minimal point of f on S. If the functional f has a directional derivative of order α at $\bar{X}$ in every direction $x - \bar{x}$ with arbitrary $x \in S$, then

$$f^{(\alpha)}(\bar{x})(x-\bar{x}) \geq 0, \qquad \textit{for all } x \in S. \tag{213}$$

(b) Let the set S be starshaped with respect to $\bar{x} \in S$ and f α-convex at $\bar{x}$. If the functional f has a directional derivative of order α at $\bar{x}$ in every direction $x - \bar{x}$ with arbitrary $x \in S$ and the inequality (213) is satisfied, then $\bar{x}$ is a minimal point of f on S.

Proof. (a) Take any $x \in S$. Since f has a directional derivative of order α at $\bar{x}$ in every direction $x - \bar{x}$, we have:

$$f^{(\alpha)}(\bar{x})(x-\bar{x}) = \lim_{\lambda \downarrow 0} \frac{f(\bar{x} - \lambda(x-\bar{x})) - f(\bar{x})}{\lambda^\alpha} \tag{214}$$

Since $\bar{x}$ is assumed to be a minimal point of f on S, we have for sufficiently small $\lambda > 0$

$$f(\bar{x} - \lambda(x-\bar{x})) \geq f(\bar{x}) \tag{215}$$

Consequently, we obtain (213).

(b) Because of the α-convexity of f at $\bar{x}$, we have for an arbitrary $x \in S$ and all $\lambda \in (\,0,1\,]$

$$f(\bar{x} - \lambda(x-\bar{x})) = f(\lambda x + (1-\lambda)\bar{x}) \leq \lambda^\alpha f(x) + (1-\lambda^\alpha) f(\bar{x}) \tag{216}$$

and hence

$$f(x) \geq f(\bar{x}) + \frac{f(\bar{x} - \lambda(x-\bar{x})) - f(\bar{x})}{\lambda^\alpha} \tag{217}$$

Taking the limit $\lambda \downarrow 0$, we obtain:

$$f(x) \geq f(\bar{x}) + f^{(\alpha)}(\bar{x})(x-\bar{x}) \tag{218}$$

With the inequality (213), we have for all $x \in S$:

$$f(x) \geq f(\bar{x}) \tag{219}$$

Consequently $\bar{x}$ is a minimal point of f on S. ∎

5.3. Clarke Fractional Derivatives

Inspired by Clarke's derivatives [8] we propose the following definition.

Definition 133. *Let X be a vector space, S a nonempty subset of X, $\alpha \in (\,0,1\,]$ and $f : S \to \mathbb{R}$ a given functional. If for $\bar{x} \in S$ and $h \in X$ the limit*

$$f_C^{(\alpha)}(\bar{x})(h) := \limsup_{x \to \bar{x}, \lambda \downarrow 0} \frac{f(x + \lambda h) - f(x)}{\lambda^\alpha} \tag{220}$$

exists, then $f_C^{(\alpha)}(\bar{x})(h)$ is called the **Clarke derivative of order** *α at $\bar{x}$ in the direction h. If this limit exists for all $h \in X$, then f is called* **Clarke differentiable to order** *α at $\bar{x}$.*

Example 134. *Let $f : \mathbb{R} \to \mathbb{R}$ be given by*

$$f(x) = |x|^\alpha \tag{221}$$

for some $\alpha \in (\,0,1\,]$. Let us prove that

$$f_C^{(\alpha)}(0)(h) = |h|^\alpha \tag{222}$$

for all $h \in \mathbb{R}$. If $h = 0$, the result is trivial. Suppose that $h \neq 0$. With the aid of the triangle and the Jensen inequalities, we have:

$$f_C^{(\alpha)}(0)(h) = \limsup_{x \to 0, \lambda \downarrow 0} \frac{|x + \lambda h|^\alpha - |x|^\alpha}{\lambda^\alpha} \leq \limsup_{x \to 0, \lambda \downarrow 0} \frac{|x|^\alpha + \lambda^\alpha |h|^\alpha - |x|^\alpha}{\lambda^\alpha} = |h|^\alpha \tag{223}$$

Now consider the sequences $\left(\frac{1}{n^2}\right)_{n \in \mathbb{N}}$ and $\left(\frac{1}{n}\right)_{n \in \mathbb{N}}$ as special cases. We have for sufficiently large n:

$$\begin{aligned} &\frac{1}{n^{-\alpha}} \left(\left|\frac{1}{n^2} + \frac{h}{n}\right|^\alpha - \left|\frac{1}{n^2}\right|^\alpha \right) = |h|^\alpha \left(\left|1 + \frac{1}{nh}\right|^\alpha - \left|\frac{1}{nh}\right|^\alpha \right) = \\ &= |h|^\alpha \left(\left(1 + \frac{1}{nh}\right)^\alpha - \left|\frac{1}{nh}\right|^\alpha \right) = |h|^\alpha \left(1 + \frac{\alpha}{nh} - \left|\frac{1}{nh}\right|^\alpha + \mathcal{O}\left(\frac{1}{nh}\right)^2 \right) \end{aligned} \tag{224}$$

As $n \to \infty$ the previous expression converges to 1. Consequently, $|h|^\alpha$ is both an accumulation point of $\lambda^{-\alpha}(|x + \lambda h|^\alpha - |x|^\alpha)$ for $x \to 0$ and $\lambda \downarrow 0$ and an upper bound of all accumulation points. Hence, we obtain (222).

Theorem 135. *Let S be a subset of a normed vector space $(X, ||\cdot||)$ with nonempty interior, let $\bar{x} \in int(S)$ be a given element, $f : S \to \mathbb{R}$ a functional which is α-Hölder continuous in a neighborhood of $\bar{x}$ for $\alpha \in (\,0,1\,]$, i.e. there exist $K, \epsilon > 0$ such that*

$$|f(x) - f(y)| \leq K ||x - y||^\alpha \tag{225}$$

for all $x, y \in B(\bar{x}, \epsilon)$. Then f is Clarke differentiable to order α at $\bar{x}$ and

$$|f_C^{(\alpha)}(\bar{x})(h)| \leq K ||h||^\alpha \qquad \text{for all } h \in X. \tag{226}$$

Proof. For arbitrary $h \in X$ we have:

$$\left| \frac{f(x+\lambda h) - f(x)}{\lambda^\alpha} \right| \leq K\lambda^{-\alpha}||x + \lambda h - x||^\alpha = K||h||^\alpha \tag{227}$$

for all $x \in S$ and $\lambda > 0$ such that $||x - \bar{x}|| < \epsilon$ and $||x+\lambda h - \bar{x}|| < \epsilon$. Because of this bound the limit superior appearing in the definition of the Clarke derivative exists. Furthermore, we have:

$$\begin{aligned} |f_C^{(\alpha)}(\bar{x})(h)| &= \left|\limsup_{x\to\bar{x},\lambda\downarrow 0} \tfrac{f(x+\lambda h)-f(x)}{\lambda^\alpha}\right| \leq \\ &\leq \limsup_{x\to\bar{x},\lambda\downarrow 0} \left|\tfrac{f(x+\lambda h)-f(x)}{\lambda^\alpha}\right| \leq K||h||^\alpha \end{aligned} \tag{228}$$

which proves the result. ■

Definition 136. *Let X be a vector space, $\alpha \in (\,0,1\,]$ and $p : X \to \mathbb{R}$ a functional. The functional p is said to be α-**sublinear**, if:*

$$\begin{aligned} &(\alpha - \textbf{positive homogeneity}) \quad p(\mu x) = \mu^\alpha p(x), \qquad \textit{for all } x \in X,\ \mu \geq 0 \\ &(\textbf{subadditivity}) \quad p(x+y) \leq p(x) + p(y), \quad \textit{for all } x, y \in X \end{aligned} \tag{229}$$

Example 137. *Let $(X; ||\cdot||)$ be a normed vector space, $\alpha \in (\,0,1\,]$. The functional $p(x) = ||x||^\alpha$ is α-sublinear. The functional $q(x) = |l(x)|^\alpha$, with l a linear functional is α-sublinear.*

Theorem 138. *Let S be a subset of a normed vector space $(X, ||\cdot||)$ with nonempty interior, $\bar{x} \in int(S)$, $\alpha \in (\,0,1\,]$ and $f : S \to \mathbb{R}$ a functional which Clarke differentiable of order α at $\bar{x}$. Then the Clarke derivative $f_C^{(\alpha)}(\bar{x})$ is an α-sublinear functional.*

Proof. First of all notice that $f_C^{(\alpha)}(\bar{x})(0_X) = 0$. For arbitrary $h \in X$ and $\mu > 0$ we have:

$$\begin{aligned} f_C^{(\alpha)}(\bar{x})(\mu h) &= \limsup_{x\to\bar{x},\lambda\downarrow 0} \tfrac{f(x+\lambda\mu h)-f(x)}{\lambda^\alpha} = \\ &= \mu^\alpha \limsup_{x\to\bar{x},\lambda\downarrow 0} \tfrac{f(x+(\lambda\mu)h)-f(x)}{(\lambda\mu)^\alpha} = \mu^\alpha f^{(\alpha)}(\bar{x})(h) \end{aligned} \tag{230}$$

This proves α-positive homogeneity. Next we prove subadditivity. For arbitrary $h_a, h_2 \in X$, we have:

$$\begin{aligned} f_C^{(\alpha)}(\bar{x})(h_1+h_2) &= \limsup_{x\to\bar{x},\lambda\downarrow 0} \tfrac{f(x+\lambda h_1+\lambda h_2)-f(x)}{\lambda^\alpha} = \\ &= \limsup_{x\to\bar{x},\lambda\downarrow 0} \tfrac{f(x+\lambda h_1+\lambda h_2)-f(x+\lambda h_2)+f(x+\lambda h_2)-f(x)}{\lambda^\alpha} \leq \\ &\leq \limsup_{x\to\bar{x},\lambda\downarrow 0} \tfrac{f(x+\lambda h_2+\lambda h_1)-f(x+\lambda h_2)}{\lambda^\alpha} + \limsup_{x\to\bar{x},\lambda\downarrow 0} \tfrac{f(x+\lambda h_2)-f(x)}{\lambda^\alpha} = \\ &= f_C^{(\alpha)}(\bar{x})(h_1) + f_C^{(\alpha)}(\bar{x})(h_2) \end{aligned} \tag{231}$$

■

Theorem 139. *Let S be a subset of a normed vector space $(X, ||\cdot||)$ with nonempty interior, $\bar{x} \in int(S)$, $\alpha \in (\,0,1\,]$ and $f : S \to \mathbb{R}$ a functional which is α-Hölder continuous and α-convex in a neighborhood of $\bar{x}$. If for a given $h \in X$, f is α-convex at $\bar{x} - h$, then the Clarke and directional derivatives of order α at $\bar{x}$ in the direction h coincide.*

Proof. Since f is α-convex at $\bar{x}$ and $\bar{x}-h$, from Theorem 131., $f^{(\alpha)}(\bar{x})(h)$ exists. Likewise, as f is α-Hölder continuous in a neighborhood of $\bar{x}$, by Theorem 135., $f_C^{(\alpha)}(\bar{x})(h)$ also exists. By the definition of these derivatives, we have automatically:

$$f^{(\alpha)}(\bar{x})(h) \leq f_C^{(\alpha)}(\bar{x})(h). \tag{232}$$

Let us now prove the converse inequality.

$$\begin{aligned} f_C^{(\alpha)}(\bar{x})(h) &= \limsup_{x \to \bar{x}, \lambda \downarrow 0} \tfrac{f(x+\lambda h)-f(x)}{\lambda^\alpha} = \\ &= \lim_{\delta \downarrow 0, \epsilon \downarrow 0} \sup_{||x-\bar{x}||<\delta} \sup_{0<\lambda<\epsilon} \tfrac{f(x+\lambda h)-f(x)}{\lambda^\alpha} \end{aligned} \tag{233}$$

Since f is α-convex in a neighborhood of $\bar{x}$, Lemma 130. leads to the equality:

$$f_C^{(\alpha)}(\bar{x})(h) = \lim_{\delta \downarrow 0, \epsilon \downarrow 0} \sup_{||x-\bar{x}||<\delta} \frac{f(x+\epsilon h)-f(x)}{\epsilon^\alpha} \tag{234}$$

For an arbitrary $\mu > 0$, we thus have:

$$f_C^{(\alpha)}(\bar{x})(h) = \lim_{\epsilon \downarrow 0} \sup_{||x-\bar{x}||<\mu\epsilon} \frac{f(x+\epsilon h)-f(x)}{\epsilon^\alpha} \tag{235}$$

Since f is α-Hölder continuous in a neighborhood of $\bar{x}$, we have for sufficiently small ϵ:

$$\begin{aligned} \left| \tfrac{f(x+\epsilon h)-f(x)}{\epsilon^\alpha} - \tfrac{f(\bar{x}+\epsilon h)-f(\bar{x})}{\epsilon^\alpha} \right| &\leq \epsilon^{-\alpha} |f(x+\epsilon h) - f(\bar{x}+\epsilon h)| + \epsilon^{-\alpha} |f(x) - f(\bar{x})| \leq \\ &\leq \tfrac{K}{\epsilon^\alpha} ||x-\bar{x}||^\alpha + \tfrac{K}{\epsilon^\alpha} ||x-\bar{x}||^\alpha \leq 2K\mu \end{aligned} \tag{236}$$

for some positive constant K. It follows that:

$$f_C^{(\alpha)}(\bar{x})(h) \leq \lim_{\epsilon \downarrow 0} \frac{f(\bar{x}+\epsilon h)-f(\bar{x})}{\epsilon^\alpha} + 2K\mu = f^{(\alpha)}(\bar{x})(h) + 2K\mu \tag{237}$$

Since $\mu > 0$ was chosen arbitrarily, we conclude that

$$f_C^{(\alpha)}(\bar{x})(h) \leq f^{(\alpha)}(\bar{x})(h). \tag{238}$$

This completes the proof. ∎

5.4. Comments

Some comments are in order.

1) The critical order of differentiability of $W_{\lambda,r}$ is $2-r$ with respect to all scales. It seems however, that the local fractional derivative of order $2-r$ exists nowhere. Can we prove that? Nevertheless, since $W_{\lambda,r}$ is $(2-r)$-Hölder continuous its Clarke derivative exists (cf. Definition 133. and Theorem 135.) everywhere.

2) If $f : X \to \mathbb{R}$ is α-Hölder continuous in some neighborhood of $\bar{x} \in X$ with $\alpha \in (\,0,1\,]$, then there exist $K, \epsilon > 0$, such that:

$$-K \leq \frac{f(x)-f(y)}{||x-y||^{\alpha}} \leq K \tag{239}$$

for all $x, y \in B(\bar{x}, \epsilon)$ with $x \neq y$. It follows that:

$$\liminf_{x\to\bar{x},\lambda\downarrow 0} \frac{f(x+\lambda h)-f(x)}{\lambda^{\alpha}} = ||h||^{\alpha} \liminf_{x\to\bar{x},\lambda\downarrow 0} \frac{f(x+\lambda h)-f(x)}{(||h||\lambda)^{\alpha}} > -K||h||^{\alpha} \tag{240}$$

so that

$$\liminf_{x\to\bar{x},\lambda\downarrow 0} \frac{f(x+\lambda h)-f(x)}{\lambda^{\alpha}} \tag{241}$$

also exists.

Acknowledgment

The authors would like to thank Rui Ferreira for drawing their attention to the Banach-Mazurkiewicz Theorem 6. This work has been supported by the grant PTDC/MAT/099880/2008 of the Portuguese Science Foundation (FCT).

References

[1] P.C. Allaart, K. Kawamura: *The Takagi function: a survey*. arxiv:math.CA/1110.1692.

[2] P.C. Allaart, K. Kawamura: The improper infinite derivatives of Takagi's nowhere differentiable function. *J. Math. Anal. Appl.* 372 (2010) 656-665.

[3] J.M. Anderson, L.D. Pitt: Probabilistic behaviour of functions in the Zygmund spaces Λ^* and λ^*. *Proc. London Math. Soc.* 59 (1989) 558-592.

[4] A. Babakhani, V. Daftardar-Gejji: On calculus of local fractional derivatives. *J. Math. Anal. and Appl.* 270 (2002) 66.

[5] P. Billingsley: Van der Waerden's continuous nowhere differentiable function. *Amer. Math. Monthly* 89 (1982) 691.

[6] F.S. Cater: Nowhere Analytic Functions. *Amer. Math. Monthly* 91 (1984) 618-624.

[7] Y. Chen, Y. Yan, K. Zhang: On the local fractional derivative. *J. Math. Anal. and Appl.* 362 (2010) 17-33.

[8] F.H. Clarke: *Optimization and nonsmooth analysis*. Wiley, New York (1983).

[9] F. Ben Adda, J. Cresson: About non-differentiable functions. *J. Math. Anal. and Appl.* 263 (2001) 721.

[10] F. Ben Adda, J. Cresson: Fractional differential equations and the Schrödinger equation. *App. Math. Comp.* 161 (2005) 323.

[11] J. Cresson: Scale calculus and the Schrödinger equation. *J. Math. Phys.* 44 (2003) 4907.

[12] K. Falconer: *Fractal Geometry.* Wiley (2003).

[13] H. Haase: Dimension functions. *Math. Nachr.* 141 (1989) 101.

[14] E. Hewitt, K. Stromberg: *Real and Abstract Analysis*. Springer-Verlag, New York (1965).

[15] S. Jaffard, Y. Meyer: Wavelet Methods for Pointwise Regularity and Local Oscillations of Functions. *Memoirs of the AMS* (1996).

[16] A.A. Kilbas, H. M. Srivastava, J.J. Trujillo: *Theory and Applications of Fractional Differential Equations.* North-Holland, Elsevier (2006).

[17] K. Kolwankar, A.D. Gangal: *Fractional differentiability of nowhere differentiable functions and dimensions.* 6 (1996) 505.

[18] K. Kolwankar, A.D. Gangal: Hölder exponents of irregular signals and local fractional derivatives. *Pramana J. Phys.* 48 (1997) 49.

[19] K. Kolwankar, A.D. Gangal: Local fractional derivatives and fractal functions of several variables. *Proc. Fractals in Eng.* (1997).

[20] K. Kolwankar, J. Lévy Véhel: Measuring functions smoothness with local fractional derivatives. *Fract. Calc. Appl. Anal.* 4 (2001) 49-68.

[21] K. Kolwankar, A.D. Gangal: Local fractional Fokker-Planck equation. *Phys. Rev. Lett.* 80 (1998) 214-217.

[22] K.B. Athreya, S.N. Lahiri: *Measure Theory and Probability Theory*. Springer (2006).

[23] M. Krüppel: On the extrema and the improper derivatives of Takagi's continuous nowhere differentiable function. *Rostock. Math.+ Kolloq.* 62 (2007) 41-59.

[24] J. Kurzweil, S. Schwabik: McShane equi-integrability and Vitali's convergence theorem. *Math. Bohem.* 129 (2004) 141-157.

[25] B.B. Mandelbrot: *The Fractal Geometry of Nature*. W.H. Freeman and Company, New York (2006).

[26] A. le Méhauté: *Fractal Geometries, Theory and Applications*. CRC Press Inc. (2000).

[27] R. Metzler, J. Klafter: The random walk's guide to anomalous diffusion: a fractional dynamics approach. *Phys. Rep.* 339 (2000) 1.

[28] K.S. Miller, B. Ross: *An Introduction to the Fractional Calculus and Fractional Differential Equations*. Wiley and Sons, New York (1993).

[29] G. Moore: *Zermelo's axiom of choide*. Springer (1982).

[30] L. Nottale: Fractal Space-Time and Microphysics; Towars a Theory of Scale Relativity. *World Scientific* (1993).

[31] K.B. Oldham, J. Spanier: The Fractional Calculus; *Theory and Applications of Differentiation and Integration to Arbitrary Order*. Dover (2002).

[32] A. Parvate, A.D. Gangal: Fractal differential equations and fractal time-dynamical systems. *Pramana J. Phys.* 64 (2005) 389-409.

[33] I. Podlubny: *Fractional Differential Equations*. Academic Press (1999).

[34] L.F. Richardson: Atmospheric diffusion shown on a distance-neighbour graph. *Proc. Ry. Soc. London A* 110 (1926) 709-737.

[35] A. Rocco, B.J. West: Fractional calculus and the evolution of fractal phenomena. *Physica A* 265 (1999) 535.

[36] S.G. Samko, A.A. Kilbas, O.I. Marichev: *Fractional Integrals and Derivatives: Theory and Applications*. Gordon and Breach Science Publishers, Switzerland (1993).

[37] K.G. Spurrier: Continuous Nowhere Differentiable Functions. *Dissertation*, South Carolina Honors College (2004).

[38] E.M. Stein, A. Zygmund: On the fractional differentiability of functions. *Proc. London Math. Soc.* 14A (1965) 249-264.

[39] T. Takagi: A simple example of a continuous function without derivative. *Phys.-Math. Soc. Japan* 1 (1903) 176-177. *The Collected Papers of Teiji Takagi*, S. Kuroda, Ed., Iwanami (1973) 5-6.

[40] C. Tricot: *Curves and Fractal Dimension*. Springer (1995).

[41] K. Weierstrass: On continuous functions of a real argument which possess a definite derivative for no value of the argument. In: *Königlisch Pressischen Akademie der Wissenschaften, Mathematische Werke von Karl Weierstrass*, Berlin, Mayee and Mueller (1895) vol.2, 71-74.

[42] B.J. West, M. Bologna, P. Grigolini: *Physics of Fractal Operators*. Springer (2003).

[43] S. Willard: *General Topology*. Dover (2004).

[44] G.M. Zaslavsky: *Hamiltonian Chaos and Fractional Dynamics*. Oxford (2005).

In: Fractional Calculus in Analysis, Dynamics ... ISBN: 978-1-62948-635-2
Editor: Jacky Cresson, pp. 65-125

Chapter 2

FRACTIONAL VARIATIONAL EMBEDDING AND LAGRANGIAN FORMULATIONS OF DISSIPATIVE PARTIAL DIFFERENTIAL EQUATIONS

Jacky Cresson*
Laboratoire de Mathématiques et de leurs Applications de Pau,
Université de Pau et des Pays de l'Adour, Pau Cedex, France

PACS: 05.45-a, 52.35.Mw, 96.50.Fm

Keywords: Fractional calculus, fractional calculus of variations, variational formulation, dissipative equations

AMS Subject Classification: 53D, 37C, 65P

1. Introduction

Many problems of physic cannot be formalized using the usual Lagrangian or Hamiltonian formalisms as described for example in [1]. This is in particular the case for dissipative systems. Many authors have tried to overcome this difficulty. The main reason behind these generalizations is that the Lagrangian formalism provides efficient tools to study the dynamics and properties of the underlying equations as well as an intrinsic structure (i.e. not depending on coordinates systems) related to a first principle of physic by the *least action principle*. As examples of previous attempts we can cite :

- Bateman ([4],[23])which constructs a mirror dynamics assuming that the description of dissipative systems is physically incomplete.
- Tveter [41] formalism which preserves the form of the Hamiltonian equations.
- Riewe's approach ([36],[37]) using fractional variational calculus.

*E-mail address: jacky.cresson@univ-pau.fr

- Stochastic differential equations whose mean behaviour corresponds to a dissipative systems.

This list is far from being exhaustive. All of these tentative are based on a particular assumption. The formalism of Tveter is dedicated to the preservation of a Hamiltonian like structure. This mathematical demand is not supported by any physic idea and the interpretation of the new structure is not clear. The stochastic approach is not sufficiently general due to mathematical difficulties in the effective computation of the mean dynamics. The idea is to see dissipative phenomenon as the trace of an irreversible dynamics conducted by a stochastic differential equation. We refer to ([31],§.2.2.3,p.59-60) for an example. Bateman's work is also supported by a physic idea related to the irreversibility of the equation but his construction appears as purely formal: the construction of the mirror dynamics is not fixed by the theory. One must construct the equation for each case. Finally, Riewe's approach is supported by physic. The dissipative phenomenon induces forces which are of fractional nature and which appear in the Lagrangian formalism via fractional derivatives. His tentative fails for technical reasons which are discussed in details in Section 2.5 and can not be used in its present form.

In this Chapter we follow Riewe's idea and provide a fractional variational framework for dissipative systems. The main advantage of this formulation is that dissipative terms are then connected to the emergence of fractional terms in the Lagrangian formulation. As a consequence, the interpretation of our formalism is clear.

Most of the material contains in this chapter can be found in many different articles during the last years. As a consequence, it is difficult for the readers to have a global view of the tools we have developed and the results we have obtained. This chapter can be considered as an overview of our contributions in the direction of fractional calculus of variations.

Our work in fractional calculus is part of a general program to study generalization of ordinary or partial differential equations called *embedding formalisms*. This point of view has evolved along the year and a first review can be found in [17]. Although this formalism is continuously developed and modified we provide in Section 3 a general introduction to this idea. We can formally resume this point of view as follows : try to find functorial or categorical ways to generalize ODEs and PDEs and classify all the existing generalizations using this formalism. Although this program is not completely fulfilled, we provide a first tentative in this direction.

The main part of this chapter is then dedicated to the asymmetric fractional embedding developed by Pierre Inizan and the author in [10] and its applications.

2. Variational Principles and Dissipative Systems

In this section we discuss the problem of constructing a variational principle for dissipative systems. We review some past issues to solve this problem. Starting from the fact that irreversibility was the main obstruction, we propose to use a doubling phase space which takes into account the evolution of the system toward past or future and different time derivatives operators for each of these variables. We also discuss Riewe's approach to dissipative systems using the fractional calculus and prove that it does not give a satisfying solution. This part can be avoided in a first lecture and must be considered as an heuristic support for the mathematical framework of asymmetric fractional calculus of variation developed in the next part.

2.1. Obstruction to Variational Principles: Helmholtz's Conditions

The classical *inverse problem of the calculus of variations* is the following: Having a given set of differential or partial differential equations is it possible to know if they derive from a variational principle, *i.e.* as a critical point of an explicit Lagrangian functional. There exists a huge literature on this subject. The Helmholtz's conditions give necessary and sufficient conditions for a given set of equations to be obtained as an Euler-Lagrange equation. We refer to [34] for more details. Helmholtz's conditions are algebro-analytic and are related to the self adjointness of the differential operator attached to the equations.

2.1.1. Second Order Differential Operators

A second order differential equation on an interval $[a,b]$ is defined by $O^{a,b}(q) = 0$ where O is a second order differential operator, *i.e.*:

$$O: \quad a<b \longmapsto O^{a,b}: \begin{array}{ccc} \mathscr{C}^2([a,b],\mathbb{R}) & \longrightarrow & \mathscr{C}^0([a,b],\mathbb{R}) \\ q & \longmapsto & O^{a,b}(q) \end{array} \tag{1}$$

with:

$$O^{a,b}(q): \begin{array}{ccc} [a,b] & \longrightarrow & \mathbb{R} \\ t & \longmapsto & O^{a,b}(q)(t) = \overline{O}\big(q(t),\dot{q}(t),\ddot{q}(t),t\big) \end{array} \tag{2}$$

where $\dot{q}$ (resp. $\ddot{q}$) is the first (resp. second) derivative of q and where:

$$\overline{O}: \begin{array}{ccc} \mathbb{R}^4 & \longrightarrow & \mathbb{R} \\ (x,v,w,t) & \longmapsto & \overline{O}(x,v,w,t). \end{array} \tag{3}$$

Hence, a second order differential equation (independently of the interval $[a,b]$) is entirely determined by the application $\overline{O}$.

2.1.2. Lagrangian Systems and Euler-Lagrange Equations

A Lagrangian system derives from a variational principle. Precisely, let us consider two reals $a<b$ and the following Lagrangian functional:

$$\mathcal{L}^{a,b}: \begin{array}{ccc} \mathscr{C}^2([a,b],\mathbb{R}) & \longrightarrow & \mathbb{R} \\ q & \longmapsto & \displaystyle\int_a^b L(q,\dot{q},t)\,dt, \end{array} \tag{4}$$

where L is a Lagrangian, *i.e.* an application of the type:

$$L: \begin{array}{ccc} \mathbb{R}^3 & \longrightarrow & \mathbb{R} \\ (x,v,t) & \longmapsto & L(x,v,t). \end{array} \tag{5}$$

Let $\mathscr{C}^2_0([a,b],\mathbb{R}) := \{w \in \mathscr{C}^2([a,b],\mathbb{R}),\ w(a) = w(b) = 0\}$ denote the set of *variations*. Then, $q \in \mathscr{C}^2([a,b],\mathbb{R})$ is said to be a *critical point* of $\mathcal{L}^{a,b}$ if for any variation w,

$D\mathcal{L}^{a,b}(q)(w) = 0$. The calculus of variations allows to characterize the critical points of $\mathcal{L}^{a,b}$ as the solutions on $[a,b]$ of the following *second order Euler-Lagrange equation*:

$$\frac{\partial L}{\partial x}(q,\dot{q},t) - \frac{d}{dt}\left(\frac{\partial L}{\partial v}(q,\dot{q},t)\right) = 0. \qquad (\mathrm{EL}^{a,b})$$

A dynamical system governed by such an Euler-Lagrange equation is called a *second order Lagrangian system*. We refer to [1, p.55-57] for more details concerning Lagrangian systems.

2.1.3. Helmholtz's Conditions

The classical Helmholtz's result can be stated as follows:

Theorem 1 (Helmholtz's condition). *Let O be a second order differential operator. The second order differential equation associated with O can be written as a second order Euler-Lagrange equation if and only if all the Frechet derivatives of $O^{a,b}$ are self-adjoint for any real $a < b$.*

We refer to [34, Theorem 5.92, p.364-365] for a detailed proof. As remarked by J-P. Olver [34, p.365], the condition of self-adjointness of all the Frechet derivatives of $O^{a,b}$ for any real $a < b$ are often referred as the *Helmholtz's condition*.

Nevertheless, the Helmholtz's condition can be more explicitly formulated:

Lemma 2. *Let O be a second order differential operator. The operator O satisfies the Helmholtz's condition if and only if*

$$\forall a < b,\ \forall q \in \mathscr{C}^2([a,b],\mathbb{R}),\ \frac{d}{dt}\left(\frac{\partial \overline{O}}{\partial w}(q,\dot{q},\ddot{q},t)\right) = \frac{\partial \overline{O}}{\partial v}(q,\dot{q},\ddot{q},t). \qquad (\mathrm{H}_{\mathrm{cont}})$$

Proof. Let $a < b$ and $q \in \mathscr{C}^2([a,b],\mathbb{R})$. One can easily obtain that:

$$\forall u \in \mathscr{C}^2([a,b],\mathbb{R}),\ DO^{a,b}(q)(u) = \frac{\partial \overline{O}}{\partial x}(q,\dot{q},\ddot{q},t)u + \frac{\partial \overline{O}}{\partial v}(q,\dot{q},\ddot{q},t)\dot{u} + \frac{\partial \overline{O}}{\partial w}(q,\dot{q},\ddot{q},t)\ddot{u}. \qquad (6)$$

Using integrations by part, it holds for any $u \in \mathscr{C}^2([a,b],\mathbb{R})$:

$$DO^{a,b}(q)^*(u) = \frac{\partial \overline{O}}{\partial x}(q,\dot{q},\ddot{q},t)u - \frac{d}{dt}\left[\frac{\partial \overline{O}}{\partial v}(q,\dot{q},\ddot{q},t)u\right] + \frac{d^2}{dt^2}\left[\frac{\partial \overline{O}}{\partial w}(q,\dot{q},\ddot{q},t)u\right]. \qquad (7)$$

Finally, we obtain that $DO^{a,b}(q) = DO^{a,b}(q)^*$ if and only if:

$$\frac{d}{dt}\left(\frac{\partial \overline{O}}{\partial w}(q,\dot{q},\ddot{q},t)\right) = \frac{\partial \overline{O}}{\partial v}(q,\dot{q},\ddot{q},t). \qquad (8)$$

This concludes the proof. □

However, they do not give an idea of the physical origin of the obstruction to the existence of a variational principle. In the next Section, we discuss the case of dissipative systems.

2.2. The Case of Dissipative Systems: Irreversibility as an Obstruction to a Lagrangian Variational Structure

For dissipative systems a classical result of Bauer, [5], in 1931 states that a linear set of differential equations with constant coefficients cannot derive from a variational principle. In order to illustrate this idea, let us consider the classical linear friction problem given by:

$$q" = \gamma\dot{q} - U'(q), \quad \gamma \neq 0. \tag{9}$$

Using the notations of the previous section, the previous equation is associated to the application

$$\overline{O}(x, v, w, t) = U'(x) - \gamma v + w. \tag{10}$$

We have

$$\frac{\partial \overline{O}}{\partial w} = 1, \quad \frac{\partial \overline{O}}{\partial v} = -\gamma, \tag{11}$$

so that the Helmholtz's condition is $\gamma = 0$. In other words, we have the following Lemma:

Lemma 3. *Equation (9) admits a Lagrangian variational formulation if and only if* $\gamma = 0$.

As a consequence, the presence of a dissipative (or damping depending on the sign of γ) is an obstruction to the existence of a Lagrangian variational formulation. Another way to interpret this result is to say that the underlying dynamics becomes irreversible in time.

2.3. Irreversibility

It must be note that the starting point of Riewe was the fact that most of classical processes observed in Nature are nonconservative and then irreversible. Precisely, Riewe writes:

> It is a strange paradox that the most advanced methods of classical mechanics deal only with conservative systems, while almost all classical processes observed in the physical world are nonconservative. Conservative systems are time reversible by definition, while nonconservative systems exhibit the familiar arrow of time due to irreversible dissipative effects such as friction.

As a consequence, most of the classical systems do not possess a Lagrangian variational structure due to the loss of reversibility. However, one can manage to overcome this difficulty:

- Bateman [4], pointed out that this obstruction is only valid if one understand that the variational principle does not produce additional equations. In particular, Bateman constructs a complementary set of equations which enables him to find a variational formulation. The main idea behind Bateman's approach is that a dissipative system must be seen as physically incomplete due to its underlying irreversibility.
- Contrary to dissipative systems, for a classical reversible system, the time evolution in the two directions (past and future) of time is well described by a single differential equation. In fact, we do not know what is the dynamical evolution if we reverse the arrow of time for a dissipative system. This is for example the case of the diffusion equation or the convection-diffusion equation. Actually, the diffusion phenomenon is non reversible in time.

2.4. Dealing with Irreversibility: Doubling Phase Space and Operators

Bauer's theorem and Bateman construction give two distinct ways to overcome the obstruction to a variational formulation:

- To avoid Bauer's obstruction to the existence of a variational principle, a natural idea is to deal with a different kind of differential calculus. This is precisely what we will do in the next part, following a previous attempt of Riewe ([36],[37]) where the Riemann-Liouville and Caputo fractional calculus are used. However, as we will see, the attempt of Riewe does not work and an appropriate modification of his formalism has to be done.

- The lost of reversibility is interpret by Bateman as an incomplete description of a system. We suggest to say that we have two different dynamical variables (x_+, x_-) which represent the evolution of the systems toward future or past. In experiments we have only access to the time evolution x_+, i.e. that a physical process can be modelled by a curve $t \mapsto x_+(t)$. The evolution of the same physical process if we reverse the arrow of time is not known. But this doubling of phase space is only one part of the dynamical modelization of a physical process. In experiment, we need an operator acting on the phase variables which controls the dynamical behavior. In classical mechanics, this is done using the ordinary differential calculus and in particular the classical derivative with respect to time d/dt. In the new setting, one must look for two time operators d_+ and d_- acting on x_+ and x_- which we call left and right time operators in the sequel.

These simple remarks tell us that if we look for a variational formulation of a general set of equations then the functional must at least be defined on the doubled phase space $X = (x_+, x_-)$ with a differential operator D given by $DX = (\mathcal{D}_+^\alpha x_+, \mathcal{D}_-^\alpha x_-)$ where $\mathcal{D}_+^\alpha$ and $\mathcal{D}_-^\alpha$ are right and left Riemann-Liouville or Caputo derivatives.

2.5. Riewe's Approach to Dissipative Systems: Fractional Mechanics

In 1996-1997, Riewe ([36],[37]) defined a *fractional Lagrangian* framework to deal with dissipative systems. Riewe's theory follows from a simple observation: "*If the Lagrangian contains terms proportional to* $\left(\frac{d^n x}{dt^n}\right)^2$, *then the Euler-Lagrange equation will have a term proportional to* $\frac{d^{2n}x}{dt^{2n}}$. *Hence a frictional force* $\gamma \frac{dx}{dt}$ *should derive from a Lagrangian containing a term proportional to the fractional derivative* $\left(\frac{d^{1/2}x}{dt^{1/2}}\right)^2$" where the notation $\frac{d^{1/2}}{dt^{1/2}}$ represents formally an operator satisfying the composition rule $\frac{d^{1/2}}{dt^{1/2}} \circ \frac{d^{1/2}}{dt^{1/2}} = \frac{d}{dt}$. He then studied fractional Lagrangian functional using the left and right Riemann-Liouville fractional derivatives which satisfy the composition rule formula. Let L a Lagrangian defined by:

$$
\begin{aligned}
L : \mathbb{R} \times \mathbb{R} \times \mathbb{R} \times \mathbb{R} &\longrightarrow \mathbb{R}, \\
(x, v_+, v_-, w) &\longmapsto L(x, v_+, v_-, w)
\end{aligned}
$$

and $x : [a, b] \to \mathbb{R}$ a trajectory. With this notation the functional studied by Riewe is given by

$$\mathcal{L}(x) = \int_a^b L(x, \mathcal{D}_+^{1/2} x, \mathcal{D}_-^{1/2} x, \frac{dx}{dt})dt.$$

He proved that critical points x of this functional correspond to the solutions of the generalised fractional Euler-Lagrange equation

$$\frac{d}{dt}\left(\frac{\partial L}{\partial w}(\star^{1/2})\right) + \mathcal{D}_-^{1/2}\left(\frac{\partial L}{\partial v_+}(\star^{1/2})\right) + \mathcal{D}_+^{1/2}\left(\frac{\partial L}{\partial v_-}(\star^{1/2})\right) = \frac{\partial L}{\partial x}(\star^{1/2}),$$

where $\star^{1/2} := (x, \mathcal{D}_+^{1/2} x, \mathcal{D}_-^{1/2} x, \frac{dx}{dt})$.

Riewe derived such a generalised Euler-Lagrange equation for more general functionals depending on left and right Riemann-Liouville derivatives $\mathcal{D}_-^{\alpha}$ and $\mathcal{D}_+^{\alpha}$ with arbitrary $\alpha > 0$ (see [36], equation (45) p. 1894).

The main drawback of this formulation is that the dependence of L with respect to $\mathcal{D}_+^{1/2}$ (resp. $\mathcal{D}_-^{1/2}$) induces a derivation with respect to $\mathcal{D}_-^{1/2}$ (resp. $\mathcal{D}_+^{1/2}$) in the equation. As a consequence, we will always obtain mixed terms of the form $\mathcal{D}_-^{1/2} \circ \mathcal{D}_+^{1/2} x$ or $\mathcal{D}_+^{1/2} \circ \mathcal{D}_-^{1/2} x$ in the associated Euler-Lagrange equation. For example, if we consider the Lagrangian

$$L(x, v_+, v_-, w) = \frac{1}{2}mw^2 - U(x) + \frac{1}{2}\gamma v_+^2,$$

we obtain as a generalised Euler-Lagrange equation

$$m\frac{d^2x}{dt^2} + \gamma\mathcal{D}_-^{1/2} \circ \mathcal{D}_+^{1/2} x + U'(x) = 0.$$

However, in general

$$\mathcal{D}_-^{1/2} \circ \mathcal{D}_+^{1/2} x \neq \frac{dx}{dt},$$

so that this theory cannot be used in order to provide a variational principle for the linear friction problem. This problem of the mixing between the left and right derivatives in the fractional calculus of variations is well known (see for example Agrawal [2]). It is due to the *integration by parts* formula which is given for f and g in $C_0^0([a, b])$ by

$$\int_a^b f(t)\, \mathcal{D}_-^{\alpha} g(t)dt = \int_a^b \mathcal{D}_+^{\alpha} f(t)\, g(t)dt.$$

In ([36], p.1897) Riewe considered the limit $a \to b$ while keeping $a < b$. He then approximated $\mathcal{D}_-^{\alpha}$ by $\mathcal{D}_+^{\alpha}$. However, this approximation is not justified in general for a large class of functions so that Riewe's derivation of a variational principle for the linear friction problem is not valid.

In [8], Cresson tried to overcome this problem by modifying the underlying set of variations in the fractional calculus. The set of variations is made of functions h satisfying $\mathcal{D}_-^{\alpha} h = \mathcal{D}_+^{\alpha} h$. A critical point of the fractional functional under this restriction is called a weak critical point. In that case, we obtain that solutions of the linear friction problem corresponds to weak critical point of the functional but not an equivalence as in the usual case.

The main problem is that the set of variations is too small to derive a Dubois-Raymond result. We refer in particular to the work of Klimek, [27], for more details.

In the next section, we review the main result of [10] allowing us to obtain an equivalence between solutions of the linear friction problem and critical points of a fractional functional.

2.6. A Remark about Irreversibility, Causality and the Least Action Principle

The link between the least action principle and the causalilty principle has always been ambiguous and it will be more present in the fractional calculus context. Poincaré [17] sumed it up as follows :

> L'énoncé du principe de moindre action a quelque chose de choquant pour l'esprit. Pour se rendre d'un point à un autre, une molécule matérielle, soustraite à l'action donnée de toute force, mais assujettie à se mouvoir sur une surface, prendra la ligne géodésique, c'est-à-dire le chemin le plus court. Cette molécule semble connaître le point où on veut la mener, prévoir le temps qu'elle mettra à l'atteindre en suivant tel ou tel chemin, et choisir ensuite le chemin le plus convenable. L'énoncé nous la présente pour ainsi dire comme un être animé et libre. Il est clair qu'il vaudrait mieux le remplacer par un énoncé moins choquant, et où, comme diraient les philosophes, les causes finales ne sembleraient pas se substituer aux causes efficientes.

Henri Poincaré, La science et l'hypothèse, 1902.

In the fractional calculus, the derivatives are non-local, which makes the past and future of the functions appear explicitly. Because of the simultaneous presence of left and right derivatives, none of those equations respect the causality principle. This difficulty may have been seen as a definite failure, and alternatives have notably been proposed in [7, 22] to get around this problem. However, because we believe that the least action principle should remain fundamental in any formalism, we choose to keep this approach - particularly the one in [5], and we prove in this paper that causality may be respected. To this purpose, we formulate the following remark: when one observe a phenomena which one wants to describe using a differential equation, one only have access to the left derivatives of the functions, i.e. to the differential operators dependent on the past values of the function. Thus, this characteristic represents a trace of the time arrow, and the differential equation becomes attached to the forward temporal direction. If we assume the existence of a similar differential equation, but related to the backward temporal evolution, we obtain a formal approach of irreversibility. Difficulties about causality inherent to the least action principle can be solved using this formalism. More precisely, we show that using a new embedding, termed asymmetric embedding, it is possible to obtain causal Euler-Lagrange equations. In doing so, we observe that the information on the future lies in fact in the variations used by this variational method. The virtual status of these could hence moderate the finalist aspect of the least action principle. In addition, we prove that the asymmetric embedding is coherent, i.e. that this procedure is globally compatible with the least action principle.

If a differential equation has a physical content, it should only involve left derivatives. Indeed, the state of a system at time t should be fixed by its past states at times τ, $\tau < t$. We also note that if we study the reversibility of a system, equations describing the backward evolution should only contain right derivatives [11]. This motivates the following definition.

Definition 4 (Causality). *A fractional differential equation is said to be* causal *if it involves fractional derivatives of a single type.*

3. Embeddings Formalisms of ODEs and PDEs

3.1. Introduction

This part is an introduction to the formalism of embeddings for ordinary or partial differential equations developed in ([13],[14],[8],[6],[9],[10],[21],[19]) in different contexts. The formalism of embedding takes its origin in the following, not exhaustive list of problems:

- *Bath of a given biological or physical model* : in many problems of Biology or Physics, the model is first constructed by neglecting or forgetting the particular environment of the experiment which is called a *bath*. For example, if one deals with the Davidov's model of energy transfer in proteins there exists a phonon bath. In general the model without the bath is a given ordinary or partial differential equation. A natural question is then to look for the new equation of the dynamics when the bath is taken into account.

- *Turbulence* : Fluid dynamics is modeled by partial differential equations. Solutions of these equations must be sufficiently smooth. However, there exists *turbulent* behaviour which correspond to very irregular trajectories. If the underlying equation has a physical meaning, then one must give a sense to this equation on irregular functions. Why ? Because the partial differential equation is derived using first principles of Physics. These principles are always valid despite the fact that the mathematical formulation makes initially some regularity assumptions. This is the case for the turbulent solutions of the Navier-Stokes equations. The previous remark has then leaded many Physicist and Mathematicians to define "generalized" solutions of PDEs. For example, Jean Leray [29] has introduced what he has called a *quasi-derivation* and the notion of *weak solutions* for the Navier-Stokes equations. This first work has a long history and we refer to [3] for an overview's of Jean Leray's work going through for example Laurent Schwartz's distribution theory.

- *Deformation quantization problems*: The problem is to go from classical mechanics to quantum mechanics trough a deformation involving the Planck constant. Roughly speaking, we have a one parameter family of spaces and operators such that they reduce to usual spaces and operators in classical mechanics, i.e. Riemannian manifolds and classical derivatives.

- *Long term behaviour of Chaotic systems*: Chaotic systems play an important role in Physics. As an example, the dynamics of the Solar system, modeled by a n-body problem, is known to be Chaotic over 100 Myr (see [28],[33]). The main characteristic of these systems is that the long term behaviour can be considered as random, i.e.

can be associated to a stochastic process. The characterization of such stochastic processes is of course difficult. However, it leads naturally to the problem of describing the dynamics satisfied by these stochastic processes knowing that at the beginning we have a given differential equations.

- *Discretisation of ODEs or PDEs*: This is a classical problem in Numerical Analysis : For a given ODE or PDE one want to construct a numerical scheme, based on discrete data, which can be used to study the behaviour of solutions. The main problem is to keep trace of some intrinsic properties of the equation at the discrete level, like symmetries, maximum principle, variational structures, etc.

These problems although completely distinct have a common core : we need to define what is the natural or canonical analogue of a given ordinary or partial differential equation over a new vector space. This space can be formed for example of stochastic processes, non-differentiable functions, discrete data, etc. Such a generalization of the initial equation can be done using an *embedding formalism*.

An embedding formalism can be considered as the minimal amount of operators and maps that one must defined in order to derive such a generalization. Indeed, an embedding being fixed, we have defined three ways to extend a given equation : differential, integral and variational. In a given embedding framework, each of these generalization are uniquely determined.

The differential embedding use the differential operator point of view. As an example of such a generalization, one can think to extension of PDEs in Schwartz's distribution sense. The integral embedding is based on the integral representation of the solutions of a given equation. As an example of such generalization, one can consider Itô's approach to stochastic differential equations. The last generalization is based on the variational formulation, when it exists, of the equation. An example of such a procedure is given by the construction of variational integrators in numerical analysis.

3.2. Embedding Formalism

An embedding formalism is a collection of operators and mappings defined over a vectorial set V. A given embedding formalism can be used to define various analogue of a differential or partial differential equations, as well as other classical objects like functionals over V. Many different generalizations of ODEs and PDEs can be included in this framework as we will see in the next chapter. The interest to introduce such a general setting is to understand on which structures and how many a generalization does depend. This is in general not clear. Knowing where we have made a choice, we see what are the possibilities we have in a given construction. As an embedding formalism is a finite number of data, we see that all existing generalizations can be improved in a limited number of ways, the improvement depending on the problem.

3.2.1. Definition

In this Section we define the notion of *abstract embedding*. We denote by $C([a,b];\mathbb{R}^d)$ the set of functions $x : t \in [a,b] \to \mathbb{R}^d$, $d \in \mathbb{N}^*$ and by $C^i([a,b],\mathbb{R}^d)$ the set of i-th differentiable functions.

Definition 5 (Asbtract embedding). *Let V be a given vector space. We assume that there exists :*

- *A mapping $\iota : C([a,b],\mathbb{R}^d) \to V$.*
- *A linear operator $D : V \to V$ which takes place for a generalized derivative on V.*
- *A linear operator $J : V \to \mathbb{R}$ which takes place for a generalized integral on V.*
- *An operator $\star : V \times V \to V$ such that*

$$\iota(f.g) = \iota(f) \star \iota(g). \tag{12}$$

The operator J and D are related by the generalized fundamental theorem of differential calculus

$$J \circ D = D \circ J = \mathrm{Id}, \tag{13}$$

where Id *is the identity map* Id $: V \to V$, $f \mapsto f$.

As we will see in the next Section, these elements are always sufficient in order to provide an extension of an ordinary or partial differential equations over V. In order to illustrate the previous definition, we give in the following Section an example of embedding for which the operator D, J and the product $\star$ can be explicitly defined assuming the existence of the map ι and two lift operators $\tau : V \to C^1$ and $\tau_D : V \to C^0$ satisfying $\tau_D \circ \iota = \mathrm{Id}$. In that case, we derive all the properties of the operator D and J as for example the generalized Leibniz formula or product formula. This example is important as many important class of embedding formalisms can be put in this framework or in a close analogue.

3.3. Example: Basic Embeddings

3.3.1. Construction of a Generalized Derivative

The operator d/dt acts naturally on the set of differentiable functions $C^1([a,b],\mathbb{R}^d)$. Then, a first idea is to define the operator generalized operator D on the subset $\iota(C^1) = V_1$ by

$$D \circ \iota = \iota \circ \frac{d}{dt}, \tag{14}$$

meaning that we are looking to the following diagram :

$$\begin{array}{ccc} f & \xrightarrow{d/dt} & \dfrac{d}{dt}(f) \\ \big\downarrow \iota & & \big\downarrow \iota \\ \iota(f) & \xrightarrow{D} & D(\iota(f)) \end{array} \tag{15}$$

The problem is that in general the quantity $\iota \circ \dfrac{d}{dt}(f)$ can not be written in term of $\iota(f)$. Even if this is the case, then one must also extend the previously define operator to V.

Another way to construct such an operator is to assume that there exists a lift operator $\tau : V \to C^1([a,b],\mathbb{R}^d)$. In this case, we can define D as follow :

Definition 6. *A generalized derivative on V denoted by D is defined for all $\mathbf{F} \in V$ by*

$$D(\mathbf{F}) = \iota\left(\frac{d}{dt}\tau(\mathbf{F})\right), \tag{16}$$

As a consequence, the following diagram commutes:

$$\begin{array}{ccc} \tau(\mathbf{F}) & \xrightarrow{\frac{d}{dt}} & \frac{d}{dt}(\tau(\mathbf{F})) \\ \uparrow \tau & & \downarrow \iota \\ \mathbf{F} & \xrightarrow{D} & D(\mathbf{F}) \end{array} \tag{17}$$

We have not in general that τ and ι are inverse one each others. So that D does not reduce to d/dt over the image of $\iota(C^1)$.

3.3.2. Construction of a Generalized Integral

The mapping J must be constructed in such a way that we obtain an analogue of the fundamental theorem of differential calculus. As a consequence, we must take care of the following diagram :

$$\begin{array}{ccccc} \tau(\mathbf{F}) & \xrightarrow{\frac{d}{dt}} & \frac{d}{dt}(\mathbf{F}) & \xrightarrow{\int_a^t} & \int_a^t \tau_D(\frac{d}{dt}(\mathbf{F})) \\ \uparrow \tau & & \uparrow \tau_D & & \downarrow \iota \\ \mathbf{F} & \xrightarrow{D} & D(\mathbf{F}) & \xrightarrow{J_D} & J_D(D(\mathbf{F})). \end{array} \tag{18}$$

It is easy to see that we need a new lifting operator τ_D is we want to satisfy the commutativity. Indeed, as $\tau : V^1 \to C^1$ but $\frac{d}{dt} : C^1 \to C^0$, we must define for all $\mathbf{F} \in V^0$ where $V^0 = D(V^1)$ a natural element in C^0 such that

$$\iota \circ \tau_D = \iota \circ \tau = \mathrm{Id}, \tag{19}$$

and

$$\tau_D \circ D = \frac{d}{dt} \circ \tau. \tag{20}$$

As a consequence, we are lead to the following definition:

Definition 7. *A generalized integral $J_D : V \to V$ is defined for all $\mathbf{F} \in V$ by*

$$J_D(\mathbf{F}) = \iota\left(\int_a^t \tau_D(\mathbf{F})\right). \tag{21}$$

Using the properties of ι, τ and τ_D, we verify that:

Lemma 8. *For all* $\mathbf{F} \in V^1$, $J_D(D(\mathbf{F})) = \mathbf{F} + \mathbf{C}$, *where* $\mathbf{C}$ *is a constant for* D, *i.e.* $D(\mathbf{C}) = 0$.

Proof. We have by definition of J_D and D that

$$\begin{aligned} J_D(D(\mathbf{F})) &= \iota\left(\int_a^t \tau_D(\Delta(\mathbf{F}))\right), \\ &= \iota\left(\int_a^t \tau_D\left(\iota\frac{d}{dt}(\tau(\mathbf{F}))\right)\right). \end{aligned} \tag{22}$$

As $\tau_D \circ \iota = \mathrm{Id}$ over C^0 we deduce that

$$\begin{aligned} J_D(D(\mathbf{F})) &= \iota\left(\int_a^t \frac{d}{dt}(\tau(\mathbf{F}))\right), \\ &= \iota\left(\tau(\mathbf{F}) + c\right), \end{aligned} \tag{23}$$

where $c \in \mathbb{R}$ is a constant. As $\iota \circ \tau = \mathrm{Id}$ over V, we deduce finally

$$J_D(D(\mathbf{F})) = \mathbf{F} + \mathbf{C}, \tag{24}$$

where $\mathbf{C} = \iota(c)$ is a constant for D. □

The generalized integral operator J_D is then construct with respect to the generalized derivative D and we then verify that we have the generalized fundamental result of differential calculus. However, as we are dealing with linear operators, one can define the integral operator J_D in a pure algebraic way as a solution of the equation $J_D \circ D = \mathrm{Id}$. We must also note that we can reverse the construction by defining first the generalized integral and then the derivative.

3.3.3. Product Rule Formula

We keep the notation of the previous Section. We want to discuss what is preserve in the new setting from the classical differential calculus. As an example, what can be said about the product rule formula

$$\int_a^t \frac{df}{dt} \cdot g = -\int_a^t f \cdot \frac{dg}{dt} \ ? \tag{25}$$

We define the "product" on V compatible with the classical product of functions.

Definition 9. *We denote by* $\star$ *the product on* V *defined by*

$$\star : \begin{array}{ccc} V \times V & \longrightarrow & V, \\ (\mathbf{F}, \mathbf{G}) & \longmapsto & \mathbf{F} \star \mathbf{G} = \iota(\tau(\mathbf{F}) \cdot \tau(\mathbf{G})). \end{array} \tag{26}$$

By definition of the map ι we verify that

$$\begin{array}{ccc} (f, g) & \xrightarrow{\ \cdot\ } & f \cdot g \\ \big\downarrow \iota & & \big\downarrow \iota \\ (\mathbf{F}, \mathbf{G}) & \xrightarrow{\ \star\ } & \mathbf{F} \star \mathbf{G}. \end{array} \tag{27}$$

As a consequence, the formal data of the maps ι, τ and τ_D provide an example of embedding.

Definition 10. *Let* $\iota : C(\mathbb{R}, \mathbb{R}^d) \to V$, $\tau : V \to C^1$ *and* $\tau_D : V \to C^0$ *such that* $\tau_D \circ \iota = \mathrm{Id}$ *over* C^0 *and* $\iota \circ \tau = \iota \circ \tau_D = \mathrm{Id}$ *over* V. *We call basic embedding and we denote by* $\mathcal{E}_(\iota, \tau, \tau_D)(V)$ *the embedding defined by* $D = \iota\left(\frac{d}{dt}(\tau(\mathbf{F}))\right)$, $J_D(\mathbf{F}) = \iota\left(\int_a^t \tau_d(\mathbf{D})\right)$ *and* $\mathbf{F} \star \mathbf{G} = \iota(\tau(\mathbf{F}) \cdot \tau(\mathbf{G}))$.

In the next Section, we study some properties of a basic embedding.

3.4. Properties of Basic Embeddings

In the following, we assume that we have fixed a basic embedding $\mathcal{E}_(\iota, \tau, \tau_D)(V)$.

3.4.1. Leibniz Formula

Using the previous definition, we can prove that the operator D satisfied a Leibniz type formula:

Lemma 11. *Let* $\mathbf{F} \in V^1$, $\mathbf{G} \in V^1$, *then*

$$D(\mathbf{F} \star \mathbf{G}) = D(\mathbf{F}) \star \mathbf{G} + \mathbf{F} \star D(\mathbf{G}). \tag{28}$$

Proof. By definition of D and $\star$ we have

$$D(\mathbf{F} \star \mathbf{G}) = \iota\left(\frac{d}{dt}(\tau(\mathbf{F}) \cdot \tau(\mathbf{G}))\right). \tag{29}$$

By the classical Leibnitz formula, we obtain

$$D(\mathbf{F} \star \mathbf{G}) = \iota\left(\frac{d}{dt}(\tau(\mathbf{F})) \cdot \tau(\mathbf{G}) + \tau(\mathbf{F}) \cdot \frac{d}{dt}(\tau(\mathbf{G}))\right). \tag{30}$$

By linearity of D and the fact that $\iota(f \cdot g) = \iota(f) \cdot \iota(g)$, we deduce

$$D(\mathbf{F} \star \mathbf{G}) = \iota\left(\frac{d}{dt}(\tau(\mathbf{F}))\right) \cdot \iota(\tau(\mathbf{G})) + \iota(\tau(\mathbf{F})) \cdot \left(\frac{d}{dt}(\tau(\mathbf{G}))\right). \tag{31}$$

As $\iota \circ \tau = \mathrm{Id}$, the definition of D and the definition of $\star$ this formula reduces to

$$D(\mathbf{F} \star \mathbf{G}) = D(\mathbf{F}) \cdot \mathbf{G} + \mathbf{F} \cdot D(\mathbf{G}). \tag{32}$$

□

In such an embedding the Leibniz relation is then preserved, meaning that the linear operator D is a derivation on the set V.

3.4.2. Generalized Scalar Product

The classical scalar product on functions denoted by $\langle \cdot, \cdot \rangle$ is defined by

$$\langle f, g \rangle = \int f \cdot g. \tag{33}$$

The natural analogue of the scalar product which is given by the previous embedding is:

Definition 12. *We call generalized scalar product and we denote by* $(\cdot, \cdot) : V \times V \to \mathbb{R}$ *the quantity*

$$(\mathbf{F}, \mathbf{G}) = J_D(\mathbf{F} \star \mathbf{G}). \tag{34}$$

It must be noticed that the previous definition make sense in general. As a consequence, one can look for the adjoint operator associated to D and denoted by D_*, satisfying

$$(D(\mathbf{F}), \mathbf{G}) = (\mathbf{F}, D_*(\mathbf{G})). \tag{35}$$

In the setting of the $\mathcal{E}_{\iota,\tau,\tau_D}(V)$ embedding, we have:

Lemma 13. *The adjoint operator associated to* D *is* $D_* = -D$.

Proof. By Lemma 11 we have for all $(\mathbf{F}, \mathbf{G}) \in V^1 \times V^1$ that

$$D(\mathbf{F} \star \mathbf{G}) = D(\mathbf{F}) \star \mathbf{G} + \mathbf{F} \star D(\mathbf{G}).$$

Using Lemma 8, we obtain

$$\mathbf{F} \star \mathbf{G} = J_D(D(\mathbf{F}) \star \mathbf{G}) + J_D(\mathbf{F} \star D(\mathbf{G})). \tag{36}$$

□

3.4.3. Generalized Functionals

In this section, using the previous definitions, we give a natural analogue of an evaluation functional and of a Lagrangian functional. We denote by $\mathcal{F}$ the set of functionals defined over $C(\mathbb{R}, \mathbb{R}^d)$.

Evaluation Functionals: We introduce the following particular class of functional:

Definition 14. *Let* $a : \mathbb{R}^m \to \mathbb{R}^d$ *be a function. The evaluation functional associated to* a *is denoted by a capital letter* A *and defined by*

$$A : \begin{array}{ccl} C(\mathbb{R}^m, \mathbb{R}^d) & \longrightarrow & \mathcal{F}, \\ x & \mapsto & A(x) := \{t \in \mathbb{R}, A(x)(t) = a(x(t))\}. \end{array} \tag{37}$$

A generalized evaluation functional over V is then defined by :

Definition 15. *Let* a *be a given function and* A *its associated evaluation functional. Then the embed evaluation functional denoted by* $\mathcal{A}$ *is defined for all* $\mathbf{F} \in V$ *by*

$$\mathcal{A}(\mathcal{F}) = \iota(A(\tau(\mathcal{F}))) = \iota(a(\tau(\mathbf{F})(t))). \tag{38}$$

Most of the time we will write the same letter a for the evaluation functional A associated to a and its embedded version.

Lagrangian Functionals: Let $L : \mathbb{R} \times \mathbb{R}^d \times \mathbb{R}^d \to \mathbb{R}$ be a function of class C^2 in all its variables. The Lagrangian functional associated to L and defined over $x \in C^1(\mathbb{R}, \mathbb{R}^d)$ is given by

$$\mathcal{L}(x) = \int_a^b L\left(t, x(t), \frac{dx}{dt}\right) dt. \tag{39}$$

Using the previous construction we define the generalized Lagrangian functional as:

Definition 16. *The generalized Lagrangian functional associated to L, and denoted by $\mathcal{L}_\mathcal{E}$ is defined by for all $\mathbf{X} \in V^1$ by*

$$\mathcal{L}_\mathcal{E}(\mathbf{X}) = \mathcal{L}(\tau(\mathbf{X})). \tag{40}$$

3.5. Differential, Integral and Variational Embedding of Differential Equations

In this Section, we describe three different ways to generalize an ordinary differential equation. This reasoning will of course extend to partial differential equation.

An ordinary differential equation is usually given in differential form, i.e.

$$\frac{dx}{dt} = f(x,t), \quad x \in \mathbb{R}^n, \ t \in \mathbb{R}. \tag{41}$$

However, one can also consider the integral form of the equation

$$x(t) = x(0) + \int_0^t f(x(s), s) ds. \tag{42}$$

The differential form is related to dynamics via the time derivative. The integral form is useful for proving existence and uniqueness of solutions or to study analytical properties of solutions.

In order to give a meaning to a differential equation over a new set (stochastic processes, non-differentiable functions, discrete sets) one can used the differential or the integral form. In general, these two generalizations do not give the same object. In the differential case, we need to extend first he time derivative. As an example of such a procedure, we can look to Schwartz's distributions [38] or backward/forward finite differences in the discrete case. Using these new derivative one can then generalize differential operators and then differential equations of arbitrary order. In the integral case, one need to give a meaning to the integral over the new set. This strategy is for example used by K. Ito [30] in order to define stochastic differential equations, defining first stochastic integrals.

In general, the integral form imposes less constraints on the underlying objects. This is already the case in the classical case, where we need a differentiable function to write the differential form but only continuity or weaker regularity to give a meaning to the integral form.

Of particular importance for many applications in physics and mathematics is the case Lagrangian systems governed by an Euler-Lagrange equation. The main property of these equations is that there solutions corresponds to critical point of a Lagrangian functional, i.e. is associated to a variational principle. Using a given embedding formalism we provide an analogous of the classical functional. By developing the corresponding calculus of variations, we have obtained an embedded Euler-Lagrange equation. This extension of the original Euler-Lagrange equation passing by the embedding of the functional is called a *variational embedding*. Such constructions are well known in Numerical Analysis in the study of discretisation of Lagrangian systems and give rise to *variational integrators* ([24],[32]).

We then have three ways to extend a given ordinary differential equation: differential, integral or variational embedding. All these extensions are a priori different. The coherence problem introduced in [14] in the context of the stochastic embedding consider the problem of finding conditions under which these extensions coincide. We refer to [6],[21] and [14] for such example.

3.6. Differential Embedding

An abstract embedding can be used to define an analogous of a given ordinary differential equation on V; Precisely, we consider a differential oeprator defined by

$$P = \sum_{i=1}^{k} a_i(\cdot).\frac{d^i}{dt^i},$$

acting on $C^k(I, \mathbb{R}^d)$ functions.

Assume that the functional $a_i(.)$ keep sense over V. We denote by V^k the subset of V such that $D^i = D \circ \cdots \circ D$, i-th times, is well defined. Then the abstract embedding of P is given by

$$\text{Emb}(P) = \sum_{i=0}^{k} a_i(\cdot) \star D^i, \tag{43}$$

over V^k.

If we look for the classical Euler-Lagrange equation associated to a Lagrangian function and given by

$$\frac{d}{dt}\left(\frac{\partial L}{\partial v}(t, x, \frac{dx}{dt})\right) - \frac{\partial L}{\partial x}(t, x, \frac{dx}{dt}) = 0, \tag{44}$$

where $x : t \in \mathbb{R} \to x(t) \in \mathbb{R}$, the embedding is formally given by

$$D\left(\frac{\partial L}{\partial v}(T, X, DX)) - \frac{\partial L}{\partial x}(T, X, DX)\right) = 0, \tag{45}$$

Of course, we have implicitely assumed that all the quantities here are well defined.

As an example, in the discretisation case, we obtain as a discrete embedded Euler-Lagrange equation

$$\Delta_+\left(\frac{\partial L}{\partial v}(\mathbf{T}, \mathbf{X}, \Delta_+\mathbf{X}) - \frac{\partial L}{\partial x}(\mathbf{T}, \mathbf{X}, \Delta_+\mathbf{X})\right) = 0, \tag{46}$$

3.7. Integral Embedding

Let us consider the following integral equation

$$x(t) = x(0) + \int_a^t f(s, x(s))\, ds. \tag{47}$$

Using an embedding formalism we define the following integral embedding of a differential equation:

Definition 17. *Let $\mathcal{E}$ be a fixed embedding formalism. The integral embedding of (47) is defined by*

$$\mathbf{X} = \iota(x(0)) + J(f(\mathbf{T}, \mathbf{X})). \tag{48}$$

The integral embedding is a priori different from the differential embedding although in some cases the two procedures gives the same result. This is the case in the finite differences embedding formalism.

The integral form of the Euler-Lagrange equations is given by

$$\frac{\partial L}{\partial v}(t, x(t), \dot{x}(t)) = \int_a^t \frac{\partial L}{\partial x}(\tau, x(\tau), \dot{x}(\tau))\, d\tau + c, \tag{49}$$

for some constant c and all $t \in [a, b]$. Assume that a basic embedding $\mathcal{E}_{\iota,\tau,\tau_D}$ is fixed. The integral embedding of the Euler-Lagrange equation is given by

$$\frac{\partial L}{\partial v}(\mathbf{T}, \mathbf{X}, D(\mathbf{X})) = J_D\left(\frac{\partial L}{\partial x}(\mathbf{T}, \mathbf{X}, D(\mathbf{X}))\right) + \mathbf{C}, \tag{50}$$

where $\mathbf{C} = \iota(c)$.

It is not a priori trivial that the integral and differential embedding gives the same result. This problem is discussed in Section 3.9.

3.8. Variational Embedding

3.8.1. Lagrangian Systems

In this section, we recall classical definitions and theorems concerning Lagrangian systems. We refer to [1] for a detailed study and for the proof of theorem 19.

Definition 18. *A Lagrangian functional is an application defined by:*

$$\mathcal{L} \quad \begin{array}{ccc} C^2([a,b], \mathbb{R}^d) & \to & \mathbb{R}, \\ q & \mapsto & \displaystyle\int_a^b L(q(t), \dot{q}(t), t)dt. \end{array} \tag{51}$$

where L is a Lagrangian i.e. a C^2 application defined by:

$$L : \begin{array}{ccc} [a,b] \times \mathbb{R}^d \times \mathbb{R}^d & \to & \mathbb{R}, \\ (t, x, v) & \mapsto & L(t, x, v). \end{array}$$

The functional (51) is also called **action functional**.

Let $\mathcal{L}$ be a Lagrangian functional, we denote by $D\mathcal{L}(q)(w)$ the Fréchet derivative of $\mathcal{L}$ in q along the direction w in $C^2([a,b],\mathbb{R}^d)$, i.e.

$$D\mathcal{L}(q)(w) = \lim_{\varepsilon\to 0} \frac{\mathcal{L}(q+\varepsilon w) - \mathcal{L}(q)}{\varepsilon}.$$

An *extremal* (or *critical point*) of a Lagrangian functional $\mathcal{L}$ is a trajectory q such that $D\mathcal{L}(q)(w) = 0$ for any *variations* w (i.e. $w \in C^2([a,b],\mathbb{R}^d)$, $w(a) = w(b) = 0$). Extremals of a Lagrangian functional can be characterized as solution of a differential equation of order 2 given by:

Theorem 19 (Variational principle)**.** *Let $\mathcal{L}$ be a Lagrangian functional associated to the Lagrangian L and let $q \in C^2([a,b],\mathbb{R}^d)$. Then, q is an extremal of $\mathcal{L}$ if and only if q is solution of the* ***Euler-Lagrange equation*** *given by:*

$$\forall t \in [a,b],\ \frac{\partial L}{\partial x}(q(t),\dot{q}(t),t) - \frac{d}{dt}\left(\frac{\partial L}{\partial v}(q(t),\dot{q}(t),t)\right) = 0. \tag{EL}$$

3.8.2. Embedded Lagrangian Functionals

An abstract embedding $\mathcal{E}$ being fixed, we can define the analogous of a Lagrangian functional over V.

Definition 20. *Let L be an admissible Lagrangian system and $\mathcal{L}$ the associated functional. The embedded Lagrangian functional of $\mathcal{L}$ under $\mathcal{E}$, denoted by $\mathcal{L}_{\mathcal{E}}$ is given by*

$$\mathcal{L}_{\mathcal{E}}(\mathbf{X}) = J(L(\mathbf{T},\mathbf{X},D\mathbf{X})), \tag{52}$$

for all $\mathbf{X} \subset V^1$.

As V is a linear space, we can define the Frechet derivative of $\mathcal{L}_{\mathcal{E}}$ along a given direction $\mathbf{H} \in V$ and denoted by

$$\mathcal{D}\mathcal{L}_{\mathcal{E}}(\mathbf{X})(\mathbf{H}) = \lim_{\epsilon\to 0}\frac{1}{\epsilon}\left(\mathcal{L}_{\mathcal{E}}(\mathbf{X}+\epsilon\mathbf{H}) - \mathcal{L}_{\mathcal{E}}(\mathbf{X})\right). \tag{53}$$

The corresponding notion of critical points is again given by:

Definition 21. *A vector $\mathbf{X} \in V$ is a critical point of the embedded Lagrangian functional $\mathcal{L}_{\mathcal{E}}$ if $\mathcal{D}\mathcal{L}_{\mathcal{E}}(\mathbf{X})(\mathbf{H}) = 0$ for all $\mathbf{H} \in V$.*

The characterization of the critical points of an embedded functional is based on a calculus of variations that one must construct.

3.8.3. Formal Embedded Calculus of Variations

Lemma 22. *Let $\mathbf{H} \in V_0^1$ and $\mathbf{X} \in V^1$, then the Frechet derivative of $\mathcal{L}_{\mathcal{E}}$ in $\mathbf{X}$ in direction $\mathbf{H}$ is given by*

$$\mathcal{D}\mathcal{L}_{\mathcal{E}}(\mathbf{X})(\mathbf{H}) = J\left(\frac{\partial L}{\partial x}(T,X,DX)\star H + D_*\left(\frac{\partial L}{\partial v}(T,X,DX)\right)\star H\right). \tag{54}$$

Proof. Formally, we have for all $\epsilon > 0$ sufficiently small and $H \in V_0^1$, $V_0^1 = V^1 \cap V_0$,

$$\begin{aligned}\mathcal{L}_{\mathcal{E}}(\mathbf{X}+\epsilon\mathbf{H}) &= J(\mathbf{T},\mathbf{X},D(\mathbf{X})) \\ &\quad +\epsilon J\left(\frac{\partial L}{\partial x}(\mathbf{T},\mathbf{X},D(\mathbf{X}))\star\mathbf{H}+\frac{\partial L}{\partial v}(T,X,DX)\star D(\mathbf{H})\right), \\ &= \mathcal{L}_{\mathcal{E}}(\mathbf{X}) \\ &\quad +\epsilon J\left(\frac{\partial L}{\partial x}(\mathbf{T},\mathbf{X},D(\mathbf{X}))\star\mathbf{H}+D_*\left(\frac{\partial L}{\partial v}(\mathbf{T},\mathbf{X},D(\mathbf{X}))\right)\star\mathbf{H}\right)+\dots\end{aligned} \tag{55}$$

up to terms of order ϵ^2. □

In order to obtain the embedded Euler-Lagrange equation we need an analogue of the Dubois-Raymond lemma.

Definition 23. *An embedding $\mathcal{E}$ is said to satisfy the Dubois-Raymond property if*

$$J(\mathbf{A}\star\mathbf{H}) = 0\ \forall\mathbf{H}\in V_0 \text{ then } \mathbf{A}=0. \tag{56}$$

As a consequence, we can say that formally the expected Euler-Lagrange equation associated to the embedded functional is given by

Theorem 24. *Let $\mathcal{E}$ be an embedding satisfying the Dubois-Raymond property. Let L be an admissible Lagrangian and $\mathcal{L}_{\mathcal{E}}$ the associated embedded Lagrangian functional. An element $\mathbf{X}\in V$ is a critical point of $\mathcal{L}_{\mathcal{E}}$ if and only if*

$$\frac{\partial L}{\partial x}(\mathbf{T},\mathbf{X},D(\mathbf{X}))+D_*\left(\frac{\partial L}{\partial v}(\mathbf{T},\mathbf{X},D(\mathbf{X}))\right)=0 \tag{57}$$

As an example, in the basic embedding case, the Euler-Lagrange equation obtained is

$$\frac{\partial L}{\partial x}(\mathbf{T},\mathbf{X},D\mathbf{X})-D\left(\frac{\partial L}{\partial v}(\mathbf{T},\mathbf{X},D\mathbf{X})\right)=0, \tag{58}$$

which coincides with the differential embedding of the Euler-Lagrange equation.

3.9. Coherence between Differential, Integral and Variational Embeddings

The previous formalism gives three different ways a priori to generalize a differential equation. A natural question is then the problem of coherence between these generalized versions of the same equation:

Coherence problem. *Let $\mathcal{E}$ be a given embedding formalism. Do we have equivalence between the differential, integral and variational embedding of a given equation ?*

A natural answer can be given for what concerns the differential and variational embedding:

Lemma 25. *The differential and variational embedding coincide if and only if $D_* = -D$.*

This follows from the previous computations.

This is of course the case in the classical differential case, as well as in the Schwartz's distribution case. However, this is not satisfied in general. This multiplicity of associated natural generalized equations can be considered as a serious difficulty. However, we can also think that for some problems, we have some choices which can be justified depending on the problem and what we are interested in.

Remark 1. *In fact, it is also possible to obtain coherence by modifying the underlying calculus of variations. Precisely, making some assumptions on the set of variations. For example, assume that a variation* $\mathbf{H} \in V$ *satisfies*

$$D\mathbf{H} = D_*\mathbf{H}, \tag{59}$$

we recover also a coherent embedding. However, this assumption has a main drawback : The set of variations satisfying this symmetry property can be reduced to zero or too small in order to ensure the validity of a Dubois-Raymond lemma. As a consequence, we have no equivalence between critical point of the Lagrangian functional and the Euler-Lagrange equation.

3.10. Example: Finite Differences Embedding

This Section deals with a simple example of embedding over discrete sets using finite differences. We recover several results of the classical literature in Numerical Analysis in a more compact and conceptual form.

3.10.1. Notations

Let $I \subset \mathbb{R}$ be a closed interval. We denote by $C(I, \mathbb{R})$ the set of functions with value in $\mathbb{R}$ defined over I. Let $N \in \mathbb{N}$ and a $\mathbb{R}$, $b > a$. We pose $h = b - a/N$ and we denote by $\mathbb{T} = \{t_i\} \in \mathbb{R}^{N+1}$ the vector defined by $t_k = a + k.h$, $h = (b-a)/N$, $k = 0, \ldots, N$.

We denote by $L([a,b])$ the set of continuous functions which are piecewise linear and by $S_+([a,b])$ (resp. $S_-([a,b])$) the set of step functions which are defined over subintervals $I_i = [t_i, t_{i+1}]$ such that $\bigcup_i I_i = [a,b]$, constant over $[t_i, t_{i+1}[$ (resp. $]t_i, t_{i+1}]$) and right (resp. left) continuous.

An element of $\mathbb{R}^{N+1}$ is always denoted by a capital boldface letter. For $\mathbf{F} \in \mathbb{R}^{N+1}$, we denote by small letter the components $\mathbf{F} = \{f_i\}$.

3.10.2. Discretisation of Functions

We keep the notations of the previous chapter about abstract embedding . A basic way to make a *discretisation* of a function is to associate a finite number of its values over a finite interval. In the following, we choose a uniform distribution over the interval $[a,b]$ but our arguments and construction can be immediately extend to more general setting like a given time scales in time scales calculus.

Definition 26. *Let* $\mathbb{T} \subset [a,b]$ *be a finite discrete set with cardinal* $N+1$. *The element of* $\mathbb{T}$ *are denoted by* t_i, $i = 0, \ldots, N$. *Let* f *be a function* $f : [a,b] \to \mathbb{R}$ *and* $Df \subset [a,b]$ *its* *domain of definition. The discrete version of the function f over $\mathbb{T}_f = Df \cap \mathbb{T}$, denoted by $\mathbf{F} : \mathbb{T}_f \to \mathbb{R}$ is given by $\mathbf{F}(t_i) = f(t_i)$.*

In general, we identify $\mathbf{F}$ with the element of $\mathbb{R}^{N+1}$ given by

$$\iota(f) = \mathbf{F} = \{f_k\}_{k=0,\dots,N}. \tag{60}$$

The discrete analogue of the functional space can then be chosen as $\mathcal{F}(\mathbb{T},\mathbb{R})$ or $\mathbb{R}^{N+1}$.

3.10.3. Backward and Forward Discrete Derivatives

The discrete analogue of the classical derivative can be derived in many ways. In this section, we make the simplest construction leading to the classical backward and forward derivatives.

The lifting map $\tau : \mathbb{R}^{N+1} \to L([a,b])$ can be obtain by linear interpolation and is defined by

$$\tau(\mathbf{F}) = f_k + \frac{f_k - f_{k-1}}{h}(t - t_k), \;\; t_k \leq t \leq t_{k+1}, \;\; k = 0,\dots,N-1. \tag{61}$$

The function $\tau(\mathbf{F})$ is not derivable but we can always define a left and right derivative in the points t_k. As a consequence, we can define Δ_+ and Δ_- the two discrete operators corresponding to the left and right derivative denoted by $\frac{d^\sigma}{dt}$ with $\sigma = \pm$, by a commutative diagram :

Definition 27. *The forward (resp. backward) discrete derivative denoted by Δ_+ (resp. Δ_-) are defined by*

$$\begin{array}{ccc} \tau(\mathbf{F}) & \xrightarrow{d^\sigma/dt} & \dfrac{d^\sigma}{dt}(\tau(\mathbf{F})) \\ \uparrow \tau & & \downarrow \iota \\ \mathbf{F} & \xrightarrow{\sigma} & \Delta_\sigma(\mathbf{F}) \end{array} \tag{62}$$

for $\sigma = \pm$.

From the point of view of embedding two strategies can be considered. The first one is to consider the set $V = \mathcal{F}(\mathbb{T},\mathbb{R})$. As $d_+(L([a,b])$ is only defined over $[a,b[$, we have $\mathbb{T}_{d^+(L([a,b]))} = \mathbb{T}_+$ where

$$\mathbb{T}_+ = \{t_0,\dots,t_{N-1}\}. \tag{63}$$

As a consequence, the operator Δ_+ is not defined on $\mathbb{T}$ as $\mathbf{F}$ but on $\mathbb{T}_+$.

The same can be done for Δ_- and we define $\mathbb{T}_- = \{t_1,\dots,t_N\}$.

Definition 28. *The forward (resp. backward) discrete derivative Δ_+ (resp. Δ_-) defined over $\mathbb{T}_+$ (resp. $\mathbb{T}_-$) is given for all $\mathbf{F} \in \mathcal{F}(\mathbb{T},\mathbb{R})$ by*

$$\Delta_+ = \iota\left(d^+(\tau(\mathbf{F}))\right) \;\; \left(\text{resp. } \Delta_- = \iota\left(d^-(\tau(\mathbf{F}))\right)\right). \tag{64}$$

Another point of view is to extend the two operators over $\mathbb{T}$ by choosing an ad-hoc extension of $\tau(\mathbf{F})$ outside of $[a, b]$. For example, we can extend the map τ by continuity to f_0 when $t \leq a$ and by f_N when $t \geq b$. In that case, simple computations lead to the following expression for these linear operators over $\mathbb{R}^{N+1}$:

Lemma 29. *The discrete derivatives* $\Delta_\sigma : \mathbb{R}^{N+1} \to \mathbb{R}^{N+1}$ *are given by for all* $\mathbf{F} \in \mathbb{R}^{N+1}$ *by*

$$\Delta_+(\mathbf{F}) = \left\{ \frac{f_{k+1} - f_k}{h}, \, 0 \leq k \leq N-1 \, , \, 0 \right\}, \tag{65}$$

and

$$\Delta_-(\mathbf{F}) = \left\{ 0, \frac{f_k - f_{k-1}}{h}, \, 1 \leq k \leq N \right\}. \tag{66}$$

It must be note that in general the backward and forward derivatives are not defined as operators acting on $\mathbb{R}^{N+1}$ but directly by components. As a consequence, most of Textbooks in numerical analysis loose the differential character of this map.

An important property of these operators is the following characterisation of the kernel:

Lemma 30. *Let* $\sigma = \pm$. *We have* $\Delta_\sigma \mathbf{F} = 0$ *if and only if* $\mathbf{F} = f_0 \mathbf{1}$, *where* $\mathbf{1} = (1, \ldots, 1)$.

In the following a vector of the form $\lambda \mathbf{1}$ will be called a *constant*.

3.10.4. Discrete Integration Operator

The discrete integration operator can be obtained by two different methods. The first one follows the previous construction of a discrete derivative. The second one is algebraic and only based on the fact that we want to construct a linear mapping respecting an analogue of the fundamental result of differential calculus.

The Analytic Method: Assume that we have fixed as a discrete analogue of the derivative the forward (resp. backward) discrete derivative Δ_+ (resp. Δ_-). We want to define a linear operator $J_+ : \mathbb{R}^{N+1} \times \mathbb{R}^{N+1}$ (resp. $J_- : \mathbb{R}^{N+1} \to \mathbb{R}^{N+1}$) such that

$$J_+ \circ \Delta_+ = \mathrm{Id} \;\; (resp. \;\; J_- \circ \Delta_- = \mathrm{Id}). \tag{67}$$

As a consequence, we are looking the following diagram

$$\begin{array}{ccccc} \kappa(\mathbf{F}) & \xrightarrow{d_+} & d_+(\mathbf{F}) & \xrightarrow{\int_a^t} & \int_a^t d_+(\mathbf{F}) \\ \uparrow \kappa & & \uparrow ? & & \downarrow \iota \\ \mathbf{F} & \xrightarrow{\Delta_+} & \Delta_+(\mathbf{F}) & \xrightarrow{J_+} & J_+(\Delta_+(\mathbf{F})). \end{array} \tag{68}$$

Where we must defined correctly the map denoted by ? in the previous diagram. Indeed, the operator d_+ satisfies

$$d_+(L([a, b]) \subset S_+([a, b]). \tag{69}$$

Or the image of $\mathbb{R}^{N+1}$ by the lift operator τ is $L([a,b])$ and not $S_+([a,b])$. As a consequence, we must defined a new lift operator $\kappa_+ : \mathbb{R}^{N+1} \to S_+([a,b])$ in order that the previous diagram becomes commutative and satisfying

$$\iota \circ \tau_+ = \iota \circ \tau. \tag{70}$$

A natural mapping from $\mathbb{R}^{N+1}$ to $S_+([a,b])$ is given by

$$\kappa_+ : \begin{array}{lcl} \mathbb{R}^{N+1} & \longrightarrow & M([a,b]), \\ \mathbf{F} = \{f_k\} & \longmapsto & f_k,\ t_k \leq t < t_{k+1},\ k = 0,\dots,N-1. \end{array} \tag{71}$$

We then obtain:

Definition 31. *The discrete integral $J_+ : \mathbb{R}^{N+1} \to \mathbb{R}^{N+1}$ is defined for all $\mathbf{F} \in \mathbb{R}^{N+1}$ by*

$$J_+(\mathbf{F}) = \iota \left(\int_a^t \kappa^+(\mathbf{F}) \right). \tag{72}$$

A computation leads to the following explicit formula:

Lemma 32. *Let $\mathbf{F} \in \mathbb{R}^{N+1}$. We have*

$$J_+(\mathbf{F}) = \{0, f_0.h, (f_0 + f_1).h, \dots, (f_0 + \dots + f_{N-1}).h\}. \tag{73}$$

We deduce the discrete version of the fundamental result of differential calculus:

Lemma 33. *For all $\mathbf{F} \in \mathbb{R}^{N+1}$ we have*

$$J_+(\Delta_+(\mathbf{F})) = \mathbf{F} - f_0\mathbf{1}, \tag{74}$$

where

$$\mathbf{1} = (1,\dots,1). \tag{75}$$

The Algebraic Method: As Δ_+ and J_+ are linear map, we can search directly J_+ as:

$$J_+(\mathbf{F}) = M_+.\mathbf{F},\ M_+\mathcal{M}(N+1,N+1), \tag{76}$$

where $\mathcal{M}(N+1,N+1)$ is the set of $(N+1)\times(N+1)$ matrices.

The forward derivative Δ_+ can be written as

$$\Delta_+(\mathbf{F}) = \frac{1}{h} A_+.\mathbf{F}, \tag{77}$$

where the matrix $A_+ \in \mathcal{M}(N+1,N+1)$ is given by

$$A_+ = \begin{pmatrix} -1 & 1 & 0 & \dots & 0 \\ 0 & \ddots & \ddots & \ddots & 0 \\ \vdots & & & & \vdots \\ 0 & \dots & & -1 & 1 \\ 0 & \dots & & & 0 \end{pmatrix} \tag{78}$$

The problem is then to define the matrix M_+ such that

$$M_+.A_+.\mathbf{F} = \mathbf{F} + \lambda.\mathbf{1},\ \lambda \in \mathbb{R}. \tag{79}$$

The matrix A_+ is non invertible. The problem has then not a unique solution.

We first rewrite the matrix A_+ as

$$A_+ = I_+ + N_+, \tag{80}$$

where N_+ is a almost null matrices given by

$$N_+ = \begin{pmatrix} 0 & \dots & & 0 \\ \dots & & \ddots & \dots \\ 0 & \dots & & 0 \\ 0 & \dots & 0 & 1 \end{pmatrix} \tag{81}$$

and I_+ is an invertible matrix defined by

$$I_+ = \begin{pmatrix} -1 & 1 & 0 & \dots & 0 \\ 0 & \ddots & \ddots & \ddots & 0 \\ \vdots & & & & \vdots \\ 0 & \dots & & -1 & 1 \\ 0 & \dots & & & -1 \end{pmatrix} \tag{82}$$

As a consequence, we have

$$J_+(\Delta_+(\mathbf{F})) = \frac{1}{h}\left(M_+.I_+ + M_+.N_+\right).\mathbf{F}. \tag{83}$$

Choosing $M_+ = h.I_+^{-1}$, we obtain

$$J_+(\Delta_+(\mathbf{F})) = \mathbf{F} + I_+^{-1}.N^+.\mathbf{F}. \tag{84}$$

It is then sufficient to prove that $I_+^{-1}.N^+.\mathbf{F}$ is a constant for Δ_+. A simple computation leads to

$$I_+^{-1}.N^+.\mathbf{F} = -f_N.\mathbf{1}. \tag{85}$$

The matrix I_+^{-1} is given by

$$I_+^{-1} = \begin{pmatrix} -1 & \dots & & -1 \\ 0 & \ddots & & \vdots \\ \vdots & & & \vdots \\ 0 & \dots & 0 & -1 \end{pmatrix}. \tag{86}$$

As a consequence, the algebraic characterization of J_+ leads to the following operator J_+^{alg}:

Definition 34. *The operator* $J_+^{\text{alg}} : \mathbb{R}^{N+1} \to \mathbb{R}^{N+1}$ *is given by*

$$J_+^{\text{alg}}(\mathbf{F}) = \left(-(f_0 + \cdots + f_N), -(f_1 + \cdots + f_N), \dots, -f_N\right). \tag{87}$$

We have the following relation

$$J_+^{\text{alg}}(\Delta_+(\mathbf{F})) = \mathbf{F} - f_N.\mathbf{1}. \tag{88}$$

3.10.5. The Product Formula

We define the "product" on $\mathbb{R}^{N+1}$ compatible with the classical product of functions. We denote by $\star$ the product on $\mathbb{R}^{N+1}$ defined by

$$\star : \begin{array}{lcl} \mathbb{R}^{N+1} \times \mathbb{R}^{N+1} & \longrightarrow & \mathbb{R}^{N+1}, \\ (\mathbf{F}, \mathbf{G}) & \longmapsto & \mathbf{F} \star \mathbf{G} = \{f_k \cdot g_k\}_{k=0,\dots,N}, \end{array} \tag{89}$$

where $\mathbf{F} = \{f_k\}_{k=0,\dots,N}$ and $\mathbf{G} = \{g_k\}_{k=0,\dots,N}$.

By definition of the map ι we verify that

$$\begin{array}{ccc} (f, g) & \xrightarrow{\ \cdot\ } & f \cdot g \\ \downarrow \iota & & \downarrow \iota \\ (\mathbf{F}, \mathbf{G}) & \xrightarrow{\ \star\ } & \mathbf{F} \star \mathbf{G}. \end{array} \tag{90}$$

3.10.6. Discrete Scalar Product

The classical scalar product on functions denoted by $\langle \cdot, \cdot \rangle$ is defined by

$$\langle f, g \rangle = \int_a^b f \cdot g. \tag{91}$$

The natural analogue of the scalar product which is given by the previous embedding is:

Definition 35. *We call discrete scalar product and we denote by* $(\cdot, \cdot) : \mathbb{R}^{N+1} \times \mathbb{R}^{N+1} \to \mathbb{R}$ *the quantity*

$$(\mathbf{F}, \mathbf{G}) = J_{+,N}(\mathbf{F} \star \mathbf{G}). \tag{92}$$

This scalar product is in fact degenerate as the equation $(\mathbf{F}, \mathbf{F}) = 0$ induces only

$$\mathbf{F} = (0, \dots, 0, f_N), \quad f_N \in \mathbb{R}. \tag{93}$$

3.10.7. Discrete Integration by Part

The generalized integration by part formula can also be seen as the fact that the operator $-\Delta_-$ and Δ_+ are adjoint with respect to this scalar product structure.

The left and right derivatives are linked by the following integration by part formula

$$\int_a^t d^+ f \cdot g \, dt = -\int_a^t f \cdot d^- g \, dt + (f(t)g(t) - f(a)g(a)), \tag{94}$$

over the set of functions f and g admitting right and left derivatives respectively and such that $f(a)g(a) = f(b)g(b) = 0$.

This integration by part formula has a direct analogue over $\mathbb{R}^{N+1}$. Indeed, we have for all $(\mathbf{F}, \mathbf{G}) \in \mathbb{R}^{N+1} \times \mathbb{R}^{N+1}$

$$\begin{array}{ccccc} (\kappa(\Delta_+(\mathbf{F})), \kappa(\mathbf{G})) & \xrightarrow{\ \cdot\ } & \kappa(\Delta_+(\mathbf{F})) \cdot \kappa(\mathbf{G}) & \xrightarrow{\ \int\ } & \int (\kappa(\Delta_+(\mathbf{F})) \cdot \kappa(\mathbf{G}) \\ \uparrow \kappa & & \uparrow \kappa & & \downarrow \iota \\ (\Delta_+(\mathbf{F}), \mathbf{G}) & \xrightarrow{\ \star\ } & \Delta_+(\mathbf{F}) \star \mathbf{G} & \xrightarrow{\ J_+\ } & J_+(\Delta_+(\mathbf{F}) \star \mathbf{G}). \end{array} \tag{95}$$

It is easy to prove that

$$\kappa \circ \iota = \text{id over } S_+([a,b]). \tag{96}$$

As $d_+(L([a,b])) \in S_+([a,b])$ we deduce that

$$\kappa \circ \iota \circ d^+ \circ \tau(\mathbf{F}) = d^+ \circ \tau(\mathbf{F}),\ \forall\, \mathbf{F} \in \mathbb{R}^{N+1}. \tag{97}$$

We deduce that

$$\begin{aligned} \int_a^t (\kappa(\Delta_+(\mathbf{F})) \cdot \kappa(\mathbf{G}) &= \int_a^t (\kappa \circ \iota \circ d^+ \circ \tau(\mathbf{F})) \cdot \kappa(\mathbf{G}), \\ &= \int_a^t (d^+ \circ \tau(\mathbf{F})) \cdot \kappa(\mathbf{G}), \\ &= \int_a^t (-d^- \circ \tau(\mathbf{F})) \cdot \kappa(\mathbf{G}) + \tau(\mathbf{F}) \cdot \kappa(\mathbf{G}) \\ &- \tau(\mathbf{F})(a) \cdot \kappa(\mathbf{G})(a). \end{aligned} \tag{98}$$

The only problem that we have is that $d^-(L([a,b]) \subset S_-([a,b])$ so that we can not use the definition of J_+ in order to obtain the result. Indeed we have in general

$$\kappa(-\Delta_-(\mathbf{F})) \neq -d^- \circ \tau(\mathbf{F}). \tag{99}$$

However, up to a set of zero measure, we have

$$\kappa(-\sigma(\Delta_-(\mathbf{F}))) = -d^- \circ \tau(\mathbf{F}), \tag{100}$$

where

$$\sigma : \begin{array}{ccc} \mathbb{R}^{N+1} & \longrightarrow & \mathbb{R}^{N+1} \\ \mathbf{F} = \{F_k\} & \longmapsto & \sigma(\mathbf{F}) = \mathbf{H} = \{H_k = F_{k+1}\}. \end{array} \tag{101}$$

As a consequence, we obtain the natural object associated to $-d^- \circ \tau(\mathbf{F})) \cdot \kappa(\mathbf{G})$ is

$$\int_a^t (\kappa(\Delta_+(\mathbf{F})) \cdot \kappa(\mathbf{G}) = \int_a^t (\kappa(-\sigma(\Delta_-(\mathbf{F}))) \cdot \kappa(\mathbf{G}) + \mathbf{F} \star \mathbf{G} - f_0 g_0 \mathbf{1}. \tag{102}$$

Finally, we obtain the following generalized integration by part formula:

Lemma 36. *For all* $(\mathbf{F},\mathbf{G}) \in \mathbb{R}^{N+1} \times \mathbb{R}^{N+1}$*, we have*

$$J_+(\Delta_+(\mathbf{F}) \star \mathbf{G}) = J_+(-\mathbf{F} \star \sigma(\Delta_-(\mathbf{G}))) + \mathbf{F} \star \mathbf{G} - f_0 g_0 \mathbf{1}. \tag{103}$$

3.10.8. Discrete Dubois-Raymond Lemma

The discrete version of the Dubois-Raymond lemma is valid. We first introduce the set $V_0^1 \subset \mathbb{R}^{N+1}$ defined by

$$V_0^1 = \left\{\mathbf{H} \in \mathbb{R}^{N+1},\ h_0 = h_N = 0\right\}. \tag{104}$$

Lemma 37. *Let* $\mathbf{F} \in \mathbb{R}^{N+1}$ *such that* $J_{+,N}(\mathbf{F} \star \mathbf{G}) = 0$ *for all* $\mathbf{G} \in V_0^1$ *then* $f_k = 0$ *for* $k = 1, \ldots, N-1$.

Proof. We take $\mathbf{H} = \{0, f_1, \ldots, f_{N-1}\} \in V_0^1$. Hence, we obtain

$$J_{+,N}(\mathbf{F} \star \mathbf{G}) = \sum_{k=1}^{N-1} f_k^2 h = 0. \tag{105}$$

This concludes the proof. □

A stronger result can be derived when $\mathbf{F}$ is replaced by $\Delta_+(\mathbf{F})$:

Lemma 38. *Let* $\mathbf{F} \in \mathbb{R}^{N+1}$ *such that* $J_{+,N}(\Delta_+(\mathbf{F}) \star \mathbf{G}) = 0$ *for all* $\mathbf{G} \in V$ *then* $\Delta_+(\mathbf{F}) = 0$.

Proof. The proof follows from a simple computation. As

$$J_{+,N}(\Delta_+(\mathbf{F}) \star \mathbf{H}) = \sum_{k=0}^{N-1} \frac{f_{k+1} - f_k}{h} h_k h, \tag{106}$$

we obtain, taking $\mathbf{H} = \Delta_+(\mathbf{F}) \in \mathbb{R}^{N+1}$ that $f_{k+1} - f_k = 0$ for all $k = 0, \ldots, N-1$, so that $\mathbf{F} = f_0.\mathbf{1}$. As a consequence, $\mathbf{F}$ is a constant for Δ_+ which concludes the proof. □

3.10.9. Discrete Differential Embedding

Let $x \in \mathbb{R}^d$, we consider the ordinary differential equations

$$\frac{dx}{dt} = f(x, t). \tag{107}$$

Using the finite differences embedding the discrete version of this equation is

$$\Delta_+ \mathbf{X} = f(\mathbf{X}, \mathbf{T}), \;\; \mathbf{X} \in (\mathbb{R}^{N+1})^d, \;\; \mathbf{T} \in \mathbb{R}^{N+1}. \tag{108}$$

As $\Delta_+ \mathbf{X}$ is defined on $\mathbb{T}_+$, we obtain for each $i = 0, \ldots, N-1$

$$\frac{x_{i+1} - x_i}{h} = f(x_i, t_i), \tag{109}$$

where $t_i = a + i\dfrac{b-a}{h}$.

3.10.10. Discrete Integral Embedding

The integral formulation of the previous ordinary differential equation is given by

$$x(t) = x(0) + \int_0^t f(s, x(s))\, ds. \tag{110}$$

The forward finite difference integral embedding of this equation is then given by

$$\mathbf{X} = x_0 \mathbf{1} + J_+(f(\mathbf{T}, \mathbf{X})). \tag{111}$$

By definition, this equation is equivalent to

$$x_i = x_0 + h\sum_{j=0}^{i-1} f(t_j, x_j), \ \ i = 1, \dots, N. \tag{112}$$

As a consequence, we obtain

$$x_{i+1} - x_i = hf(t_i, x_i), \ \ i = 0, \dots, N-1, \tag{113}$$

which is the classical *one step forward Euler scheme* in Numerical Analysis.

The forward finite differences differential embedding of the equation is equivalent to the differential case. As a consequence, we see that in this simple case, we have coherence between the two discrete version of the equation.

3.10.11. Discrete Variational Embedding

The discrete calculus of variations is defined over discrete Lagrangian functionals which are obtained using the discrete embedding that we have fixed.

Definition 39. *Let L be an admissible Lagrangian function and $\mathcal{L}$ the associated functional. Let $\sigma = \pm$. The discrete forward (resp. backward) Lagrangian functional associated to L is defined by*

$$\mathcal{L}_\sigma(\mathbf{X}) = J_{+,N}(L(\mathbf{T}, \mathbf{X}, \Delta_\sigma(\mathbf{X}))). \tag{114}$$

A discrete variation is a vector $\mathbf{H} \in \mathbb{R}^{N+1}$ such that $h_0 = h_N = 0$.

Definition 40. *A discrete critical point $\mathbf{X} \in \mathbb{R}^{N+1}$ is a vector satisfying $D\mathcal{L}_\sigma(\mathbf{X})(\mathbf{H}) = 0$ for all $\mathbf{H} \in V_0^1 \subset \mathbb{R}^{N+1}$.*

As usual the main difficulty is to characterize the critical point of a discrete Lagrangian functional by explicit conditions on L and the components of $\mathbf{X}$. The previous results lead easily to :

Theorem 41 (Discrete Euler-Lagrange equation). *Let L be an admissible Lagrangian function. A vector $\mathbf{X} \in \mathbb{R}^{N+1}$ is a discrete critical point of the discrete Lagrangian functional associated to L if and only if it satisfies the system*

$$\left[-\sigma\Delta_-\left(\frac{\partial L}{\partial v}(\mathbf{T}, \mathbf{X}, \Delta_+(\mathbf{X}))\right) + \frac{\partial L}{\partial x}(\mathbf{T}, \mathbf{X}, \Delta_+(\mathbf{X}))\right]_{k=1\dots,N-1} = 0. \tag{115}$$

Proof. Using a Taylor expansion of L, we prove easily that

$$\mathcal{DL}_+(\mathbf{X})(\mathbf{H}) = J_{+,N}\left(\frac{\partial L}{\partial v}(\mathbf{T}, \mathbf{X}, \Delta_+(\mathbf{X})).\Delta_+(\mathbf{H}) + \frac{\partial L}{\partial x}(\mathbf{T}, \mathbf{X}, \Delta_+(\mathbf{X})).\mathbf{H}\right). \tag{116}$$

As $\mathbf{H} \in V_0^1$, we have by Lemma 36 that

$$J_{+,N}(\frac{\partial L}{\partial v}(\mathbf{T}, \mathbf{X}, \Delta_+(\mathbf{X})).\Delta_+(\mathbf{H})) = J_{+,N}(-\sigma\Delta_-\left(\frac{\partial L}{\partial v}(\mathbf{T}, \mathbf{X}, \Delta_+(\mathbf{X}))\right).\mathbf{H}), \tag{117}$$

and

$$\mathcal{D}\mathcal{L}_+(\mathbf{X})(\mathbf{H}) = J_{+,N}\left(-\sigma\Delta_-\left(\frac{\partial L}{\partial v}(\mathbf{T},\mathbf{X},\Delta_+(\mathbf{X}))\right).\mathbf{H} + \frac{\partial L}{\partial x}(\mathbf{T},\mathbf{X},\Delta_+(\mathbf{X})).\mathbf{H}\right). \tag{118}$$

Using Lemma 37 we conclude the proof. □

We refer to [24] for a different formulation of the same result. As we will see in the following, the discrete embedding formalism can be used as a guiding line to generalized classical theorem. This point of view is illustrated in the discrete version of Noether's theorem [7].

4. Asymmetric Fractional Embedding

4.1. Notations and Assumptions

4.1.1. Domain

Let $d \in \mathbb{N}$. We consider a smooth d-dimensional bounded convex domain Ω with boundary $\partial\Omega$. Let $(e_1, \ldots, e_d)$ be the canonical basis for $\mathbb{R}^d$. For any $x \in \mathbb{R}^d$, we denote by x_i the i-th component of x in the canonical basis of $\mathbb{R}^d$. Let $1 \leq i \leq d$ and $x \in \overline{\Omega}$. We denote by $\delta_{i,x}$ the straight line of $\mathbb{R}^d$ defined by

$$\delta_{i,x} = x + \mathrm{Span}(e_i)$$

and $\Omega_{i,x} = \overline{\Omega} \cap \delta_{i,x}$. As Ω is bounded and convex, $\Omega_{i,x}$ is a segment. Then, it exists $a_{i,x} \leq b_{i,x}$ such that

$$\Omega_{i,x} := \{(x_1, \ldots, x_{i-1}, t, x_{i+1}, \ldots, x_d) \mid t \in [a_{i,x}, b_{i,x}]\}. \tag{119}$$

4.1.2. Functional Sets

For two sets A and B, $\mathcal{F}(A, B)$ denotes the vector space of functions $f : A \to B$. Let $a, b \in \mathbb{R}$, $a < b$. Let $m, n \in \mathbb{N}^*$ and $p, q \in \mathbb{N}$. Let $\mathcal{U}$ be an open subset of $\mathbb{R}^m$ or the finite interval $[a, b]$. The vector space of functions $\mathcal{U} \to \mathbb{R}^n$ of class C^p is denoted by $C^p(\mathcal{U})$. For $f \in \mathcal{F}([a, b] \times \Omega, \mathbb{R})$, we denote by:

$$\forall t \in [a,b] \quad f_t : \begin{array}{ll} \Omega & \longrightarrow \mathbb{R} \\ x & \mapsto f(t,x) \end{array} \quad \text{and} \quad \forall x \in \Omega \quad f_x : \begin{array}{ll} [a,b] & \longrightarrow \mathbb{R}. \\ t & \mapsto f(t,x) \end{array}$$

Let $C^{p,q}([a, b] \times \Omega)$ and $C^p([a, b] \times \Omega)$ the functional spaces defined as follow:

$$C^{p,q}([a,b]\times\Omega) := \{f \in \mathcal{F}([a,b]\times\Omega,\mathbb{R}) |\, \forall t \in [a,b],\ f_t \in C^p(\Omega), \forall x \in \Omega,\ f_x \in C^q([a,b])\},$$

and $C^p([a, b] \times \Omega) := C^{p,p}([a, b] \times \Omega)$ when $q = p$. Let $C^p_0(\Omega) := \{f \in C^p(\Omega) \mid f = 0 \text{ on } \partial\Omega\}$.
For $p = 0$, we introduce the following vector spaces:

$$C^0_+([a,b]) := \{f \in C^0([a,b]) \mid f(a) = 0\},$$

$$C^0_-([a,b]) := \{f \in C^0([a,b]) \mid f(b) = 0\},$$

and for $p \geq 1$, let

$$C^p_+([a,b]) := \{f \in C^p([a,b]) \mid f^{(k)}(a) = 0,\ 0 \leq k \leq p-1\},$$

$$C^p_-([a,b]) := \{f \in C^p([a,b]) \mid f^{(k)}(b) = 0,\ 0 \leq k \leq p-1\}.$$

$$C^p_0([a,b]) := C^p_+([a,b]) \cap C^p_-([a,b]).$$

The set of absolutely continuous functions over $[a,b]$ is denoted by $AC([a,b])$ and $AC^{p+1}([a,b])$ is the set defined by

$$AC^{p+1}([a,b]) := \{f \in C^p([a,b]),\ f^{(p)} \in AC([a,b])\}.$$

Then $C^p([a,b]) \subset AC^p([a,b])$. A natural functional space for the study of classical PDEs is

$$F^{p,q}([a,b]\times\overline{\Omega}) := \{f \in \mathcal{F}([a,b]\times\overline{\Omega},\mathbb{R}) | \forall t \in [a,b],\ f_t \in C^p(\overline{\Omega}), \forall x \in \overline{\Omega},\ f_x \in AC^q([a,b])\}.$$

Let $F^p([a,b]\times\overline{\Omega}) := F^{p,p}([a,b]\times\overline{\Omega})$ when $q = p$. For any integer $m \geq 0$, $H^m(\Omega)$ denotes the Sobolev space of order m.

4.1.3. Fields

In this paper we are interested in fields u depending on time $t \in [a,b]$ and space $x \in \Omega$:

$$\begin{array}{rccc} u: & [a,b]\times\overline{\Omega} & \longrightarrow & \mathbb{R}. \\ & (t,x) & \longmapsto & u(t,x) \end{array}$$

The notation $\nabla u(t,x) \in \mathbb{R}^d$ is the gradient of $x \mapsto u(t,x)$ and $\partial_t u(t,x) \in \mathbb{R}$ the partial derivative of u according to t and $\partial_{x_i} f$ the derivative of f in the i-th space-variable. The divergence of a vector field $F = (F_1,\ldots,F_d)$ is $\text{div}F = \sum_{i=1}^{d} \partial_{x_i} F_i$. Let $v : \Omega \to \mathbb{R}$, $1 \leq i \leq d$ and $x \in \overline{\Omega}$. We denote by $v_{i,x}$ the function defined by

$$\begin{array}{rccc} v_{i,x}: & \Omega_{i,x} & \longrightarrow & \mathbb{R} \\ & y & \longmapsto & v(x_1,\ldots,x_{i-1},y,x_{i+1},\ldots,x_d). \end{array} \tag{120}$$

If $v : \overline{\Omega} \to \mathbb{R}$, we denote also by v its extension to $\mathbb{R}^d$ such that $v(x) = 0$ if $x \in \mathbb{R}^d \backslash \overline{\Omega}$. For $x, y \in \mathbb{R}^d$, $x \times y$ denotes the vector in $\mathbb{R}^d$ defined by

$$x \times y := (x_1 y_1, \ldots, x_d y_d)^t. \tag{121}$$

4.2. Fractional Operators: The One-Dimensional Case

For a general overview of the fractional calculus and more details we refer to the classical book of Samko, Kilbas and Marichev, [39].

4.2.1. Fractional Integrals

Let $\beta > 0$, and $f, g : [a, b] \to \mathbb{R}^n$.

Definition 42. *The left and right Riemann-Liouville fractional integrals of f are respectively defined by*

$$\begin{aligned} (\mathcal{I}_+^\beta f)(t) &= \frac{1}{\Gamma(\beta)} \int_a^t (t-\tau)^{\beta-1} f(\tau) d\tau, \\ (\mathcal{I}_-^\beta f)(t) &= \frac{1}{\Gamma(\beta)} \int_t^b (\tau-t)^{\beta-1} f(\tau) d\tau, \end{aligned}$$

for $t \in [a, b]$, where Γ is the Gamma function.

4.2.2. Fractional Derivatives

Let $\alpha > 0$. Let $p \in \mathbb{N}$ such that $p - 1 \leq \alpha < p$.

Definition 43. *Let $t \in [a, b]$, the function f is left (resp. right) Riemann-Liouville differentiable of order α at t if the quantity*

$$\mathcal{D}_+^\alpha f(t) = \left(\frac{d^p}{dt^p} \circ \mathcal{I}_+^{p-\alpha} \right) f(t) \quad \textit{and} \quad \mathcal{D}_-^\alpha f(t) = \left((-1)^p \frac{d^p}{dt^p} \circ \mathcal{I}_-^{p-\alpha} \right) f(t),$$

exist respectively. The value $\mathcal{D}_+^\alpha f(t)$ (resp. $\mathcal{D}_-^\alpha f(t)$) is called the left (resp. right) Riemann-Liouville fractional derivative of order α of f at t.

Exchanging the order of composition we obtain the left and right Caputo fractional derivatives:

Definition 44. *Let $t \in [a, b]$, the function f is left (resp. right) Caputo differentiable of order α at t if the quantity*

$${}^c\mathcal{D}_+^\alpha f(t) = \left(\mathcal{I}_+^{p-\alpha} \circ \frac{d^p}{dt^p} \right) f(t) \quad \textit{and} \quad {}^c\mathcal{D}_-^\alpha f(t) = \left(\mathcal{I}_-^{p-\alpha} \circ (-1)^p \frac{d^p}{dt^p} \right) f(t),$$

exist respectively. The value ${}^c\mathcal{D}_+^\alpha f(t)$ (resp. ${}^c\mathcal{D}_-^\alpha f(t)$) is called the left (resp. right) fractional derivative of order α of f at t.

In a general setting the subscript ^{c} refers to Caputo derivative, whereas no subscript is used to specify the Riemann-Liouville fractional derivative.

4.2.3. Properties

Let us mention the following relations linking both Riemann-Liouville and Caputo fractional derivatives:

Lemma 45. *Let $0 \leq \alpha < 1$ and $f \in AC^1([a, b])$ then $\mathcal{D}_+^\alpha$ and $\mathcal{D}_-^\alpha$ exist almost everywhere and*

$$\begin{aligned} \mathcal{D}_+^\alpha f(t) &= {}^c\mathcal{D}_+^\alpha f(t) + \frac{(t-a)^{-\alpha}}{\Gamma(1-\alpha)} f(a) \\ \mathcal{D}_-^\alpha f(t) &= {}^c\mathcal{D}_-^\alpha f(t) + \frac{(b-t)^{-\alpha}}{\Gamma(1-\alpha)} f(b) \,. \end{aligned}$$

The proof can be found in ([39], thm.2.2, p.39). The following lemma concerns the semi-group property and the integration by parts formula for the fractional derivatives of both Riemann-Liouville and Caputo types:

Lemma 46. *1. Composition rule formula: Let $f \in AC^2([a,b])$, then*

$$ {}^c\mathcal{D}_+^{1/2} \circ {}^c\mathcal{D}_+^{1/2} = \frac{d}{dt} \qquad \mathcal{D}_+^{1/2} \circ {}^c\mathcal{D}_+^{1/2} = \frac{d}{dt}. $$

2. *Integration by parts formula: Let $0 \leq \alpha < 1$. Let $f \in AC^1([a,b])$ and $g \in C_0^1([a,b])$ then*

$$ \begin{aligned} \int_a^b ({}^c\mathcal{D}_+^\alpha f)(t)g(t)dt &= \int_a^b f(t)({}^c\mathcal{D}_-^\alpha g)(t)dt \\ \int_a^b (\mathcal{D}_+^\alpha f)(t)g(t)dt &= \int_a^b f(t)({}^c\mathcal{D}_-^\alpha g)(t)dt. \end{aligned} $$

We refer to [39], p.46 for a proof.

Lemma 47 (Regularity). *Let $\alpha > 0$ and $p \in \mathbb{N}$.*

1. *If $f \in C^1([a,b])$, then ${}^c\mathcal{D}_+^\alpha f \in C_+^0([a,b])$.*
2. *If $f \in C_+^1([a,b])$, then $\mathcal{D}_+^\alpha f = {}^c\mathcal{D}_+^\alpha f$, and $\mathcal{D}_+^\alpha f \in C_+^0([a,b])$.*
3. *If $f \in C_+^{p+1}([a,b])$, then ${}^c\mathcal{D}_+^\alpha f \in C_+^p([a,b])$.*

The previous lemma is stated under more general but implicit assumptions (as $f \in \mathcal{I}_+^\alpha(L^1)$) in [39]. However, as we need explicit conditions on the functional spaces in order to develop the fractional calculus of variations, we give a less general result but with explicit functional spaces.

Proof. 1. As $\frac{df}{dt} \in C^0([a,b])$, we have $\mathcal{I}_+^\alpha(\frac{df}{dt}) \in C_+^0([a,b])$ using (Thm. 3.1 of [39], p.53 with $\lambda = 0$).

2. It results from 1. and lemma 45.

3. Let $1 \leq k \leq p$. As $f \in C_+^{p+1}([a,b])$, then $\frac{df}{dt} \in C_+^k([a,b])$

$$ \frac{d^k}{dt^k}{}^c\mathcal{D}_+^\alpha f = \frac{d^k}{dt^k}\mathcal{I}_+^{1-\alpha}\frac{df}{dt} = \mathcal{I}_+^{1-\alpha} f^{(k+1)}. $$

As $f^{(k+1)} \in C^0([a,b])$, we have $\mathcal{I}_+^{1-\alpha} f^{(k+1)} \in C_+^0([a,b])$ using (Thm. 3.1 of [39], p.53 with $\lambda = 0$). Then ${}^c\mathcal{D}_+^\alpha f \in C^k([a,b])$ and $\frac{d^k}{dt^k}{}^c\mathcal{D}_+^\alpha f(a) = 0$ for $1 \leq k \leq p$. Moreover ${}^c\mathcal{D}_+^\alpha f(a) = 0$ by 1., so that ${}^c\mathcal{D}_+^\alpha f \in C_+^p([a,b])$.

□

A similar result holds for the right derivative.

Lemma 48. *1. If $f \in AC([a,b])$, then $\mathcal{I}_+^\beta f \in AC([a,b])$ and $\mathcal{I}_-^\beta f \in AC([a,b])$.*

2. If $f \in C^0([a,b])$, then $\mathcal{I}_+^\beta f \in C_+^0([a,b])$ and $\mathcal{I}_-^\beta f \in C_-^0([a,b])$.

Proof. 1. We refer to Lemma 2.1 in [39, p.32].

2. We refer to Theorem 3.1 in [39, p.53], with $\lambda = 0$.

□

Lemma 48 and Lemma 45 directly provide the following results.

Lemma 49. *1. If $f \in C^p([a,b])$, then ${}^c\mathcal{D}_+^\alpha f \in C_+^0([a,b])$.*

2. If $f \in C_+^p([a,b])$, then $\mathcal{D}_+^\alpha f = {}^c\mathcal{D}_+^\alpha f$. In particular, $\mathcal{D}_+^\alpha f \in C_+^0([a,b])$.

Similar results hold for the right derivatives.

For $x > 0$ the integers $\overline{x}$ and $\underline{x}$ are defined by $\overline{x} - 1 \leq x < \overline{x}$ and $\underline{x} - 1 < x \leq \underline{x}$. A preliminary lemma will be useful for the following ones.

Lemma 50. *Let $\beta > 0$ and $p \in \mathbb{N}^*$. If $f \in C_+^p([a,b])$, then $\dfrac{d^p}{dt^p} \circ \mathcal{I}_+^\beta f = \mathcal{I}_+^\beta \circ \dfrac{d^p}{dt^p} f$.*

Proof. We prove it by induction. For $p = 1$ and $f \in C_+^1([a,b])$,

$$\begin{aligned} \frac{d}{dt}\mathcal{I}_+^\beta f(t) &= \frac{(t-a)^{\beta-1}}{\Gamma(\beta)} f(a) + \frac{1}{\Gamma(\beta)} \int_a^t (t-\tau)^{\beta-1} f'(\tau)\, d\tau \\ &= \mathcal{I}_+^\beta f'(t), \end{aligned}$$

since $f(a) = 0$.

Now let $p \in \mathbb{N}^*$ and $f \in C_+^{p+1}([a,b])$. Since $f' \in C_+^p([a,b])$, we may apply the induction hypothesis:

$$\frac{d^p}{dt^p} \circ \mathcal{I}_+^\beta f' = \mathcal{I}_+^\beta \circ \frac{d^p}{dt^p} f'. \tag{122}$$

From case $p = 1$, $\mathcal{I}_+^\beta f' = \dfrac{d}{dt}\mathcal{I}_+^\beta f$. Hence (122) becomes

$$\frac{d^{p+1}}{dt^{p+1}} \circ \mathcal{I}_+^\beta f = \mathcal{I}_+^\beta \circ \frac{d^{p+1}}{dt^{p+1}} f,$$

which concludes the proof. □

4.3. Fractional Operators: The Multidimensional Case

We generalise the previous definitions and properties on Ω by introducing the multidimensional fractional operators.

For a function $v : \Omega \to \mathbb{R}$, we denote by $v_{i,x}$ the function defined on $\Omega_{i,x}$, as in (119) acting in the i-th component of v. The i-th partial fractional derivatives are given by:

$$\partial_i^\alpha v(x) := \mathcal{D}_+^\alpha v_{i,x}(x_i), \quad {}^c\partial_i^\alpha v(x) := {}^c\mathcal{D}_+^\alpha v_{i,x}(x_i),$$

the right fractional Riemann-Liouville and Caputo partial derivatives with respect to the i-th component of v. In a same way the left fractional Riemann-Liouville and Caputo partial derivatives are given by

$$\overline{\partial}_i^\alpha v(x) := \mathcal{D}_-^\alpha\, v_{i,x}(x_i), \quad {}^c\overline{\partial}_i^\alpha v(x) := {}^c\mathcal{D}_-^\alpha\, v_{i,x}(x_i).$$

The associated Riemann-Liouville and Caputo *fractional gradient* of u, denoted by $\nabla^\alpha u$ and ${}^c\nabla^\alpha u$, are defined by

$$\nabla^\alpha u(t,x) := \begin{pmatrix} \partial_1^\alpha u(t,x) \\ \vdots \\ \partial_d^\alpha u(t,x) \end{pmatrix}, \qquad {}^c\nabla^\alpha u(t,x) := \begin{pmatrix} {}^c\partial_1^\alpha u(t,x) \\ \vdots \\ {}^c\partial_d^\alpha u(t,x) \end{pmatrix}.$$

The Riemann-Liouville and Caputo *fractional divergence* of $v : \Omega \to \mathbb{R}^d$ are defined by

$$\text{div}^\alpha v(x) = \sum_{i=1}^d \partial_i^\alpha v_i(x) \quad {}^c\text{div}^\alpha v(x) = \sum_{i=1}^d {}^c\partial_i^\alpha v_i(x).$$

The analogous definitions of the left fractional gradient and divergence hold adding a bar on the previous symbols.

Lemma 51. *Let $\gamma = (\gamma_1, \ldots, \gamma_d) \in \mathbb{R}^d$, $u \in AC^2(\overline{\Omega})$ and $x \in \Omega$, then we have*

$$\text{div}^{1/2}\left(\gamma \times {}^c\nabla^{1/2} u(x)\right) = \gamma \cdot \nabla u(x).$$

Proof. Since $\gamma \times {}^c\nabla^{1/2} u = \begin{pmatrix} \gamma_1\, {}^c\partial_1^{1/2} u \\ \vdots \\ \gamma_d\, {}^c\partial_d^{1/2} u \end{pmatrix}$, then

$$\text{div}^{1/2}\left(\gamma \times {}^c\nabla^{1/2} u(x)\right) = \sum_{i=1}^d \gamma_i \left(\partial_i^{1/2} \circ {}^c\partial_i^{1/2}\right) u(x).$$

Applying lemma 46 concludes the proof. □

4.3.1. Fractional Green-Riemann Formula

The following lemma is a consequence of the unidimensional integration by parts formula of lemma 45 and leads to the fractional version of the Green-Riemann theorem mixing Caputo and Riemann-Liouville derivatives given in lemma 53.

Lemma 52. *Let $u \in AC^1(\overline{\Omega})$ and $v \in C_0^1(\Omega)$ then*

$$\int_\Omega u(x)\, {}^c\overline{\partial}_i^\alpha v(x)dx = \int_\Omega \partial_i^\alpha u(x) v(x) dx.$$

Proof. We first extend u, v, $\overline{^c\partial_i^\alpha} v$ and $\partial_i^\alpha u$ over $\mathbb{R}^d$ by associating the value 0 if $x \in \mathbb{R}^d \backslash \overline{\Omega}$. In this case, we have

$$\begin{aligned}\int_\Omega u(x)\, \overline{^c\partial_i^\alpha} v(x) dx &= \int_{\mathbb{R}^d} u(x)\, \overline{^c\partial_i^\alpha} v(x) dx \\ &= \int_{\mathbb{R}^{d-1}} \left[\int_{x_i=-\infty}^{\infty} u(x)\, \overline{^c\partial_i^\alpha} v(x) dx_i \right] dx_1 \dots dx_{i-1} dx_{i+1} \dots dx_d.\end{aligned}$$

As Ω is convex, there exists $a_{i,x}$ and $b_{i,x}$ such that $a_{i,x} \leq b_{i,x}$ and

$$\int_{x_i=-\infty}^{\infty} u(x)\, \overline{^c\partial_i^\alpha} v(x) dx_i = \int_{a_{i,x}}^{b_{i,x}} u_{i,x}(x_i)^c\, \mathcal{D}_-^\alpha\, v_{i,x}(x_i).$$

As $u_{i,x} \in AC^1([a_{i,x}, b_{i,x}])$ and $v_{i,x} \in C_0^1([a_{i,x}, b_{i,x}])$, the integration by parts formula from lemma 46 leads to:

$$\begin{aligned}\int_{a_{i,x}}^{b_{i,x}} u_{i,x}(x_i)^c\, \mathcal{D}_-^\alpha\, v_{i,x}(x_i) dx_i &= \int_{a_{i,x}}^{b_{i,x}} {}^c\mathcal{D}_+^\alpha\, u_{i,x}(x_i) v_{i,x}(x_i) dx_i \\ &= \int_{a_{i,x}}^{b_{i,x}} \partial_i^\alpha u(x) v_{i,x}(x_i) dx_i \\ &= \int_{x_i=-\infty}^{+\infty} \partial_i^\alpha u(x) v(x) dx_i.\end{aligned}$$

As a consequence, we have

$$\begin{aligned}\int_{\mathbb{R}^d} u(x)\, \overline{^c\partial_i^\alpha} v(x) dx_i &= \int_{\mathbb{R}^{d-1}} \left[\int_{x_i=-\infty}^{\infty} \partial_i^\alpha u(x) v(x) dx_i \right] dx_1 \dots dx_{i-1} dx_{i+1} \dots dx_d, \\ &= \int_\Omega \partial_i^\alpha u(x)\, v(x) dx.\end{aligned}$$

□

Lemma 53 (Fractional Green-Riemann theorem)**.** *Let $\Omega \in \mathbb{R}^d$ be a smooth d-dimensional bounded convex domain, $u \in C_0^1(\overline{\Omega}, \mathbb{R})$ and $v = (v_1, \dots, v_d) \in AC^1(\Omega)^d$, then we have*

$$\int_\Omega v(x) \cdot \overline{^c\nabla^\alpha} u(x)\, dx = \int_\Omega \mathrm{div}^\alpha(v(x))\, u(x)\, dx. \tag{123}$$

Proof. The proof results from lemma 52. On the canonical basis the scalar product is given by

$$\int_\Omega v(x) \cdot \overline{^c\nabla^\alpha} u(x)\, dx = \sum_{i=1}^d \int_\Omega v_i(x)\, \overline{^c\partial_i^\alpha} u(x)\, dx.$$

Applying lemma 52, we have for any $1 \leq i \leq d$,

$$\int_\Omega v_i(x)\, \overline{^c\partial_i^\alpha} u(x)\, dx = \int_\Omega \partial_i^\alpha v_i(x)\, u(x)\, dx.$$

The definition of div^α concludes the proof. □

4.4. Differential Asymmetric Fractional Embedding

We adapt here the presentation done in [8].

Let $M, N \in \mathbb{N}^*$. If $f \in \mathcal{F}(\mathbb{R}^{M+1}, \mathbb{R}^N)$ and $y \in \mathcal{F}([a,b], \mathbb{R}^M)$, we denote by $f(y(\bullet), \bullet)$ the function defined by

$$\begin{array}{rccc} f(y(\bullet),\bullet): & [a,b] & \longrightarrow & \mathbb{R}^N \\ & t & \longmapsto & f(y(t),t). \end{array}$$

Let $p, k \in \mathbb{N}$. If $\mathbf{f} = \{f_i\}_{0\le i\le p}$ and $\mathbf{g} = \{g_j\}_{1\le j\le p}$ are two families of $\mathcal{F}(\mathbb{R}^{n(k+1)+1}, \mathbb{R}^m)$ ($\mathbf{g} = \emptyset$ if $p = 0$), with $f_j \in C^0(\mathbb{R}^{n(k+1)+1})$ and $g_j \in C^j(\mathbb{R}^{n(k+1)+1})$ for $1 \le j \le p$, we introduce the operator $\mathcal{O}_{\mathbf{f}}^{\mathbf{g}}$ defined by

$$\begin{array}{rccc} \mathcal{O}_{\mathbf{f}}^{\mathbf{g}}: & C^{k+p}([a,b]) & \longrightarrow & \mathcal{F}([a,b], \mathbb{R}^m) \\ & x & \longmapsto & \left[f_0 + \sum_{i=1}^{p} f_i \star \frac{d^i}{dt^i} g_i\right]\left(x(\bullet), \dots, \frac{d^k}{dt^k}x(\bullet), \bullet\right), \end{array} \tag{124}$$

where, for two operators $A = (A_1, \dots, A_m)$ and $B = (B_1, \dots, B_m)$, $A \star B$ is defined by

$$(A \star B)(y) = (A_1(y)B_1(y), \dots, A_m(y)B_m(y)).$$

The fractional embedding presented in [8] consists in replacing d/dt by a fractional derivative. Here we want to keep this idea, but additionnaly we want to split in two the functional space of the trajectories, in order to make the asymmetry between left and right fractional derivatives explicitly appear.

Let $0 < \alpha < 1$. For $X = (x_+, x_-) \in C^1([a,b])^2$ ($AC^2([a,b])^2$ would be sufficient), we introduce the fractional derivative ${}^c\mathcal{D}^\alpha$, defined by

$${}^c\mathcal{D}^\alpha X = \left({}^c\mathcal{D}_+^\alpha\, x_+, -\,{}^c\mathcal{D}_-^\alpha\, x_-\right).$$

The classical case is recovered for $\alpha \to 1^-$ (and not for $\alpha = 1$).

Lemma 54. *Let $X \in C^1([a,b])^2$. Then*

$$\forall t \in (a,b),\ \lim_{\alpha\to 1^-} {}^c\mathcal{D}^\alpha X(t) = \frac{d}{dt}X(t).$$

Proof. Let $X = (x_+, x_-) \in C^1([a,b])^2$. Since $x'_+ \in C^0([a,b])$, all points of (a,b) are Lebesgue points of x'. We may then apply Theorem 2.7 of [39, p.51]:

$$\forall t \in (a,b),\ \lim_{\alpha\to 1^-} \mathcal{I}_+^{1-\alpha}\, x'(t) = x'(t).$$

We proceed likewise for x_-. □

Hence for $k \in \mathbb{N}^*$ and a "suitable" function X,

$$({}^c\mathcal{D}^\alpha)^k X = \left(({}^c\mathcal{D}_+^\alpha)^k x_+, (-\,{}^c\mathcal{D}_-^\alpha)^k x_-\right).$$

The following lemma provides an example of such "suitable" functions.

Lemma 55. *Let $k \in \mathbb{N}^*$. If $f \in C^k_+([a,b])$, then we have*

$$\left({}^c\mathcal{D}^{\alpha}_+\right)^k f = {}^c\mathcal{D}^{\alpha k}_+ f.$$

A similar result holds for the right derivative.

Proof. We prove it by induction on k. For $k = 1$, the result is obvious. Now, let $k \in \mathbb{N}^*$ and $f \in C^{k+1}_+([a,b])$. Since $f \in C^k_+([a,b])$, we use the induction hypothesis: $({}^c\mathcal{D}^{\alpha}_+)^k f = {}^c\mathcal{D}^{\alpha k}_+ f = \mathcal{I}^{\overline{\alpha k}-\alpha k}_+ f^{(\overline{\alpha k})}$. We have $\overline{\alpha k} \leq k$, so $f^{(\overline{\alpha k})} \in AC([a,b])$, and from Lemma 48, $({}^c\mathcal{D}^{\alpha}_+)^k f \in AC([a,b])$. Moreover, from Lemma 49, $({}^c\mathcal{D}^{\alpha}_+)^k f(a) = 0$. We may then apply Lemma 45:

$$\mathcal{D}^{\alpha}_+ \circ ({}^c\mathcal{D}^{\alpha}_+)^k f = {}^c\mathcal{D}^{\alpha}_+ \circ ({}^c\mathcal{D}^{\alpha}_+)^k f = ({}^c\mathcal{D}^{\alpha}_+)^{k+1} f.$$

On the other hand, $\mathcal{D}^{\alpha}_+ \circ ({}^c\mathcal{D}^{\alpha}_+)^k f = \dfrac{d}{dt} \circ \mathcal{I}^{1-\alpha}_+ \circ \mathcal{I}^{\overline{\alpha k}-\alpha k}_+ f^{(\overline{\alpha k})}$. We have $f^{(\overline{\alpha k})} \in C^0([a,b])$, so we may use formula 2.21 of [39, p.34]:

$$\mathcal{I}^{1-\alpha}_+ \circ \mathcal{I}^{\overline{\alpha k}-\alpha k}_+ f^{(\overline{\alpha k})} = \mathcal{I}^{\beta}_+ f^{(\overline{\alpha k})},$$

where $\beta = 1 + \overline{\alpha k} - \alpha(k+1)$.

Since $f^{(\overline{\alpha k})} \in C^1_+([a,b])$, from Lemma 50,

$$\frac{d}{dt} \circ \mathcal{I}^{\beta}_+ f^{(\overline{\alpha k})} = \mathcal{I}^{\beta}_+ f^{(\overline{\alpha k}+1)}.$$

We have $\overline{\alpha k} + 1 \in \{\overline{\alpha(k+1)}, \overline{\alpha(k+1)} + 1\}$, so we consider two cases.

- If $\overline{\alpha k} + 1 = \overline{\alpha(k+1)}$, then

$$\begin{aligned} \mathcal{D}^{\alpha}_+ \circ ({}^c\mathcal{D}^{\alpha}_+)^k f &= \mathcal{I}^{\overline{\alpha(k+1)}-\alpha(k+1)}_+ f^{(\overline{\alpha(k+1)})}, \\ &= {}^c\mathcal{D}^{\alpha(k+1)}_+ f. \end{aligned}$$

- If $\overline{\alpha k} = \overline{\alpha(k+1)}$, then

$$\begin{aligned} \mathcal{D}^{\alpha}_+ \circ ({}^c\mathcal{D}^{\alpha}_+)^k f &= \mathcal{I}^{1+\overline{\alpha(k+1)}-\alpha(k+1)}_+ f^{(\overline{\alpha(k+1)}+1)}, \\ &= \mathcal{I}^{\overline{\alpha(k+1)}-\alpha(k+1)}_+ \circ \mathcal{I}^1_+ f^{(\overline{\alpha(k+1)}+1)}. \end{aligned}$$

  We have $\mathcal{I}^1_+ f^{(\overline{\alpha(k+1)}+1)}(t) = f^{(\overline{\alpha(k+1)})}(t) - f^{(\overline{\alpha(k+1)})}(a)$. But $\overline{\alpha(k+1)} \leq k$, so $f^{(\overline{\alpha(k+1)})}(a) = 0$.

  Consequently,

$$\begin{aligned} \mathcal{D}^{\alpha}_+ \circ ({}^c\mathcal{D}^{\alpha}_+)^k f &= \mathcal{I}^{\overline{\alpha(k+1)}-\alpha(k+1)}_+ f^{(\overline{\alpha(k+1)})}, \\ &= {}^c\mathcal{D}^{\alpha(k+1)}_+ f. \end{aligned}$$

In both cases, we have proved that $({}^c\mathcal{D}^{\alpha}_{+})^{k+1} f = {}^c\mathcal{D}^{\alpha(k+1)}_{+} f$, which concludes the proof. □

Consequently, if $X = (x_+, x_-) \in C^k_+([a,b]) \times C^k_-([a,b])$, $({}^c\mathcal{D}^{\alpha})^k X$ verifies

$$({}^c\mathcal{D}^{\alpha})^k X = ({}^c\mathcal{D}^{\alpha k}_{+} x_+, (-1)^k \, {}^c\mathcal{D}^{\alpha k}_{-} x_-),$$

and $({}^c\mathcal{D}^{\alpha})^k X \in C^0_+([a,b]) \times C^0_-([a,b])$.

Let us now precise the splitting we are interested in.

Definition 56. *Let $k \in \mathbb{N}$ and $m, n \in \mathbb{N}^*$. Let $f \in \mathcal{F}(\mathbb{R}^{n(k+1)+1}, \mathbb{R}^m)$. The asymmetric representation of f, denoted by $\tilde{f}$, is defined by*

$$\begin{array}{rccc} \tilde{f}: & \mathbb{R}^{2n(k+1)+1} & \longrightarrow & \mathbb{R}^m \\ & (x_0, y_0, \ldots, x_k, y_k, t) & \longmapsto & f(x_0 + y_0, \ldots, x_k + y_k, t). \end{array}$$

Actually, the relevant functions will be in $\mathcal{F}([a,b], \mathbb{R}^n) \times \{0\}$ or $\{0\} \times \mathcal{F}([a,b], \mathbb{R}^n)$. That is why we introduce the following "selection" matrix.

Let $\mathcal{M}_{m,2m}(\mathbb{R})$ be the set of real matrices with m rows and $2m$ columns. We note I_m the identity matrix of dimension m, and we introduce the operator σ defined by

$$\begin{array}{rccl} \sigma: & \mathcal{F}([a,b], \mathbb{R}^m)^2 & \longrightarrow & \mathcal{M}_{m,2m}(\mathbb{R}) \\ & X & \longmapsto & \begin{cases} (I_m \quad 0) & \text{if } X \in \mathcal{F}([a,b], \mathbb{R}^n) \times \{0\} \text{ and } X \neq 0, \\ (0 \quad I_m) & \text{if } X \in \{0\} \times \mathcal{F}([a,b], \mathbb{R}^n) \text{ and } X \neq 0, \\ (0 \quad 0) & \text{otherwise.} \end{cases} \end{array}$$

Now we can define the asymmetric fractional embedding of a differential operator.

Definition 57. *With the previous notations, the asymmetric fractional embedding of operator* (124)*, denoted by $\mathcal{E}_\alpha(\mathcal{O}^{\mathbf{g}}_{\mathbf{f}})$, is defined on a subset $E^\alpha \subset \mathcal{F}([a,b], \mathbb{R}^n)^2$, by*

$$\begin{array}{rccl} \mathcal{E}_\alpha(\mathcal{O}^{\mathbf{g}}_{\mathbf{f}}): & E^\alpha & \longrightarrow & \mathcal{F}([a,b], \mathbb{R}^m) \\ & X & \longmapsto & \left[\tilde{f}_0 + \sigma(X) \sum_{i=1}^{p} \begin{pmatrix} \tilde{f}_i \star \mathcal{D}^{\alpha i}_{+} \tilde{g}_i \\ \tilde{f}_i \star (-1)^i \mathcal{D}^{\alpha i}_{-} \tilde{g}_i \end{pmatrix} \right] \left(X(\bullet), \ldots, ({}^c\mathcal{D}^{\alpha})^k X(\bullet), \bullet \right). \end{array} \tag{125}$$

The definition set E^α of $\mathcal{E}_\alpha(\mathcal{O}^{\mathbf{g}}_{\mathbf{f}})$ depends on $\mathbf{f}$ and $\mathbf{g}$. We also introduce the following spaces:

$$E^\alpha_+ = E^\alpha \cap (\mathcal{F}([a,b], \mathbb{R}^n) \times \{0\}), \quad E^\alpha_- = E^\alpha \cap (\{0\} \times \mathcal{F}([a,b], \mathbb{R}^n)).$$

In particular, for $(x_+, 0) \in E^\alpha_+$, (125) becomes

$$\mathcal{E}_\alpha(\mathcal{O}^{\mathbf{g}}_{\mathbf{f}})(x_+, 0)(t) = \left[f_0 + \sum_{i=1}^{p} f_i \star \mathcal{D}^{\alpha i}_{+} g_i \right] \left(x_+(t), \ldots, ({}^c\mathcal{D}^{\alpha}_{+})^k x_+(t), t \right),$$

and for $(0, x_-) \in E^\alpha_-$, we have

$$\mathcal{E}_\alpha(\mathcal{O}^{\mathbf{g}}_{\mathbf{f}})(0, x_-)(t) = \left[f_0 + \sum_{i=1}^{p} f_i \star (-1)^i \mathcal{D}^{\alpha i}_{-} g_i \right] \left(x_-(t), \ldots, (-{}^c\mathcal{D}^{\alpha}_{-})^k x_-(t), t \right).$$

Remark 2. *There exists of course several ways to define a fractional embedding because of the different definitions of fractional derivatives. As it will be seen later, with our choice, the action of a fractional Lagrangian is well defined, we may obtain a coherent and causal embedding. This property is not shared by the others fractional embedding formalisms.*

For the sake of clarity, we will often denote by $\overline{x}$ the integer which verifies $\overline{x}-1 \leq x < \overline{x}$, where $x \in \mathbb{R}^+$. We also denote by $\underline{x}$ the integer which verifies $\underline{x}-1 < x \leq \underline{x}$.

Precisions on E^α_+ and E^α_- can be given thanks to the following lemma.

Lemma 58. *Let $\beta > 0$ and $p \in \mathbb{N}$. If $f \in C^{\underline{\beta}+p}_+([a,b])$, then ${}^c\mathcal{D}^\beta_+ f \in C^p_+([a,b])$. A similar result holds for the right derivative.*

Proof. If $\beta \in \mathbb{N}^*$, ${}^c\mathcal{D}^\beta_+ f(t) = f^{(\beta)}(t) - f^{(\beta)}(a)$, and ${}^c\mathcal{D}^\beta_+ f \in C^p_+([a,b])$. Else, let $1 \leq k \leq p$. Since $f^{(\underline{\beta})} \in C^k_+([a,b])$, from Lemma 50,

$$\begin{aligned}\frac{d^k}{dt^k}\,{}^c\mathcal{D}^\beta_+ f &= \frac{d^k}{dt^k}\,\mathcal{I}^{\underline{\beta}-\beta}_+ f^{(\underline{\beta})} \\ &= \mathcal{I}^{\underline{\beta}-\beta}_+ f^{(\underline{\beta}+k)}\end{aligned}$$

Given that $f^{(\underline{\beta}+k)} \in C^0([a,b])$, $\mathcal{I}^{\underline{\beta}-\beta}_+ f^{(\underline{\beta}+k)} \in C^0_+([a,b])$, from Lemma 48. Hence ${}^c\mathcal{D}^\beta_+ f \in C^k([a,b])$ and $\frac{d^k}{dt^k}\,{}^c\mathcal{D}^\beta_+ f(a) = 0$. Moreover, ${}^c\mathcal{D}^\beta_+ f(a) = 0$ from Lemma 49. Finally, ${}^c\mathcal{D}^\beta_+ f \in C^p_+([a,b])$. □

Corollary 59.

- *If $\frac{\partial g_i}{\partial t} = 0$ for all $1 \leq i \leq p$, $C^{p+k}_+([a,b]) \times C^{p+k}_-([a,b]) \subset E^\alpha$, and for all $X \in C^{p+k}_+([a,b]) \times C^{p+k}_-([a,b])$, $\mathcal{E}_\alpha(\mathcal{O}^{\mathbf{g}}_f)(X) \in C^0([a,b])$.*
- *If $p = 0$ ($\mathbf{g} = \emptyset$) and $k = 1$, $C^1([a,b])^2 \subset E^\alpha$ and for all $X \in C^1([a,b])^2$, $\mathcal{E}_\alpha(\mathcal{O}^{\emptyset}_f)(X) \in C^0([a,b])$.*

Proof.

- Let $X = (x_+, x_-) \in C^{p+k}_+([a,b]) \times C^{p+k}_-([a,b])$. For all $1 \leq j \leq k$, $({}^c\mathcal{D}^\alpha)^j X = ({}^c\mathcal{D}^{\alpha j}_+ x_+, (-1)^j\,{}^c\mathcal{D}^{\alpha j}_- x_-)$ and $({}^c\mathcal{D}^\alpha)^j X \in C^p_+([a,b]) \times C^p_-([a,b])$, from Lemmas 55 and 58.

  If $x_+ \neq 0$ and $x_- \neq 0$, $\mathcal{E}_\alpha(\mathcal{O}^{\mathbf{g}}_{\mathbf{f}})(X) = \tilde{f}_0\left(X(\bullet),\ldots,({}^c\mathcal{D}^\alpha)^k X(\bullet),\bullet\right) \in C^0([a,b])$.

  If $x_- = 0$, let $1 \leq i \leq p$. Since g_i is of class C^i, we have $g_i(x_+) : t \mapsto g_i(x_+(t),\ldots,{}^c\mathcal{D}^{\alpha k}_+ x_+(t),t) \in C^i([a,b])$. Moreover, $g_i(x_+)'(a) = \sum_{j=1}^{k} \partial_j g_i(x_+)(a)\cdot x_+^{(j)}(a) + \frac{\partial g_i(x_+)}{\partial t}(a)$. Since $\frac{\partial g_i}{\partial t} = 0$ and $x_+^{(j)}(a) = 0$ for all $1 \leq j \leq k$, we obtain $g_i(x_+)'(a) = 0$. By induction, $g_i(x_+)^{(l)}(a) = 0$ for all $1 \leq l \leq i$. Hence $g_i(x_+) \in C^i_+([a,b])$, and from Lemma 49, $\mathcal{D}^{\alpha i}_+ g_i(x_+) \in C^0([a,b])$.

  We proceed likewise if $x_+ = 0$.
- Let $X \in C^1([a,b])^2$. We have $\tilde{f}_0(X(\bullet), {}^c\mathcal{D}^\alpha X(\bullet),\bullet) \in C^0([a,b])$ from Lemma 49, so $\mathcal{E}_\alpha(\mathcal{O}^{\emptyset}_{\mathbf{f}})(X) = \tilde{f}_0(X(\bullet), {}^c\mathcal{D}^\alpha X(\bullet),\bullet)$ is well defined and is a function of $C^0([a,b])$.

□

In order to clarify those notations, we give here a short example.

Example. We set $n = m = p = 1$, $k = 2$, and we suppose that $0 < \alpha < 1/2$.

Let f_0, f_1, $g_1 : \mathbb{R}^3 \times \mathbb{R} \longrightarrow \mathbb{R}$ be three functions defined by

$$\begin{aligned} f_0(a,b,c,t) &= c + e^{-t}\cos b, \\ f_1(a,b,c,t) &= 1, \\ g_1(a,b,c,t) &= \cos a. \end{aligned}$$

The associated operator $\mathcal{O}_{\mathbf{f}}^{\mathbf{g}}$ verifies

$$\mathcal{O}_{\mathbf{f}}^{\mathbf{g}}(x)(t) = \frac{d^2}{dt^2}x(t) + e^{-t}\cos\left(\frac{d}{dt}x(t)\right) + \frac{d}{dt}\cos(x(t)),$$

for $x \in C^2([a,b])$ and $t \in [a,b]$.

Moreover, for any $(x_+, x_-) \in AC^2([a,b])^2$, $({}^c\mathcal{D}^\alpha)^2(x_+,x_-) = ({}^c\mathcal{D}_+^{2\alpha}x_+, {}^c\mathcal{D}_-^{2\alpha}x_-)$ as it will be shown in Lemma 78. The asymmetric fractional embedding $\mathcal{E}_\alpha(\mathcal{O}_{\mathbf{f}}^{\mathbf{g}})$ is hence given by

$$\begin{aligned} \mathcal{E}_\alpha(\mathcal{O}_{\mathbf{f}}^{\mathbf{g}})(x_+,x_-)(t) = {}&{}^c\mathcal{D}_+^{2\alpha}x_+(t) + {}^c\mathcal{D}_-^{2\alpha}x_-(t) + e^{-t}\cos({}^c\mathcal{D}_+^\alpha x_+(t) - {}^c\mathcal{D}_-^\alpha x_-(t)) \\ &+\sigma(x_+,x_-)\begin{pmatrix} \mathcal{D}_+^\alpha \cos(x_+(t)+x_-(t)) \\ -\mathcal{D}_-^\alpha \cos(x_+(t)+x_-(t)) \end{pmatrix}. \end{aligned}$$

For $(x_+,0) \in AC^2([a,b]) \times \{0\}$, the fractional embedding becomes

$$\mathcal{E}_\alpha(\mathcal{O}_{\mathbf{f}}^{\mathbf{g}})(x_+,0)(t) = {}^c\mathcal{D}_+^{2\alpha}x_+(t) + e^{-t}\cos({}^c\mathcal{D}_+^\alpha x_+(t)) + \mathcal{D}_+^\alpha\cos(x_+(t)),$$

and for $(0,x_-) \in \{0\} \times AC^2([a,b])$, we have

$$\mathcal{E}_\alpha(\mathcal{O}_{\mathbf{f}}^{\mathbf{g}})(0,x_-)(t) = {}^c\mathcal{D}_-^{2\alpha}x_-(t) + e^{-t}\cos(-{}^c\mathcal{D}_-^\alpha x_-(t)) - \mathcal{D}_-^\alpha\cos(x_-(t)).$$

The ordinary differential equations may be written by using operators $\mathcal{O}_{\mathbf{f}}^{\mathbf{g}}$. Following [8], we consider the differential equations of the form

$$\mathcal{O}_{\mathbf{f}}^{\mathbf{g}}(x) = 0, \quad x \in C^{p+k}([a,b]). \tag{126}$$

Definition 60. *With the previous notations, the asymmetric fractional embedding of differential equation* (126) *is defined by*

$$\mathcal{E}_\alpha(\mathcal{O}_{f}^{\mathbf{g}})(X) = 0, \quad X \in E^\alpha. \tag{127}$$

Consequently, if $(x_+,0) \in E_+^\alpha$, (127) becomes

$$\left[f_0 + \sum_{i=1}^{p} f_i \star \mathcal{D}_+^{\alpha i} g_i\right]\left(x_+(t),\ldots,({}^c\mathcal{D}_+^\alpha)^k x_+(t),t\right) = 0,$$

and for $(0,x_-) \in E_-^\alpha$, we obtain

$$\left[f_0 + \sum_{i=1}^{p} f_i \star (-1)^i \mathcal{D}_-^{\alpha i} g_i\right]\left(x_-(t),\ldots,({}^c\mathcal{D}_-^\alpha)^k x_-(t),t\right) = 0.$$

We verify that for these two cases, the asymmetric fractional embedding respects causality, in the sense of Definition 4.

4.5. Variational Asymmetric Fractional Embedding-ODEs Case

4.5.1. Asymmetric Fractional Embedding of Lagrangian Systems

Let $0 < \alpha < 1$. Let L be a Lagrangian.

For $X = (x_1, x_2) \in \mathbb{R}^{2n}$, $Y = (y_1, y_2) \in \mathbb{R}^{2n}$, and $t \in [a, b]$, the asymmetric representation of L, denoted by $\tilde{L}$, verifies

$$\tilde{L}(X, Y, t) = L(x_1 + x_2, y_1 + y_2, t).$$

Given that

$$\frac{\partial L}{\partial x_1}(x_1 + x_2, y_1 + y_2, t) = \frac{\partial L}{\partial x_2}(x_1 + x_2, y_1 + y_2, t) = \partial_1 L(x_1 + x_2, y_1 + y_2, t),$$

we deduce $\partial_1 \tilde{L}(X, Y, t) = \partial_1 L(x_1 + x_2, y_1 + y_2, t)$. Similarly, we note $\partial_2 \tilde{L}(X, Y, t) = \partial_2 L(x_1 + x_2, y_1 + y_2, t)$.

Theorem 61. *The asymmetric fractional embedding of* (EL) *is defined by*

$$\mathcal{E}_\alpha(EL) \qquad \partial_1 \tilde{L}(X(t), {}^c\mathcal{D}^\alpha X(t), t) - \sigma(X) \begin{pmatrix} \mathcal{D}^\alpha_+ \, \partial_2 \tilde{L}(X(t), {}^c\mathcal{D}^\alpha X(t), t) \\ -\, \mathcal{D}^\alpha_- \, \partial_2 \tilde{L}(X(t), {}^c\mathcal{D}^\alpha X(t), t) \end{pmatrix} = 0. \tag{128}$$

In particular, for $(x_+, 0) \in E^\alpha_+$, (128) *becomes*

$$\mathcal{E}_\alpha(EL)_+ \qquad \partial_1 L(x_+(t), {}^c\mathcal{D}^\alpha_+ x_+(t), t) - \mathcal{D}^\alpha_+ \, \partial_2 L(x_+(t), {}^c\mathcal{D}^\alpha_+ x_+(t), t) = 0, \tag{129}$$

and for $(0, x_-) \in E^\alpha_-$,

$$\mathcal{E}_\alpha(EL)_- \qquad \partial_1 L(x_-(t), -\,{}^c\mathcal{D}^\alpha_- x_-(t), t) + \mathcal{D}^\alpha_- \, \partial_2 L(x_-(t), -\,{}^c\mathcal{D}^\alpha_- x_-(t), t) = 0. \tag{130}$$

Proof. Equation (EL) may be written like (126) with $k = 1$, $p = 1$, $\mathbf{f} = \{\partial_1 L, 1\}$ and $\mathbf{g} = \{-\partial_2 L\}$. We conclude by using Definitions 57 and 60. □

On the other hand, the asymmetric fractional embedding of the Lagrangian L, which will be noted L_α, verifies

$$L_\alpha(X)(t) = L(x_+(t) + x_-(t), {}^c\mathcal{D}^\alpha_+ x_+(t) - {}^c\mathcal{D}^\alpha_- x_-(t), t),$$

for all $(x_+, x_-) \in C^1([a, b])^2$ and $t \in [a, b]$.

The associated action (51) now becomes

$$\begin{array}{rccc} \mathcal{A}(L_\alpha) : & C^1([a,b])^2 & \longrightarrow & \mathbb{R} \\ & X & \longmapsto & \displaystyle\int_a^b \tilde{L}\left(X(t), {}^c\mathcal{D}^\alpha X(t), t\right) dt. \end{array}$$

Remark 3. *We see here the necessity to choose the Caputo derivative inside the functions. If we had taken the Riemann-Liouville derivative, the action could be undefined even for* *regular functions. For example, if* $L(x,v,t) = \frac{1}{2}v^2 - U(x)$ *and* $x_+ \in C^1([a,b])$*, with* $x_+(a) \neq 0$*, we would have*

$$L(x_+(t),\, \mathcal{D}_+^\alpha x_+(t), t) \underset{a}{\sim} \frac{1}{2}\left(\frac{x_+(a)}{\Gamma(1-\alpha)}\right)^2 (t-a)^{-2\alpha},$$

and $\mathcal{A}(L_\alpha)(x_+, 0)$ *would not be defined for* $\alpha \geq 1/2$.

The obtention of the differential of the action first requires a formula for integration by parts with fractional derivatives.

Lemma 62. *Let* $\beta > 0$*. If* $f \in AC^{\underline{\beta}}([a,b])$ *and* $g \in C_0^{\overline{\beta}}([a,b])$*, then we have the following formula for fractional integration by parts:*

$$\int_a^b f(t) \cdot {}^c\mathcal{D}_-^\beta\, g(t)\, dt = \int_a^b \mathcal{D}_+^\beta\, f(t) \cdot g(t)\, dt.$$

Similarly, we have:

$$\int_a^b f(t) \cdot {}^c\mathcal{D}_+^\beta\, g(t)\, dt = \int_a^b \mathcal{D}_-^\beta\, f(t) \cdot g(t)\, dt.$$

Proof. If $\beta \in \mathbb{N}^*$, this is the classical formula for integration by parts. Else, since $g \in C_0^{\overline{\beta}}([a,b])$, $g^{\underline{\beta}} \in L_p([a,b])$, with $p \geq 1/\beta$. Furthermore, $f^{(\underline{\beta})} \in L_1([a,b])$, so equation 2.20 of [39, p.34] is valid:

$$\int_a^b f(t) \cdot {}^c\mathcal{D}_-^\beta\, g(t)\, dt = (-1)^{\underline{\beta}} \int_a^b \mathcal{I}_+^{\underline{\beta}-\beta}\, f(t) \cdot g^{(\underline{\beta})}(t)\, dt.$$

Moreover, for all $0 \leq k \leq \underline{\beta} - 1$, $g^{(k)}(a) = g^{(k)}(b) = 0$. Therefore, iterating the classical integration by parts $\underline{\beta}$ times leads to:

$$(-1)^{\underline{\beta}} \int_a^b \mathcal{I}_+^{\underline{\beta}-\beta}\, f(t) \cdot g^{(\underline{\beta})}(t)\, dt = \int_a^b \frac{d^{\underline{\beta}}}{dt^{\underline{\beta}}} \mathcal{I}_+^{\underline{\beta}-\beta}\, f(t) \cdot g(t)\, dt = \int_a^b \mathcal{D}_+^\beta\, f(t) \cdot g(t)\, dt.$$

We proceed likewise for the other relation. □

Lemma 63. *Let* $X \in C^1([a,b])^2$*. We suppose that* $\partial_2 \tilde{L}(X(\bullet), {}^c\mathcal{D}^\alpha X(\bullet), \bullet) \in AC([a,b])$*. Then* $\mathcal{A}(L_\alpha)$ *is* $C_0^1([a,b])^2$*-differentiable at* X *and for all* $H = (h_+, h_-) \in C_0^1([a,b])^2$*,*

$$\begin{aligned} d\mathcal{A}(L_\alpha)(X,H) = &\int_a^b \left[\partial_1 \tilde{L}\left(X(t), {}^c\mathcal{D}^\alpha X(t), t\right) + \mathcal{D}_-^\alpha\, \partial_2 \tilde{L}\left(X(t), {}^c\mathcal{D}^\alpha X(t), t\right)\right] \cdot h_+(t)\, dt \\ &+ \int_a^b \left[\partial_1 \tilde{L}\left(X(t), {}^c\mathcal{D}^\alpha X(t), t\right) - \mathcal{D}_+^\alpha\, \partial_2 \tilde{L}\left(X(t), {}^c\mathcal{D}^\alpha X(t), t\right)\right] \cdot h_-(t)\, dt. \end{aligned}$$

Proof. Let $H = (h_+, h_-) \in C_0^1([a,b])^2$ and $\varepsilon > 0$. For all $t \in [a,b]$, we have:

$$\begin{aligned}\tilde{L}(X(t)+\varepsilon H(t), {}^c\mathcal{D}^\alpha X(t) + \varepsilon\, {}^c\mathcal{D}^\alpha H(t), t) &= \tilde{L}(X(t), {}^c\mathcal{D}^\alpha X(t), t)\\ &+ \partial_1 \tilde{L}(X(t), {}^c\mathcal{D}^\alpha X(t), t) \cdot (h_+(t) + h_-(t))\\ &+ \partial_2 \tilde{L}(X(t), {}^c\mathcal{D}^\alpha X(t), t) \cdot ({}^c\mathcal{D}_+^\alpha h_+(t) - {}^c\mathcal{D}_-^\alpha h_-(t)) + o(\varepsilon).\end{aligned}$$

Integrating this relation on $[a,b]$ leads to:

$$\begin{aligned}\mathcal{A}(L_\alpha)(X+\varepsilon H) = \mathcal{A}(L_\alpha)(X) &+ \varepsilon \int_a^b \partial_1 \tilde{L}(X(t), {}^c\mathcal{D}^\alpha X(t), t) \cdot (h_+(t) + h_-(t))\, dt\\ &+ \varepsilon \int_a^b \partial_2 \tilde{L}(X(t), {}^c\mathcal{D}^\alpha X(t), t) \cdot ({}^c\mathcal{D}_+^\alpha h_+(t) - {}^c\mathcal{D}_-^\alpha h_-(t))\, dt + o(\varepsilon).\end{aligned}$$

Since $\partial_2 \tilde{L}(X(\bullet), \mathcal{D}X(\bullet), \bullet) \in AC([a,b])$, $h_+ \in C_0^1([a,b])$ and $h_- \in C_0^1([a,b])$, we can use lemma 62:

$$\int_a^b \partial_2 \tilde{L}(X(t), {}^c\mathcal{D}^\alpha X(t), t) \cdot {}^c\mathcal{D}_+^\alpha h_+(t)\, dt = \int_a^b \mathcal{D}_-^\alpha\, \partial_2 \tilde{L}(X(t), {}^c\mathcal{D}^\alpha X(t), t) \cdot h_+(t)\, dt,$$

and

$$\int_a^b \partial_2 \tilde{L}(X(t), {}^c\mathcal{D}^\alpha X(t), t) \cdot {}^c\mathcal{D}_-^\alpha h_-(t)\, dt = \int_a^b \mathcal{D}_+^\alpha\, \partial_2 \tilde{L}(X(t), {}^c\mathcal{D}^\alpha X(t), t) \cdot h_-(t)\, dt.$$

Finally, $(h_+, h_-) \mapsto \int_a^b \left[\partial_1 \tilde{L}\left(X(t), {}^c\mathcal{D}^\alpha X(t), t\right) + \mathcal{D}_-^\alpha\, \partial_2 L\left(X(t), {}^c\mathcal{D}^\alpha X(t), t\right)\right] \cdot h_+(t)\, dt + \int_a^b \left[\partial_1 \tilde{L}\left(X(t), {}^c\mathcal{D}^\alpha X(t), t\right) - \mathcal{D}_+^\alpha\, \partial_2 L\left(X(t), {}^c\mathcal{D}^\alpha X(t), t\right)\right] \cdot h_-(t)\, dt$ is linear, which concludes the proof.

□

Then we obtain a result similar to Theorem 19.

Theorem 64. *Let $X \in C^1([a,b])^2$. We suppose that $\partial_2 \tilde{L}(X(\bullet), {}^c\mathcal{D}^\alpha X(\bullet), \bullet) \in AC([a,b])$.*

Then we have the following equivalence:

X is a $C_0^1([a,b])^2$-extremal of the action $\mathcal{A}(L_\alpha)$ if and only if it verifies

$$(EL_\alpha) \qquad \forall t \in (a,b), \quad \left\{\begin{array}{lcl} \partial_1 \tilde{L}(X(t), {}^c\mathcal{D}^\alpha X(t), t) + \mathcal{D}_-^\alpha\, \partial_2 \tilde{L}(X(t), {}^c\mathcal{D}^\alpha X(t), t) &=& 0,\\ \partial_1 \tilde{L}(X(t), {}^c\mathcal{D}^\alpha X(t), t) - \mathcal{D}_+^\alpha\, \partial_2 \tilde{L}(X(t), {}^c\mathcal{D}^\alpha X(t), t) &=& 0.\end{array}\right. \tag{131}$$

Proof. Similar to Theorem 19. The only difference is that $\mathcal{D}_-^\alpha\, \partial_2 \tilde{L}(X(t), {}^c\mathcal{D}^\alpha X(t), t)$ and $\mathcal{D}_+^\alpha\, \partial_2 \tilde{L}(X(t), {}^c\mathcal{D}^\alpha X(t), t)$ may not be continuous respectively in b and a. □

Equation (131) is very restrictive since X must verify

$$(\mathcal{D}^\alpha_+ + \mathcal{D}^\alpha_-)\partial_2\tilde{L}(X(t), {}^c\mathcal{D}^\alpha X(t), t) = 0.$$

This condition may not be related to the dynamics of the system and seems too strong. For instance, for $\alpha \in (1,2)$, functions which fulfill $(\mathcal{D}^\alpha_+ + \mathcal{D}^\alpha_-)f = 0$ are given in [27] and are very specific. By restricting the set of variations, equations more relevant will now be obtained.

Euler-Lagrange equations which have been obtained so far in [36, 2, 8, 22] involve both left and right fractional derivatives. The following result provides a similar equation.

Corollary 65. *Let $x_+ \in C^1([a,b])$. We suppose that $\partial_2 L(x_+(\bullet), {}^c\mathcal{D}^\alpha_+ x_+(\bullet), \bullet) \in AC([a,b])$. Then we have the following equivalence:*

$(x_+, 0)$ is a $C^1_0([a,b]) \times \{0\}$-extremal of the action $\mathcal{A}(L_\alpha)$ if and only if x_+ verifies

$$\partial_1 L(x_+(t), {}^c\mathcal{D}^\alpha_+ x_+(t), t) + \mathcal{D}^\alpha_- \partial_2 L(x_+(t), {}^c\mathcal{D}^\alpha_+ x_+(t), t) = 0, \tag{132}$$

for all $t \in [a,b)$.

Such an equation is not causal because of the simultaneous presence of ${}^c\mathcal{D}^\alpha_+$ and $\mathcal{D}^\alpha_-$. Moreover, regarding (129), this procedure is not coherent. Those problems are solved with the following results.

Corollary 66. *Let $x_+ \in C^1([a,b])$. We suppose that $\partial_2 L(x_+(\bullet), {}^c\mathcal{D}^\alpha_+ x_+(\bullet), \bullet) \in AC([a,b])$. Then we have the following equivalence:*

$(x_+, 0)$ is a $\{0\} \times C^1_0([a,b])$-extremal of the action $\mathcal{A}(L_\alpha)$ if and only if x_+ verifies

$$(EL_\alpha)_+ \qquad \forall t \in (a,b], \quad \partial_1 L(x_+(t), {}^c\mathcal{D}^\alpha_+ x_+(t), t) - \mathcal{D}^\alpha_+ \partial_2 L(x_+(t), {}^c\mathcal{D}^\alpha_+ x_+(t), t) = 0. \tag{133}$$

Corollary 67. *Let $x_- \in C^1([a,b])$. We suppose that $\partial_2 L(x_-(\bullet), {}^c\mathcal{D}^\alpha_- x_-(\bullet), \bullet) \in AC([a,b])$ Then we have the following equivalence:*

$(0, x_-)$ is a $C^1_0([a,b]) \times \{0\}$-extremal of the action $\mathcal{A}(L_\alpha)$ if and only if x_- verifies

$$(EL_\alpha)_- \qquad \forall t \in [a,b), \quad \partial_1 L(x_-(t), {}^c\mathcal{D}^\alpha_- x_-(t), t) - \mathcal{D}^\alpha_- \partial_2 L(x_-(t), {}^c\mathcal{D}^\alpha_- x_-(t), t) = 0. \tag{134}$$

Equations (133) and (134) are causal. Moreover, they are respectively similar to (129) and (130): $(EL_\alpha)_\pm \equiv \mathcal{E}_\alpha(EL)_\pm$. With such sets of variations, the asymmetric fractional embedding is therefore coherent.

4.5.2. Derivatives of Higher Orders

Let $\alpha \in (0,1)$ and $k \geq 2$. We consider here generalized Lagrangian systems which involve derivatives up to order k.

Definition 68. *An extended Lagrangian is a function*

$$\begin{array}{rccc} L: & \mathbb{R}^{n(k+1)} \times [a,b] & \longrightarrow & \mathbb{R} \\ & (y_0, y_1, \ldots, y_k, t) & \longmapsto & L(y_0, y_1, \ldots, y_k, t), \end{array}$$

which verifies the following properties:

- $L \in C^1(\mathbb{R}^{n(k+1)} \times [a,b])$,
- $\forall\, 1 \leq i \leq k,\ \partial_{i+1} L \in C^i(\mathbb{R}^{n(k+1)} \times [a,b])$.

The action is now defined by

$$\begin{array}{rccl} \mathcal{A}(L): & C^k([a,b]) & \longrightarrow & \mathbb{R} \\ & x & \longmapsto & \displaystyle\int_a^b L\left(x(t), \frac{d}{dt}x(t), \dots, \frac{d^k}{dt^k}x(t), t\right) dt. \end{array}$$

We then obtain:

Lemma 69. *Let L be an extended Lagrangian and $x \in C^k([a,b])$.*

We suppose that for all $1 \leq i \leq k$, $\partial_{i+1} L\left(x(\bullet), \frac{d}{dt}x(\bullet), \dots, \frac{d^k}{dt^k}x(\bullet), \bullet\right) \in AC^i([a,b])$.

Then $\mathcal{A}(L)$ is $C_0^k([a,b])$-differentiable at x and for all $h \in C_0^k([a,b])$,

$$d\mathcal{A}(L)(x,h) = \int_a^b \left[\partial_1 L + \sum_{i=1}^{k} (-1)^i \frac{d^i}{dt^i} \partial_{i+1} L\right] \left(x(t), \dots, \frac{d^k}{dt^k}x(t), t\right) dt.$$

Lemma 70. *Let L be an extended Lagrangian and $x \in C^k([a,b])$.*

We suppose that for all $1 \leq i \leq k$, $\partial_{i+1} L\left(x(\bullet), \frac{d}{dt}x(\bullet), \dots, \frac{d^k}{dt^k}x(\bullet), \bullet\right) \in AC^i([a,b])$.

Then we have the following equivalence:

x is a $C_0^k([a,b])$-extremal for $\mathcal{A}(L)$ if and only if x verifies the Euler-Lagrange equation

$$(EL_k) \qquad \forall t \in [a,b], \quad \left[\partial_1 L + \sum_{i=1}^{k} (-1)^i \frac{d^i}{dt^i} \partial_{i+1} L\right] \left(x(t), \dots, \frac{d^k}{dt^k}x(t), t\right) = 0. \tag{135}$$

Concerning the asymmetric fractional embedding, we start with the embedding of the Euler-Lagrange equation.

The asymmetric fractional embedding of (135) is given by:

$$\mathcal{E}_\alpha(EL_k) \qquad \left[\partial_1 \tilde{L} + \sigma(X) \sum_{i=1}^{k} \begin{pmatrix} (-1)^i \mathcal{D}_+^{\alpha i} \partial_{i+1} \tilde{L} \\ \mathcal{D}_-^{\alpha i} \partial_{i+1} \tilde{L} \end{pmatrix}\right] \left(X(t), \dots, ({}^c\mathcal{D}^\alpha)^k X(t), t\right) = 0.$$

In particular, for $(x_+, 0)$, we have

$$\mathcal{E}_\alpha(EL_k)_+ \qquad \left[\partial_1 L + \sum_{i=1}^{k} (-1)^i \mathcal{D}_+^{\alpha i} \partial_{i+1} L\right] \left(x_+(t), \dots, ({}^c\mathcal{D}_+^\alpha)^k x_+(t), t\right) = 0, \tag{136}$$

and for $(0, x_-)$,

$$\mathcal{E}_\alpha(EL_k)_- \qquad \left[\partial_1 L + \sum_{i=1}^{k} \mathcal{D}_-^{\alpha i} \partial_{i+1} L\right] \left(x_-(t), \dots, (-{}^c\mathcal{D}_-^\alpha)^k x_-(t), t\right) = 0. \tag{137}$$

Now we consider the embedding of the extended Lagrangian. First we need to set a vector space for the trajectories, suitable for the calculus of variations. Let F_k^α be the functional space defined by

$$F^{\alpha,k}([a,b]) = \left\{X \in C^0([a,b])^2 \mid \forall 1 \le i \le k,\ ({}^c\mathcal{D}^\alpha)^i X \in C^0([a,b])^2\right\}.$$

We also introduce

$$F_+^{\alpha,k}([a,b]) = F^{\alpha,k}([a,b]) \cap (\mathcal{F}([a,b],\mathbb{R}^n) \times \{0\}),$$

$$F_-^{\alpha,k}([a,b]) = F^{\alpha,k}([a,b]) \cap (\{0\} \times \mathcal{F}([a,b],\mathbb{R}^n)).$$

The asymmetric fractional embedding of L, still denoted by L_α, is given by

$$\begin{aligned} L_\alpha(X)(t) &= \tilde{L}(X(t),\ldots,({}^c\mathcal{D}^\alpha)^k X(t),t) \\ &= L(x_+(t)+x_-(t),\ldots,({}^c\mathcal{D}_+^\alpha)^k x_+(t) + (-\,{}^c\mathcal{D}_-^\alpha)^k x_-(t),t), \end{aligned}$$

for all $X = (x_+, x_-) \in F^{\alpha,k}([a,b])$ and $t \in [a,b]$.

The associated action is now given by

$$\begin{array}{rccl} \mathcal{A}(L_\alpha): & F^{\alpha,k}([a,b]) & \longrightarrow & \mathbb{R} \\ & X & \longmapsto & \displaystyle\int_a^b \tilde{L}\left(X(t),\ {}^c\mathcal{D}^\alpha X(t),\ldots,({}^c\mathcal{D}^\alpha)^k X(t),t\right)\,dt. \end{array}$$

The variations should be of course in $F^{\alpha,k}([a,b])$ and should be suitable for the integration by parts. The space $C_0^k([a,b])$ is suitable (but may not be optimal). In particular, $C_0^k([a,b]) \subset F^{\alpha,k}([a,b])$ from Lemmas 55 and 49.

The differential of the action is given by the following result.

Lemma 71. *Let L be an extended Lagrangian and $X \in F^{\alpha,k}([a,b])$.*

We suppose that for all $1 \le i \le k$, $\partial_{i+1}\tilde{L}(X(\bullet),\ldots,({}^c\mathcal{D}^\alpha)^k X(\bullet),\bullet) \in AC^{\underline{\alpha i}}([a,b])$. Then $\mathcal{A}(L_\alpha)$ is $C_0^k([a,b])^2$-differentiable at X and for all $H = (h_+,h_-) \in C_0^k([a,b])^2$,

$$\begin{aligned} d\mathcal{A}(L_\alpha)(X,H) &= \int_a^b \left[\partial_1\tilde{L} + \sum_{i=1}^k \mathcal{D}_-^{\alpha i}\,\partial_{i+1}\tilde{L}\right](X(t),\ldots,({}^c\mathcal{D}^\alpha)^k X(t),t)\cdot h_+(t)\,dt \\ &+ \int_a^b \left[\partial_1\tilde{L} + \sum_{i=1}^k (-1)^i\,\mathcal{D}_+^{\alpha i}\,\partial_{i+1}\tilde{L}\right](X(t),\ldots,({}^c\mathcal{D}^\alpha)^k X(t),t)\cdot h_-(t)\,dt. \end{aligned}$$

Proof. Let $H = (h_+,h_-) \in C_0^k([a,b])^2$ and $\varepsilon > 0$. Similarly to Lemma 75, we have:

$$\begin{aligned} \mathcal{A}(L_\alpha)(X+\varepsilon H) &= \mathcal{A}(L_\alpha)(X) + \varepsilon \int_a^b \partial_1\tilde{L}(X(t),\ldots,({}^c\mathcal{D}^\alpha)^k X(t),t)\cdot(h_+(t)+h_-(t))\,dt \\ &+ \varepsilon \int_a^b \sum_{i=1}^k \partial_{i+1}\tilde{L}(X(t),\ldots,({}^c\mathcal{D}^\alpha)^k X(t),t)\cdot\left(({}^c\mathcal{D}_+^\alpha)^i h_+(t) + (-\,{}^c\mathcal{D}_-^\alpha)^i h_-(t)\right)\,dt + o(\varepsilon). \end{aligned}$$

Let $1 \leq i \leq k$. Since $h_+ \in C_0^i([a,b])$, it verifies $({}^c\mathcal{D}_+^\alpha)^i h_+ = {}^c\mathcal{D}_+^{\alpha i}$, from Lemma 55. By using Lemma 62, we have:

$$\int_a^b \partial_{i+1}\tilde{L}(\dots)\cdot {}^c\mathcal{D}_+^{\alpha i} h_+(t)\,dt = \int_a^b \mathcal{D}_-^{\alpha i}\,\partial_{i+1}\tilde{L}(\dots)\cdot h_+(t)\,dt.$$

A similar relation holds for h_-. Hence we obtain

$$\begin{aligned}\mathcal{A}(L_\alpha)(X+\varepsilon H) &= \mathcal{A}(L_\alpha)(X)\\ &\quad + \varepsilon\int_a^b \left[\partial_1\tilde{L} + \sum_{i=1}^k \mathcal{D}_-^{\alpha i}\,\partial_{i+1}\tilde{L}\right](X(t),\dots,({}^c\mathcal{D}^\alpha)^k X(t),t)\cdot h_+(t)\,dt\\ &\quad + \varepsilon\int_a^b \left[\partial_1\tilde{L} + \sum_{i=1}^k (-1)^i\,\mathcal{D}_+^{\alpha i}\,\partial_{i+1}\tilde{L}\right](X(t),\dots,({}^c\mathcal{D}^\alpha)^k X(t),t)\cdot h_-(t)\,dt + o(\varepsilon).\end{aligned}$$

The terms in h_+ and h_- are linear, which concludes the proof. □

We may still obtain coherent and causal embeddings, thanks to the following equivalences.

Theorem 72. *Let $(x_+,0) \in F_+^{\alpha,k}([a,b])$.*

We suppose that for all $1 \leq i \leq k$, $\partial_{i+1}L(x_+(\bullet),\dots,({}^c\mathcal{D}_+^\alpha)^k x_+(\bullet),\bullet) \in AC^{\underline{\alpha i}}([a,b])$.

Then we have the following equivalence:

$(x_+,0)$ is a $\{0\}\times C_0^k([a,b])$-extremal of the action $\mathcal{A}(L_\alpha)$ if and only if x_+ verifies

$$(EL_{k,\alpha})_+ \forall t\in(a,b],\ \left[\partial_1 L + \sum_{i=1}^k (-1)^i\,\mathcal{D}_+^{\alpha i}\,\partial_{i+1}L\right]\Big(x_+(t),\dots,({}^c\mathcal{D}_+^\alpha)^k x_+(t),t\Big) = 0. \tag{138}$$

Let $(0,x_-) \in F_-^{\alpha,k}([a,b])$.

We suppose that for all $1 \leq i \leq k$, $\partial_{i+1}L(x_-(\bullet),\dots,(-{}^c\mathcal{D}_-^\alpha)^k x_-(\bullet),\bullet) \in AC^{\overline{\alpha i}}([a,b])$.

Then we have the following equivalence:

$(0,x_-)$ is a $C_0^k([a,b])\times\{0\}$-extremal of the action $\mathcal{A}(L_\alpha)$ if and only if x_- verifies

$$(EL_{k,\alpha})_- \forall t\in[a,b),\ \left[\partial_1 L + \sum_{i=1}^k \mathcal{D}_-^{\alpha i}\,\partial_{i+1}L\right]\Big(x_-(t),\dots,(-{}^c\mathcal{D}_-^\alpha)^k x_-(t),t\Big) = 0. \tag{139}$$

Proof. From Theorem 1.2.4 of [25], the fundamental lemma in the calculus of variations is still valid for variations in $C_0^\infty([a,b])$. Since $C_0^\infty([a,b]) \subset C_0^k([a,b])$, the result is proved. □

Equations (138) and (139) are once again similar to (136) and (137): $(EL_{k,\alpha})_\pm \equiv \mathcal{E}_\alpha(EL_k)_\pm$. The following diagrams are thus valid:

$$
\begin{array}{ccc}
L & \xrightarrow{\mathcal{E}_{\alpha}^{+}} & L_{\alpha} \\
{\scriptstyle C_0^k([a,b])}\updownarrow & & \updownarrow{\scriptstyle \{0\}\times C_0^k([a,b])} \\
(EL_k) & \xrightarrow{\mathcal{E}_{\alpha}^{+}} & \mathcal{E}_{\alpha}(EL_k)_{+}
\end{array}
\qquad
\begin{array}{ccc}
L & \xrightarrow{\mathcal{E}_{\alpha}^{-}} & L_{\alpha} \\
{\scriptstyle C_0^k([a,b])}\updownarrow & & \updownarrow{\scriptstyle C_0^k([a,b])\times\{0\}} \\
(EL_k) & \xrightarrow{\mathcal{E}_{\alpha}^{-}} & \mathcal{E}_{\alpha}(EL_k)_{-}
\end{array}
$$

4.6. Variational Asymmetric Fractional Embedding-PDEs Case

In this part, we recall the classical definitions of the left and right Riemann-Liouville and Caputo derivatives in the one dimensional case. We define the multidimensional fractional analogous for partial derivatives, gradient and divergence. In particular, we prove a fractional Green-Riemann theorem. We give a simplified version of the asymmetric fractional calculus of variations introduced in [10] for which we refer for more details and weaker assumptions on functional spaces. This formalism will be used in the last part to derive the convection-diffusion equation from a variational principle.

4.6.1. Asymmetric Fractional Calculus of Variations

The fractional Euler-Lagrange equations obtained so far in [36, 2, 8, 22] involve both left and right fractional derivatives. This is a main drawback, if ones want to recover PDEs with order one derivative as composition of fractional derivatives of order 1/2. In this section, we give a simplified version of the asymmetric calculus of variations introduced in [10] which provides causal fractional Euler-Lagrange equations. We refer to [10] and [26] for more details and in particular for weaker assumptions on the functional spaces.

Asymmetric Fractional Lagrangian Here the fractional derivatives are seen as partial fractional derivatives according to t and x. With the notations u_x and u_t from section 1 4.1.2 we denote by

$$
{}^{c}\mathcal{D}_{+}^{\alpha}\, u(t,x) := {}^{c}\mathcal{D}_{+}^{\alpha}\, u_x(t), \quad {}^{c}\nabla^{\alpha} u(t,x) := {}^{c}\nabla^{\alpha} u_t(x) \quad \text{and} \quad \nabla u(t,x) := \nabla u_t(x).
$$

For a field $U = (u_+, u_-)$ we define

$$
{}^{c}\mathcal{D}^{\alpha} U(t,x) := \left({}^{c}\mathcal{D}_{+}^{\alpha}\, u_+(t,x), -{}^{c}\mathcal{D}_{-}^{\alpha}\, u_-(t,x)\right) \quad {}^{c}\nabla^{\alpha} U(t,x) := \left({}^{c}\nabla^{\alpha} u_+(t,x), -{}^{c}\overline{\nabla^{\alpha}} u_-(t,x)\right).
$$

For a generalised Lagrangian $L(t,x,y,v,w,z)$, we denote $\partial_y L$, $\partial_v L$, ∂_w and ∂_z the partial derivatives of L.

Definition 73. *Let be a Lagrangian L defined as follows:*

$$
\begin{array}{rccl}
L: & [a,b] \times \Omega \times \mathbb{R} \times \mathbb{R} \times \mathbb{R}^d \times \mathbb{R}^d & \longrightarrow & \mathbb{R} \\
 & (t,x,y,v,w,z) & \longmapsto & L(t,x,y,v,w,z)
\end{array}
$$

The asymmetric representation of L is denoted by $\tilde{L}$ and given by:

$$
\begin{array}{rccl}
\tilde{L}: & [a,b] \times \Omega \times \mathbb{R}^2 \times \mathbb{R}^2 \times \mathbb{R}^{2d} \times \mathbb{R}^{2d} & \longrightarrow & \mathbb{R} \\
 & (t,x,U,V,W,Z) & \longmapsto & \tilde{L}(t,x,U,V,W,Z)
\end{array}
$$

where

$$\tilde{L}(t,x,(u_+,u_-),(v_+,v_-),(w_+,w_-),(z_+,z_-)) := L(t,x,u_+ + u_-, v_+ + v_-, w_+ + w_-, z_+ + z_-).$$

The asymmetric fractional Lagrangian functional $\mathcal{L}_\alpha$ is defined by:

Definition 74. *Let L be a Lagrangian as defined in definition 73 of class $C^1([a,b]\times\mathbb{R}^{3d+2})$ and $\tilde{L}$ its asymmetric representation. The associated asymmetric fractional Lagrangian functional of order α is given by*

$$\begin{array}{rcl} \mathcal{L}_\alpha : C^1([a,b]\times\Omega)^2 & \longrightarrow & \mathbb{R} \\ U & \longmapsto & \displaystyle\int_a^b \int_\Omega \tilde{L}\left(t,x,U(t,x),\, {}^c\mathcal{D}^\alpha U(t), {}^c\nabla^\alpha U(t,x), \nabla U(t,x)\right) dx\, dt. \end{array}$$

For convenience, we denote by L_α the functional defined for any $U = (u_+,u_-) \in C^1([a,b]\times\Omega)^2$ by

$$\begin{array}{rcl} L_\alpha(U)(t,x) & := & \tilde{L}\left(t,x,U(t,x),\, {}^c\mathcal{D}^\alpha U(t), {}^c\nabla^\alpha U(t,x), \nabla U(t,x)\right) \\ & = & L(t,x,u_+(t,x)+u_-(t,x),\, {}^c\mathcal{D}^\alpha_+ u_+(t,x) - {}^c\mathcal{D}^\alpha_- u_-(t,x), \\ & & {}^c\nabla^\alpha u_+(t,x) - {}^c\overline{\nabla}^\alpha u_-(t,x), \nabla u_+(t,x) + \nabla u_-(t,x)), \end{array} \tag{140}$$

for any $x\in\Omega$ and $t\in[a,b]$ so that

$$\mathcal{L}_\alpha(U) = \int_a^b \int_\Omega L_\alpha(U)(t,x)dx\, dt.$$

Let us note that as $U\in C^1([a,b]\times\Omega)$ we have ${}^c\mathcal{D}^\alpha_U(t)$, ${}^c\nabla^\alpha_x U(t,x)$ and $\nabla_x U(t,x) \in C^0([a,b]\times\Omega)$. Using the fact that L is $C^1([a,b]\times\mathbb{R}^{3d+2})$ we obtain that $L_\alpha(C^1([a,b]\times\Omega) \subset C^0([a,b]\times\Omega)$ and conclude that $\mathcal{L}_\alpha$ is well defined.

Asymmetric Calculus of Variations The next lemma explicits the differential of the functional $\mathcal{L}_\alpha$ defined on $U\in(C^1([a,b]\times\Omega))^2$ in the direction $(C^1_0([a,b]\times\Omega))^2$

Lemma 75. *Let $U \in (C^1([a,b]\times\Omega))^2$. Let $\star^\alpha := (t,x,U(t,x),\, {}^c\mathcal{D}^\alpha U(t,x), {}^c\nabla^\alpha U(t,x), \nabla U(t,x))$. We assume that*

- $\forall x\in\Omega,\ t\mapsto \partial_v L(\star^\alpha) \in AC^1([a,b])$,
- $\forall t\in[a,b],\ x\mapsto \partial_w L(\star^\alpha) \in AC^1(\overline{\Omega})$,
- $\forall t\in[a,b],\ x\mapsto \partial_z L(\star^\alpha) \in C^1(\Omega)$.

Then $\mathcal{L}_\alpha$ is $(C^1([a,b]\times\Omega))^2$-differentiable at U and in any direction $H = (h_+,h_-) \in (C^1_0([a,b]\times\Omega))^2$, the differential of $\mathcal{L}_\alpha$ is given by

$$\begin{array}{rcl} D\mathcal{L}_\alpha(U,H) & = & \displaystyle\int_a^b\int_\Omega \left[\partial_y L(\star^\alpha) + \mathcal{D}^\alpha_- \partial_v L(\star^\alpha) + \overline{\mathrm{div}}^\alpha(\partial_w L(\star^\alpha)) - \mathrm{div}(\partial_z L(\star^\alpha))\right]\cdot h_+(t)\, dx\, dt \\ & & + \displaystyle\int_a^b\int_\Omega \left[\partial_y L(\star^\alpha) - \mathcal{D}^\alpha_+ \partial_v L(\star^\alpha) - \mathrm{div}^\alpha(\partial_w L(\star^\alpha)) - \mathrm{div}(\partial_z L(\star^\alpha))\right]\cdot h_-(t)\, dx\, dt. \end{array}$$

Proof. Using a Taylor expansion of the Lagrangian L, we obtain:

$$\begin{aligned} D\mathcal{L}_\alpha(U,H) &= \int_a^b \int_\Omega \big[\partial_y L(\star)\cdot(h_+ + h_-) + \partial_v L(\star)\cdot({}^c\mathcal{D}_+^\alpha h_+ - {}^c\mathcal{D}_-^\alpha h_-) \\ &\quad + \partial_w L(\star)\cdot\left({}^c\nabla^\alpha h_+ - {}^c\overline{\nabla^\alpha} h_-\right) + \partial_z L(\star)\cdot(\nabla h_+ + \nabla h_+)\big]\, dx\, dt. \end{aligned}$$

As $\partial_v L(\star) \in AC^1([a,b])$ by assumption, using the integration by parts formula of lemma 46 with h_+ (resp. h_-) in $C_0^1([a,b])$ leads to

$$\int_a^b \int_\Omega \partial_v L(\star)\cdot({}^c\mathcal{D}_+^\alpha h_+ - {}^c\mathcal{D}_-^\alpha h_-)\, dx\, dt = \int_a^b \int_\Omega \left(\mathcal{D}_-^\alpha\, \partial_v L(\star)\cdot h_+ - \mathcal{D}_+^\alpha\, \partial_v L(\star)\cdot h_-\right) dx\, dt.$$

As $\partial_w L(\star) \in AC^1(\overline{\Omega})$ and h_+ (resp. h_-) is in $C_0^1([a,b]\times\Omega)$, we can apply the fractional Green-Riemann formula from lemma 53 to obtain

$$\int_a^b \int_\Omega \partial_w L(\star)\cdot\left({}^c\nabla^\alpha h_+ - {}^c\overline{\nabla^\alpha} h_-\right) dx\, dt = \int_a^b \int_\Omega \left(\overline{\text{div}}^\alpha(\partial_w L(\star))\cdot h_+ - \text{div}^\alpha(\partial_w L(\star))\cdot h_-\right) dx\, dt.$$

The last part of the formula comes from the usual Green-Riemann theorem. This completes the proof. □

A consequence of the previous lemma is the following characterisation of the extremals of $\mathcal{L}_\alpha$ for asymmetric fractional functional as solutions of two fractional Euler-Lagrange equations:

Theorem 76. *Let $U \in (C^1([a,b]\times\Omega))^2$.*
Let $\star^\alpha := (t, x, U(t,x), {}^c\mathcal{D}^\alpha U(t,x), {}^c\nabla^\alpha U(t,x), \nabla U(t,x))$. We assume that

- $\forall x \in \Omega,\ t \mapsto \partial_v L(\star^\alpha) \in AC^1([a,b])$,
- $\forall t \in [a,b],\ x \mapsto \partial_w L(\star^\alpha) \in AC^1(\overline{\Omega})$,
- $\forall t \in [a,b],\ x \mapsto \partial_z L(\star^\alpha) \in C^1(\Omega)$.

Then $(C_0^1([a,b]\times\Omega))^2$ extremals of $\mathcal{L}_\alpha$ correspond to solutions of the following set of fractional Euler-Lagrange equations:

$$\begin{aligned} \partial_u L(\star^\alpha) + \mathcal{D}_-^\alpha\left(\partial_v L(\star^\alpha)\right) + \overline{\text{div}}^\alpha\left(\partial_w L(\star^\alpha)\right) - \text{div}\left(\partial_z L(\star^\alpha)\right) &= 0, \\ \partial_u L(\star^\alpha) - \mathcal{D}_+^\alpha\left(\partial_v L(\star^\alpha)\right) - \text{div}^\alpha\left(\partial_w L(\star^\alpha)\right) - \text{div}\left(\partial_z L(\star^\alpha)\right) &= 0. \end{aligned}$$

4.6.2. Specialisation

This theorem is not applicable as it is for classical PDEs since as it provides a system of PDEs. However, by restricting our attention to extremals over $C^1([a,b]\times\Omega)\times\{0\}$ over variations in $\{0\}\times C_0^1([a,b]\times\Omega)$, we obtain a more interesting version:

Theorem 77. *Let $u_+ \in C^1([a,b]\times\Omega)$.*
Let $\star_+^\alpha := \left(t, x, u_+(t,x), {}^c\mathcal{D}_+^\alpha u_+(t,x), {}^c\nabla u_+(t,x), \nabla u_+(t,x)\right)$. We assume that

- $\forall x \in \Omega,\ t \mapsto \partial_v L(\star_+^\alpha) \in AC^1([a,b])$,
- $\forall t \in [a,b],\ x \mapsto \partial_w L(\star_+^\alpha) \in AC^1(\overline{\Omega})$,

- $\forall t \in [a,b],\ x \mapsto \partial_z L(\star_+^\alpha) \in C^1(\Omega)$.

Then $(u_+, 0)$ *is a* $\{0\} \times C_0^1([a,b] \times \Omega)$*-extremal of the action* $\mathcal{L}_\alpha$ *if and only if* u_+ *satisfies*

$$\partial_y L(\star_+^\alpha) - \mathcal{D}_+^\alpha\left(\partial_v L(\star_+^\alpha)\right) - \mathrm{div}^\alpha\left(\partial_w L(\star_+^\alpha)\right) - \mathrm{div}\left(\partial_z L(\star_+^\alpha)\right) = 0,$$

for any $x \in \Omega$, $t \in [a,b]$.

The proof is a consequence of lemma 75.

Remark 4. *In the previous part, we have heuristically justified the introduction of a doubled phase space by saying that irreversibility induces a natural arrow of time. Our idea to focus only on curves in* $C^1([a,b] \times \Omega) \times \{0\}$ *is precisely to say that we are interested in one direction of time (here the future). However, the selection of a transverse set for variations to the underlying phase space is not so clear. It means heuristically that the future depends mostly on the virtual variations in the past. More work are needed in this direction.*

5. Fractional Variational Formulation of Dissipative Ordinary Differential Equations

5.1. Linear Friction

The differential equation of linear friction is

$$m\frac{d^2}{dt^2}x(t) + \gamma\frac{d}{dt}x(t) - \nabla U(x(t)) = 0, \tag{141}$$

where $t \in [a,b]$, $m, \gamma > 0$ and $U \in C^1(\mathbb{R}^n)$.

Even if $U(x)$ is quadratic, it has been shown in [5] that this equation cannot be derived from a variational principle with classical derivatives. But this can be done by using fractional derivatives, since $\frac{d}{dt} = {}^c\mathcal{D}_+^{1/2} \circ {}^c\mathcal{D}_+^{1/2}$, which is proved in the following lemma.

Lemma 78. *If* $f \in AC^2([a,b])$*, we have:*

- *if* $0 < \alpha < 1/2$, ${}^c\mathcal{D}_+^\alpha \circ {}^c\mathcal{D}_+^\alpha f = {}^c\mathcal{D}_+^{2\alpha} f$,
- *if* $\alpha = 1/2$, ${}^c\mathcal{D}_+^{1/2} \circ {}^c\mathcal{D}_+^{1/2} f = f'$,
- *if* $1/2 < \alpha < 1$*, for all* $t \in (a,b]$, ${}^c\mathcal{D}_+^\alpha \circ {}^c\mathcal{D}_+^\alpha f(t) = {}^c\mathcal{D}_+^{2\alpha} f(t) + \frac{f'(a)}{\Gamma(2-2\alpha)}(t-a)^{1-2\alpha}$.

Proof. We prove it by induction on k. For $k = 1$, the result is obvious. Now, let $k \in \mathbb{N}^*$ and $f \in C_+^{k+1}([a,b])$. Since $f \in C_+^k([a,b])$, we use the induction hypothesis: $({}^c\mathcal{D}_+^\alpha)^k f = {}^c\mathcal{D}_+^{\alpha k} f = \mathcal{I}_+^{\overline{\alpha k}-\alpha k} f^{(\overline{\alpha k})}$. We have $\overline{\alpha k} \le k$, so $f^{(\overline{\alpha k})} \in AC([a,b])$, and from Lemma 48, $({}^c\mathcal{D}_+^\alpha)^k f \in AC([a,b])$. Moreover, from Lemma 49, $({}^c\mathcal{D}_+^\alpha)^k f(a) = 0$. We may then apply Theorem 45:

$$\mathcal{D}_+^\alpha \circ ({}^c\mathcal{D}_+^\alpha)^k f = {}^c\mathcal{D}_+^\alpha \circ ({}^c\mathcal{D}_+^\alpha)^k f = ({}^c\mathcal{D}_+^\alpha)^{k+1} f.$$

On the other hand, $\mathcal{D}_+^\alpha \circ ({}^c\mathcal{D}_+^\alpha)^k f = \dfrac{d}{dt} \circ \mathcal{I}_+^{1-\alpha} \circ \mathcal{I}_+^{\overline{\alpha k}-\alpha k} f^{(\overline{\alpha k})}$. We have $f^{(\overline{\alpha k})} \in C^0([a,b])$, so we may use formula 2.21 of [39, p.34]:

$$\mathcal{I}_+^{1-\alpha} \circ \mathcal{I}_+^{\overline{\alpha k}-\alpha k} f^{(\overline{\alpha k})} = \mathcal{I}_+^{\beta} f^{(\overline{\alpha k})},$$

where $\beta = 1 + \overline{\alpha k} - \alpha(k+1)$.

Since $f^{(\overline{\alpha k})} \in C_+^1([a,b])$, from Lemma 50,

$$\frac{d}{dt} \circ \mathcal{I}_+^{\beta} f^{(\overline{\alpha k})} = \mathcal{I}_+^{\beta} f^{(\overline{\alpha k}+1)}.$$

We have $\overline{\alpha k}+1 \in \{\overline{\alpha(k+1)}, \overline{\alpha(k+1)}+1\}$, so we consider two cases.

- If $\overline{\alpha k}+1 = \overline{\alpha(k+1)}$, then

$$\begin{aligned}\mathcal{D}_+^\alpha \circ ({}^c\mathcal{D}_+^\alpha)^k f &= \mathcal{I}_+^{\overline{\alpha(k+1)}-\alpha(k+1)} f^{(\overline{\alpha(k+1)})}, \\ &= {}^c\mathcal{D}_+^{\alpha(k+1)} f.\end{aligned}$$

- If $\overline{\alpha k} = \overline{\alpha(k+1)}$, then

$$\begin{aligned}\mathcal{D}_+^\alpha \circ ({}^c\mathcal{D}_+^\alpha)^k f &= \mathcal{I}_+^{1+\overline{\alpha(k+1)}-\alpha(k+1)} f^{(\overline{\alpha(k+1)}+1)}, \\ &= \mathcal{I}_+^{\overline{\alpha(k+1)}-\alpha(k+1)} \circ \mathcal{I}_+^{1} f^{(\overline{\alpha(k+1)}+1)}.\end{aligned}$$

  We have $\mathcal{I}_+^1 f^{(\overline{\alpha(k+1)}+1)}(t) = f^{(\overline{\alpha(k+1)})}(t) - f^{(\overline{\alpha(k+1)})}(a)$. But $\overline{\alpha(k+1)} \leq k$, so $f^{(\overline{\alpha(k+1)})}(a) = 0$.

  Consequently,

$$\begin{aligned}\mathcal{D}_+^\alpha \circ ({}^c\mathcal{D}_+^\alpha)^k f &= \mathcal{I}_+^{\overline{\alpha(k+1)}-\alpha(k+1)} f^{(\overline{\alpha(k+1)})}, \\ &= {}^c\mathcal{D}_+^{\alpha(k+1)} f.\end{aligned}$$

In both cases, we have proved that $({}^c\mathcal{D}_+^\alpha)^{k+1} f = {}^c\mathcal{D}_+^{\alpha(k+1)} f$, which concludes the proof.

□

We consider the function $L(x,v,w,t) = \dfrac{m}{2}w^2 - \dfrac{\gamma}{2}v^2 - U(x)$, which is an extended Lagrangian.

The variations should be chosen in $C_0^2([a,b])$, but the space

$$AC_0^2([a,b]) = \{f \in AC^2([a,b]) \mid f(a) = f(b) = 0\}$$

is actually sufficient.

Theorem 79. *Let* $x \in C^2([a,b])$. *Then* x *is solution of* (141) *if and only if* $(x,0)$ *is a* $\{0\} \times AC_0^2([a,b])$*-extremal of the action* $\mathcal{A}(L_{1/2})$.

Proof. From Lemma 78 and its proof, ${}^c\mathcal{D}^{1/2}_+ \circ {}^c\mathcal{D}^{1/2}_+ x(t) = \mathcal{D}^{1/2}_+ \circ {}^c\mathcal{D}^{1/2}_+ x(t) = \frac{d}{dt}x(t)$, for all $t \in [a,b]$. Hence $L(x(\bullet), {}^c\mathcal{D}^{1/2}_+ x(\bullet), x'(\bullet), \bullet) \in C^0([a,b])$ and the action $\mathcal{A}(L_{1/2})$ is well defined.

Let $t \in [a,b]$. The partial derivatives of L verify:

- $\partial_1 L(x(t), {}^c\mathcal{D}^{1/2}_+ x(t), x'(t), t) = -\nabla U(x(t))$,
- $\partial_2 L(x(t), {}^c\mathcal{D}^{1/2}_+ x(t), x'(t), t) = -\gamma\, {}^c\mathcal{D}^{1/2}_+ x(t)$,
- $\partial_3 L(x(t), {}^c\mathcal{D}^{1/2}_+ x(t), x'(t), t) = m\, x'(t)$.

Since $x' \in AC([a,b])$, ${}^c\mathcal{D}^{1/2}_+ x = \mathcal{I}^{1/2}_+ x' \in AC([a,b])$, from Lemma 48. Consequently, for all $1 \le i \le 2$, $\partial_{i+1} L(x(\bullet), {}^c\mathcal{D}^{1/2}_+ x(\bullet), x'(\bullet), \bullet) \in AC^{\underline{\alpha_i}}([a,b])$. Conditions of application of Theorem 72 are hence fulfilled. Since $C^\infty_0([a,b]) \subset AC^2_0([a,b])$, the choice of $\{0\} \times AC^2_0([a,b])$ for the variations is valid and Theorem 72 may be applied :

$(x,0)$ is a $\{0\} \times AC^2_0([a,b])$-extremal of the action $\mathcal{A}(L_{1/2})$ if and only if x verifies

$$\left[\partial_1 L - \mathcal{D}^{1/2}_+ \partial_2 L + \mathcal{D}^1_+ \partial_3 L\right](x(t), {}^c\mathcal{D}^{1/2}_+ x(t), x'(t), t) = 0. \tag{142}$$

Given that $\mathcal{D}^1_+ x'(t) = x''(t)$, (142) is exactly (141). □

We see here the necessity of having a causal Euler-Lagrange equation. Indeed, an equation similar to (132) would have provide $\mathcal{D}^\alpha_- \circ {}^c\mathcal{D}^\alpha_+$ which is never equal to $\frac{d}{dt}$.

Furthermore, the choice of $({}^c\mathcal{D}^\alpha)^k$ instead of ${}^c\mathcal{D}^{\alpha k}$ in the asymmetric fractional embedding is justified here. If we had taken ${}^c\mathcal{D}^{\alpha k}$, the evaluation of the Lagrangian in this example would have been $L(x(t), {}^c\mathcal{D}^{1/2}_+ x(t), x'(t) - x'(a), t)$, since ${}^c\mathcal{D}^1_+ x(t) = x'(t) - x'(a)$. Hence the initial condition $x'(a) = 0$ should have been added to obtain (141), which is too restrictive for the solutions of (141).

6. Variational Formulation of Dissipative Partial Differential Equations

6.1. The Diffusion Equation

We are now interested in the diffusion equation

$$\frac{\partial}{\partial t}u(x,t) = c\,\Delta u(x,t), \tag{143}$$

where $t \in [a,b]$, $x \in \Omega$, $c > 0$, and Δ is the Laplace operator.

We consider the generalized Lagrangian $L(u,v,w,x,t) = \frac{1}{2}w^2 - \frac{c}{2}v^2$.

Theorem 80. *Let* $u \in \mathcal{F}(\Omega \times [a,b], \mathbb{R})$ *such that*

- $\forall x \in \Omega,\ t \mapsto u(x,t) \in AC^2([a,b])$,

- $\forall t \in [a,b],\ x \mapsto u(x,t) \in C^2(\Omega)$.

Then u is solution of (143) *if and only if* $(u,0)$ *is a* $\{0\} \times C_0^2(\Omega \times [a,b])$*-extremal of the action* $\mathcal{A}(L_{1/2})$.

Proof. Let $x \in \Omega$ and $t \in [a,b]$. The partial derivatives of L verify:

- $\partial_w L(u(x,t), \nabla u(x,t), {}^c\mathcal{D}_+^{1/2}\, u(x,t), x, t) = {}^c\mathcal{D}_+^{1/2}\, u(x,t)$,
- $\forall 1 \le i \le n,\ \partial_{v_i} L(u(x,t), \nabla u(x,t), {}^c\mathcal{D}_+^{1/2}\, u(x,t), x, t) = -c\, \partial_{x_i} u(x,t)$,

so conditions of Theorem 77 are fulfilled, and we have:

$(u,0)$ is a $\{0\} \times V^\alpha(\Omega \times [a,b])$-extremal of the action $\mathcal{A}(L_{1/2})$ if and only if u verifies

$$\left[\partial_u L - \sum_{i=1}^{n} \partial_{x_i} \partial_{v_i} L - \mathcal{D}_+^{1/2}\, \partial_w L\right] (u(x,t), \nabla u(x,t), {}^c\mathcal{D}_+^{1/2}\, u(x,t), x, t) = 0. \quad (144)$$

Given that $\sum_{i=1}^{n} \partial_{x_i} \partial_{x_i} u(x,t) = \Delta u(x,t)$ and $\mathcal{D}_+^{1/2} \circ {}^c\mathcal{D}_+^{1/2}\, u(x,t) = \frac{\partial}{\partial t} u(x,t)$, (144) is exactly (143). □

Once again, causality is essential so as to obtain the term $\frac{\partial}{\partial t} u(x,t)$.

6.2. The Fractional Wave Equation

In this section, we derive the *fractional wave equation* defined by Schneider and Wyss [40] as the extremals of a fractional continuous Lagrangian systems.

The equation describing waves propagating on a stretched string of constant linear mass density ρ under constant tension T is

$$\rho \frac{\partial^2 u(t,x)}{\partial t^2} = T \frac{\partial^2 u(t,x)}{\partial x^2}, \quad (145)$$

where $u(t,x)$ denotes the amplitude of the wave at position x along the string at time t. The wave equation corresponds to the extremals of the generalized functional associated to the Lagrangian systems

$$L(t,x,y,v,w) = \frac{\rho}{2} v^2 - \frac{T}{2} w^2. \quad (146)$$

In [40], the authors define the fractional analogue of the wave equation by changing the classical derivative by a fractional one. Using our notations, the definition of the fractional wave equation is:

Definition 81. *The fractional wave equation of order α is the fractional partial differential equation*

$$-\rho\, {}^c\mathcal{D}_-^{2\alpha}\, u = T \frac{\partial^2 u}{\partial x^2}. \quad (147)$$

A natural demand with respect to this generalization which is just a formal manipulation on equations, is to keep a more structural property of the wave equation, namely the fact that it derives from a least-action principle.

It is easy to prove that Theorem 80 is also valid replacing $\alpha = 1/2$ by $0 < \alpha < 1/2$. We then deduce :

Theorem 82. *Let $u \in \mathcal{F}(\Omega \times [a,b], \mathbb{R})$ such that*

- $\forall x \in \Omega,\ t \mapsto u(x,t) \in AC^2([a,b])$,
- $\forall t \in [a,b],\ x \mapsto u(x,t) \in C^2(\Omega)$.

Then u is solution of (147) *if and only if $(u,0)$ is a $\{0\} \times C_0^2(\Omega \times [a,b])$-extremal of the action $\mathcal{A}(L_\alpha)$ with L defined by (146).*

6.3. The Convection-Diffusion Equation

The convection-diffusion equation occurs in many physical problems such as porous media, engineering, geophysics. It could model the dispersion of a pollutant in a river estuary, or groundwater transport, atmospheric pollution, concentration of electron inducing an electric current, heat transfer in a heated body. As many PDEs, the solution exists under conditions, but is often not known explicitly. Even for linear convection-diffusion equation numerical schemes are not always well understood. It is still a challenging problem to obtain efficient and robust numerical schemes to solve the convection-diffusion equation due in particular to the mixing between two different types of behavior, namely the convective and diffusive regimes. Let us first recall the equation. Let $\Omega \subset \mathbb{R}^d$ be an open subset with a Lipschitz-continuous boundary $\partial\Omega$. The time domain $[a,b]$, $0 \leq a \leq b$ is arbitrary, but fixed. Let us consider the general linear parabolic equation of second order:

$$\begin{array}{ll} u_t + (\gamma \cdot \nabla)u - \mathrm{div}(K \cdot \nabla u) + \beta u = f(t,x) & \text{in } (a,b] \times \Omega, \\ u(t,x) = 0 \quad \text{in } (a,b] \times \partial\Omega, & \\ u(a,x) = u_0(x) \quad \text{in } \Omega\,, & \end{array} \tag{148}$$

where $\gamma \in \mathbb{R}^d, K \in \mathbb{R}^{d \times d}, \beta \in \mathbb{R}$.

As an example u is the concentration of a pollutant, transported by a flow of velocity $\gamma \in \mathbb{R}^d$. The tensor K represents the diffusivity of the pollutant specie. The creation or destruction of the specie can be taken into account via βu, and f is the source term. The unknown u is both depending on time and space.

We assume the coefficients are smooth, bounded and satisfying the following properties:

- K is symmetric, uniformly positive definite s.t.

$$K \in C^0(a,b; L^\infty(\Omega)^{d \times d}) \quad and \quad \exists \lambda_1, \lambda_2 > 0 : \lambda_1 |\xi|^2 < \xi K \xi^T < \lambda_2 |\xi|^2$$

- the convection γ is such that

$$\gamma \in C^0(a,b; W^{1,\infty}(\Omega)^d) \quad and \quad \mathrm{div}\gamma = 0$$

- the reaction β is non-negative

$$\beta \in C^0(a,b;L^\infty(\Omega)) \quad and \quad \exists \beta_0 \quad \beta \geq \beta_0 \,.$$

These assumptions guarantee that the problem (148) is well posed for $f \in L^2(a,b;H^{-1}(\Omega))$, and every $u_0 \in L^2(\Omega)$.
Let us notice that in the special case of absence of convection, when $\gamma = 0$, the stationary convection-diffusion equation is simply the Poisson equation. It is well known that the Poisson equation derives from a variational principle also called least-action principle. This means that the solution u of the Poisson equation is a minimizer of the following Lagrangian functional

$$\begin{aligned} \mathcal{L} \; : \; & H_0^1(\Omega) \longrightarrow \mathbb{R} \\ & v \mapsto \mathcal{L}(v) = \int_\Omega \frac{1}{2}(K\cdot\nabla v)\cdot\nabla v\,dx - \int_\Omega f v\,dx \,. \end{aligned}$$

This is not the case of the convection-diffusion equation. For instance, the stationary convection-diffusion equation admits the following weak formulation:

$$\int_\Omega (K\cdot\nabla v)\cdot\nabla\phi\,dx + \int_\Omega (\gamma\cdot\nabla)u\,\phi + \beta u\phi\,dx = \int_\Omega f\phi\,dx\,, \qquad \text{for any } \phi\in H_0^1(\Omega). \tag{149}$$

Nevertheless the advective term is not symmetric in u and ϕ. As a consequence, the weak formulation (149) does not derive from a potential, [42]. This can also be seen as the convection-diffusion equation does not satisfy the so-called Helmholtz conditions, i.e. that the Fréchet derivative of the Euler-Lagrange expression is not self-adjoint. We refer to ([34], Thm. 5.92, p.364) for more details.

Let us note that there were some attempt to construct variational formulation for the convection-diffusion equation by Ortiz [35], where he resorts to a local transformation of the solution by use of a "dual" problem. In this Section, we apply the previous fractional formalism to obtain a fractional Lagrangian variational formulation.

Let us consider the reaction convection diffusion equation defined on $[a,b]\times\Omega$ by (148):

$$\frac{\partial}{\partial t}u(t,x) + \gamma\cdot\nabla u(t,x) - \text{div}\,(K\cdot\nabla u(t,x)) + \beta u(t,x) = f(t,x). \tag{150}$$

with constant coefficients $\gamma\in\mathbb{R}^d$, $K\in\mathbb{R}^{d\times d}$ and $\beta\in\mathbb{R}$. As we already mentioned, this equation does not derive from a variational principle in the classical sense. Nevertheless the result of the previous section allows us to overcome this difficulty and obtain a variational formulation of the convection-diffusion equation by mean of the asymmetric fractional Lagrangian. Let us defined the extended Lagrangian L given by

$$\begin{aligned} L: \quad [a,b]\times\Omega\times\mathbb{R}\times\mathbb{R}\times\mathbb{R}^d\times\mathbb{R}^d \quad & \longrightarrow \quad \mathbb{R} \\ (t,x,y,v,w,z) \quad & \longmapsto \quad f(t,x)y - \frac{1}{2}\beta y^2 + \frac{1}{2}v^2 + \frac{1}{2}(\gamma\times w)\cdot w - \frac{1}{2}(K\cdot z)\cdot z. \end{aligned}$$

The direct application of theorem 77 provides that the solutions of the convection-diffusion equation are $\{0\}\times C_0^1([a,b]\times\Omega)$-extremals of the following asymmetric fractional functional $\mathcal{L}_{1/2}$ defined for $U=(u_+,u_-)$ by

$$\begin{aligned} \mathcal{L}_{1/2}(U) = \quad & \int_a^b\int_\Omega L\big(t,x,u_+(t,x)+u_-(t,x), {}^c\mathcal{D}_+^{1/2}\,u_+(t,x) - {}^c\mathcal{D}_-^{1/2}\,u_-(t,x), \\ & {}^c\nabla^{1/2}u_+(t,x) - \overline{{}^c\nabla}^{1/2}u_-(t,x),\, \nabla u_+(t,x)+\nabla u_-(t,x)\big)\,dx\,dt. \end{aligned}$$

Namely, the following result holds:

Theorem 83. *Let* $u \in F^2(\overline{\Omega} \times [a,b])$. *Then* u *is a solution of the convection-diffusion equation* (150) *if and only if* $(u,0)$ *is a* $\{0\} \times C_0^1([a,b] \times \Omega)$ *critical point of* $\mathcal{L}_{1/2}$.

Proof. Let $x \in \Omega$ and $t \in [a,b]$. Let $t \in [a,b]$. The partial derivatives of L verify:

- $\partial_y L(t,x,u(t,x), {}^c\mathcal{D}_+^{1/2}\, x(t), {}^c\nabla^{1/2}u(t,x), \nabla u(t,x)) = f(t,x) - \beta u(t,x)$,
- $\partial_v L(t,x,u(t,x), {}^c\mathcal{D}_+^{1/2}\, x(t), {}^c\nabla^{1/2}u(t,x), \nabla u(t,x)) = {}^c\mathcal{D}_+^{1/2}\, u(t,x)$,
- $\partial_w L(t,x,u(t,x), {}^c\mathcal{D}_+^{1/2}\, x(t), {}^c\nabla^{1/2}u(t,x), \nabla u(t,x)) = \gamma \times {}^c\nabla^{1/2}u(t,x)$,
- $\partial_z L(t,x,u(t,x), {}^c\mathcal{D}_+^{1/2}\, x(t), {}^c\nabla^{1/2}u(t,x), \nabla u(t,x)) = -K \cdot \nabla u(t,x)$.

As $u \in F^2([a,b] \times \overline{\Omega})$ we have that $u_x \in AC^2([a,b])$ and as a consequence ${}^c\mathcal{D}_+^{1/2}\, u \in AC^1([a,b])$. We have also $u_t \in C^2(\overline{\Omega})$ so that using lemma 47, we deduce ${}^c\nabla^{1/2}u \in C^1(\Omega)$. Moreover $\nabla u \in C^1([a,b])$ so that the conditions of theorem 77 are fulfilled. From lemma 46, as $u_x \in AC^2([a,b])$ we have $\mathcal{D}_+^{1/2} \circ {}^c\mathcal{D}_+^{1/2}\, u = \frac{d}{dt}u$. Moreover, as $u_t \in C^2(\overline{\Omega})$, lemma 51 applies and we have

$$\mathrm{div}^{1/2}\left(\gamma \times {}^c\nabla^{1/2}u(x)\right) = \gamma \cdot \nabla u(x).$$

This concludes the proof. □

References

[1] Vladimir I. Arnold. *Mathematical methods of classical mechanics*, volume 60 of *Graduate Texts in Mathematics*. Springer-Verlag, New York, second edition, 1989. Translated from the Russian by K. Vogtmann and A. Weinstein.

[2] Om Prakash Agrawal. Formulation of Euler-Lagrange equations for fractional variational problems. *J. Math. Anal. Appl.*, 272(1):368–379, 2002.

[3] Andler M., Jean Leray (1906-1998), *Proceedings of the American Philosophical Society*, Vol. 144, no. 4, 2000.

[4] Harry Bateman. On dissipative systems and related variational principles. *Physical Review*, 38(1):815–819, 1931.

[5] P. S. Bauer. Dissipative dynamical systems. *Proc. Nat. Acad. Sci.*, 17:311–314, 1931.

[6] L. Bourdin, J. Cresson, I. Greff, P. Inizan, *Variational integrators for fractional Lagrangian systems in the framework of discrete embeddings*, Preprint 2011.

[7] L. Bourdin, J. Cresson, I. Greff, A Continuous/discrete fractional Noether's theorem, *Communication in Nonlinear Science and Numerical Simulation*, 2012.

[8] J. Cresson. Fractional embedding of differential operators and Lagrangian systems. *J. Math. Phys.*, 48(3):033504, 34, 2007.

[9] J. Cresson, I. Greff, and C. Pierre. *Coherent discrete embeddings for Lagrangian and Hamiltonian systems.* Arxiv-1107.0894, pages 1–24, 2011.

[10] J. Cresson and P. Inizan. Variational formulations of differential equations and asymmetric fractional embedding. *Journal of Mathematical Analysis and Applications*, Volume 385, Issue 2, 15 January 2012, Pages 975-997.

[11] J. Cresson and P. Inizan. *Irreversibility, least action principle and causality.* arXiv, 0812.3529v2, 2008. Unpublished.

[12] J. Cresson, Non-differentiable deformations of R^n, *International Journal of Geometric Methods in Modern Physics*, Vol. 3, no. 7 (2006) 1395-1415.

[13] J. Cresson and S. Darses (2006), Plongement stochastique des systèmes lagrangiens, *C.R. Acad. Sci. Paris Ser. I* 342 (5), 333-336.

[14] J. Cresson and S. Darses, Stochastic embedding of dynamical systems. *J. Math. Phys.*, 48(7):072703, 54, 2007.

[15] J. Cresson, S. Darses, Théorème de Noether Stochastique, *C.R. Acad. Sci. Paris, Ser. I* 344 (2007) 259-264

[16] J. Cresson, Introduction to embedding of Lagrangian systems, *International Journal of Biomathematics and Biostatistics*, Vol. 1, no. 1, 23-31, 2010

[17] J. Cresson, *Introduction to embedding formalisms and Lagrangian PDEs*, Lecture Notes TU München, 2011.

[18] J. Cresson, I. Greff, A non-differentiable Noether's theorem, *J. Math. Phys.* 52, 023513 (2011); doi:10.1063/1.3552936 (10 pages).

[19] J. Cresson and I. Greff, Non-differentiable embedding of Lagrangian systems and PDEs, 30.p, *Journal of Mathematical Analysis and Applications* Volume 384, Issue 2 (2011) Pages 626-646.

[20] J. Cresson and P. Inizan. About fractional Hamiltonian systems. *Physica Scripta*, T136:014007, 2009.

[21] J. Cresson, D. F-M. Torres, A. Malinovska, Differential, integral and variational time scales embeddings of Lagrangian systems, *Computers Mathematics with applications*, 2012.

[22] Frederico, Gastão S. F. and Torres, Delfim F. M., A formulation of Noether's theorem for fractional problems of the calculus of variations, *J. Math. Anal. Appl.*, 334, 2007,2, 834–846.

[23] Sk. Golam Ali, Benoy Talukdar, and Umapada Das. Inverse problem of variational calculus for nonlinear evolution equations. *Acta Phys. Polon. B*, 38(6):1993–2002, 2007.

[24] Ernst Hairer, Christian Lubich, and Gerhard Wanner. *Geometric numerical integration*, volume 31 of *Springer Series in Computational Mathematics*. Springer-Verlag, Berlin, second edition, 2006.

[25] L. Hörmander. *The Analysis of Linear Partial differential Operators I*. Springer-Verlag, 1983.

[26] Pierre Inizan. *Dynamique fractionnaire pour le chaos hamiltonien.* PhD thesis. 2010.

[27] Malgorzata Klimek. On analogues of exponential functions for antisymmetric fractional derivatives. *Comput. Math. Appl.*, 59(5):1709–1717, 2010.

[28] Laskar J., Large-scale chaos in the solar system, *Astron. Astrophys.* 287, L9-L12 (1994).

[29] J. Leray, Sur le mouvement d'un liquide visqueux emplissant l'espace, *Acta Math.* 63 (1934), p. 193-248.

[30] K. Itô, On stochastic differential equations, *Mem. Amer. Math. Soc.* 4 (1951), 1-51.

[31] Làzaro Cami J.A., *Stochastic geometric mechanics*, Ph.D., University of Zaragoza, 2008.

[32] Jerrold E. Marsden and Matthew West. Discrete mechanics and variational integrators. *Acta Numer.*, 10:357–514, 2001.

[33] Marmi, S., *Chaotic behaviour in the solar system following J. Laskar*, Séminaire Bourbaki, 51ème année, 1998-1999, Paper No. 854.

[34] Peter J. Olver. *Applications of Lie groups to differential equations*, volume 107 of *Graduate Texts in Mathematics*. Springer-Verlag, New York, second edition, 1993.

[35] Michael Ortiz. A variational formulation for convection-diffusion problems. *Internat. J. Engrg. Sci.*, 23(7):717–731, 1985.

[36] Fred Riewe. Nonconservative Lagrangian and Hamiltonian mechanics. *Phys. Rev. E* (3), 53(2):1890–1899, 1996.

[37] Fred Riewe. Mechanics with fractional derivatives. *Phys. Rev. E* (3), 55(3, part B):3581–3592, 1997.

[38] L. Schwartz, *Théorie des distributions*, Publications de l'Institut de Mathématique de l'Université de Strasbourg, No. IX-X. Nouvelle édition, entiérement corrigée, refondue et augmentée, Hermann, Paris, 1966.

[39] Stefan G. Samko, Anatoly A. Kilbas, and Oleg I. Marichev. *Fractional integrals and derivatives*. Gordon and Breach Science Publishers, Yverdon, 1993. Theory and applications, Edited and with a foreword by S. M. Nikol′skiĭ, Translated from the 1987 Russian original, Revised by the authors.

[40] Schneider W.R., Wyss W., Fractional diffusion and wave equations, *J. Math. Phys.* 30 (1989) 134-144.

[41] F.T. Tveter, Deriving the Hamiltonian equations of motion for a nonconservative system using a variational principle, *Journal of Mathematical Physics*, Vol. 39, no. 3 (1998), 1495-1500.

[42] M. M. Vainberg. *Variational methods for the study of nonlinear operators*. Holden-Day Inc., San Francisco, Calif., 1964. With a chapter on Newton's method by L. V. Kantorovich and G. P. Akilov. Translated and supplemented by Amiel Feinstein.

[43] G.M. Zaslavsky. *Hamiltonian Chaos & Fractional Dynamics*. Oxford University Press, Oxford, 2005.

In: Fractional Calculus in Analysis, Dynamics ... ISBN: 978-1-62948-635-2
Editor: Jacky Cresson, pp. 127-177

Chapter 3

A Class of Fractional Optimal Control Problems and Fractional Pontryagin's Systems. Variational Integrator and Existence of Continuous/Discrete Noether's Theorems

Loïc Bourdin*
Laboratoire de Mathématiques et de leurs Applications de Pau,
Université de Pau et des Pays de l'Adour, Pau Cedex, France

Introduction

The control theory is the analysis of controlled dynamical systems. These systems are varied: they can be differential, stochastic or discrete. The optimal control theory concerns the determination of a control optimizing a certain cost. Consequently, this theory is strongly connected to the 18th century classical mechanic (variational principles, Euler-Lagrange equations, etc., see [6, 32, 47]). Since the second world war, this theory has a considerable development and one can find applications in many domains: celestial mechanic [11], biology [15], hydroelectricity [16], economy [20, 22, 26], etc. The subject is widely treated and one can find a lot of references on the subject, see for example [14, 24, 37, 57].

The fractional calculus, *i.e.* the mathematical field dealing with the generalization of the derivative to any real order, plays an increasing role in many varied domains as economy [17] or probability [44, 56]. Fractional derivatives also appear in many fields of Physics (see [38]): wave mechanic [5], viscoelasticity [7], thermodynamics [39], fluid mechanic in heterogeneous media [33, 59, 60], etc. A natural question then arises: *can we develop optimal control theories for fractional differential systems?*

Recently, a subtopic of the fractional calculus gains importance: it concerns the variational principles on functionals involving fractional derivatives. This leads to the statement

*E-mail address: bourdin.l@univ-pau.fr

of fractional Euler-Lagrange equations, see [1, 9, 54]. A direct consequence is the emergence of works concerning a particular class of fractional optimal control problems, see [2, 3, 28, 29, 41] and references therein. These studies usually use a Lagrange multiplier technique allowing to write these problems as problems of optimization without constraint of *augmented functionals*. With a calculus of variations, authors then obtain a necessary condition for the existence of an optimal control. This condition is commonly given as the existence of a solution of a system of fractional differential equations called *fractional Pontryagin's system*.

In this chapter, we first give a new presentation of this result. Precisely, making an additional assumption (see Condition (f_x lip)), we rewrite *directly* these fractional optimal control problems as simpler problems of optimization without constraint of functionals *depending only on the control*. Although the method used is considerably inspired by the Lagrange multiplier technique, it allows us to give a complete proof of this result using only classical mathematical tools adapted to the fractional case: *calculus of variations*, *Gronwall's Lemma*, *Cauchy-Lipschitz Theorem* and *stability of differential equations under perturbations*.

Nevertheless, the explicit computation of controls satisfying the above necessary condition needs the resolution of a fractional Pontryagin's system which is a main drawback. Indeed, solving a fractional differential equation is in general very difficult. Consequently, in this chapter, we suggest two deviously ways in order to get informations on the solutions of a fractional Pontryagin's system.

Firstly, we study the existence of classical conservation laws, *i.e.* functions which are constant on each solution. Indeed, constants of motion, generally associated to physical quantities, give strong informations on the solutions in the phase space for example. Moreover, they also can be used in order to reduce or integrate the equation by quadrature. Previous results in this direction have been obtained by Torres and Frederico in [28, 29]. However, in each of these papers, the conservation law is not explicit but implicitly defined by a functional relation. In this chapter, inspired by a recent result obtained in [12], we prove a fractional Noether's theorem providing an explicit conservation law for fractional Pontryagin's systems exhibiting a symmetry.

As it is done in [3], the second idea is to suggest a numerical approach. In this chapter, we construct a numerical scheme preserving the variational structure of the fractional Pontryagin's systems. Indeed, this variational structure is intrinsic and induces strong constraints on the qualitative behaviour of the solutions. It seems then important to preserve it at the discrete level. A *variational integrator* is a numerical scheme preserving the variational structure at the discrete level. We refer to Section 2 for more details concerning the construction of a variational integrator and let us remind that the variational integrators are well-developed in [36, 46] for classical Euler-Lagrange equations and in [13] for fractional ones. In this chapter, we construct a variational integrator for fractional Pontryagin's systems and it is called *shifted discrete fractional Pontryagin's system*.

Finally, adapting the strategy from the continuous level to the discrete one, we prove a discrete fractional Noether's theorem providing an explicit discrete conservation law for shifted discrete fractional Pontryagin's systems exhibiting a discrete symmetry. This result is widely inspired from the discrete fractional Noether's theorem proved in [12] for discrete fractional Euler-Lagrange equations introduced in [13].

In this chapter, we also suggest:

- a link between classical optimal control problems and their fractional versions via the Stanislavsky's formalism, see [40, 56];
- a solved fractional example allowing to test numerical schemes in the strict fractional case.

The chapter is organized as follows. In Section 1, we first remind classical definitions and some results concerning fractional calculus (Section 1.1). Then, we present the class of fractional optimal control problems studied and the usual strategy leading to the fractional Pontryagin's system (Section 1.2). Making an additional assumption on the problem (see Condition (f_x lip)), we develop a new proof leading to the same result with a different presentation and using only classical mathematical tools adapted to the fractional case (Section 1.3). In Section 1.5, we suggest a way in order to transit from a classical optimal control problem to its fractional version via the Stanislavsky's formalism. Subsequently, we suggest some examples with, in particular, a solved fractional example allowing to test numerical schemes in the strict fractional case (Section 1.6). Finally, in Section 1.7, we state a fractional Noether's theorem for fractional Pontryagin's systems admitting a symmetry. Section 2 is focused on a variational integrator for fractional Pontryagin's systems. Indeed, Section 2.1 reminds usual definitions and proves some results on discrete fractional derivatives. Then, in Section 2.2, we construct a variational integrator for fractional Pontryagin's systems and it is called shifted discrete fractional Pontryagin's system. Finally, after some numerical tests in Section 2.5, we state a discrete fractional Noether's theorem for shifted discrete fractional Pontryagin's systems admitting a discrete symmetry (Section 2.5). Appendices A and B contain technical results and proofs non developed in Sections 1 and 2.

1. A Class of Fractional Optimal Control Problems

1.1. Definitions and Results about Fractional Calculus

Let us introduce the following notations available in the whole chapter. Let $a < b$ be two reals, let $d, m \in \mathbb{N}^*$ denote two dimensions and let $\|\cdot\|$ be the euclidean norm of $\mathbb{R}^d$ and $\mathbb{R}^m$.

1.1.1. Fractional Operators of Riemann-Liouville and Caputo

The fractional calculus concerns the extension of the usual notion of derivative from non-negative integer orders to any real order. Since 1695, numerous notions of fractional derivatives emerge over the year, see [42, 52, 55]. In this chapter, we only use the notions of fractional integrals and derivatives in the sense of Riemann-Liouville (1847) and Caputo (1967) whose definitions are recalled in this section. We refer to [42, 52, 55] for more details.

Let $g \in \mathscr{C}^0([a,b],\mathbb{R}^d)$ and $\alpha > 0$. The left (resp. right) fractional integral in the sense of Riemann-Liouville with inferior limit a (resp. superior limit b) of order α of g is defined by:

$$\forall t \in]a,b],\ I^{\alpha}_{-}g(t) := \frac{1}{\Gamma(\alpha)} \int_a^t (t-y)^{\alpha-1} g(y)\, dy \tag{1}$$

respectively:

$$\forall t \in [a,b[,\ I^{\alpha}_{+}g(t) := \frac{1}{\Gamma(\alpha)} \int_t^b (y-t)^{\alpha-1} g(y)\, dy, \tag{2}$$

where Γ denotes the Euler's Gamma function. Let us remind that $I^{\alpha}_{-}g$ (resp. $I^{\alpha}_{+}g$) is continuous and can be continuously extended by 0 in $t=a$ (resp. $t=b$). Let us note that $I^1_{-}g$ (resp. $-I^1_{+}g$) coincides with the anti-derivative of g vanishing in $t=a$ (resp. $t=b$). For $\alpha=0$, let $I^0_{-}g = I^0_{+}g = g$.

Now, let us consider $0 < \alpha \leq 1$. The left (resp. right) fractional derivative in the sense of Riemann-Liouville with inferior limit a (resp. superior limit b) of order α of g is then given by:

$$\forall t \in]a,b],\ D^{\alpha}_{-}g(t) := \frac{d}{dt}(I^{1-\alpha}_{-}g)(t) \quad \Big(\text{resp.} \quad \forall t \in [a,b[,\ D^{\alpha}_{+}g(t) := -\frac{d}{dt}(I^{1-\alpha}_{+}g)(t)\Big), \tag{3}$$

provided that the right side terms are defined.

In the Riemann-Liouville sense, the strict fractional derivative of a constant is not zero. Caputo then suggests the following definition. For $0 < \alpha \leq 1$, the left (resp. right) fractional derivative in the sense of Caputo with inferior limit a (resp. superior limit b) of order α of g is given by:

$$\forall t \in]a,b],\ {}_{c}D^{\alpha}_{-}g(t) := D^{\alpha}_{-}\big(g - g(a)\big)(t) \tag{4}$$

$$\Big(\text{resp.} \quad \forall t \in [a,b[,\ {}_{c}D^{\alpha}_{+}g(t) := D^{\alpha}_{+}\big(g - g(b)\big)(t)\Big),$$

provided that the right side terms are defined. Let us note that if $g(a)=0$ (resp. $g(b)=0$), then ${}_{c}D^{\alpha}_{-}g = D^{\alpha}_{-}g$ (resp. ${}_{c}D^{\alpha}_{+}g = D^{\alpha}_{+}g$).

In the *classical case* $\alpha = 1$, the fractional derivatives of Riemann-Liouville and Caputo both coincide with the classical derivative. Precisely, modulo a (-1) term in the right case, we have $D^1_{-} = {}_{c}D^1_{-} = -D^1_{+} = -{}_{c}D^1_{+} = d/dt$.

Finally, let us remind the following important result. If $g \in \mathscr{C}^1([a,b],\mathbb{R}^d)$ and $0 < \alpha \leq 1$, then we have:

$$\forall t \in]a,b],\ {}_{c}D^{\alpha}_{-}g(t) = I^{1-\alpha}_{-}\dot{g}(t) \quad \text{and} \quad \forall t \in [a,b[,\ {}_{c}D^{\alpha}_{+}g(t) = -I^{1-\alpha}_{+}\dot{g}(t), \tag{5}$$

where $\dot{g}$ denotes the derivative of g. Consequently, in this case, we have ${}_{c}D^{\alpha}_{-}g$ (resp. ${}_{c}D^{\alpha}_{+}g$) is continuous and can be continuously extended in $t=a$ (resp. $t=b$).

1.1.2. Some Properties and Results about the Fractional Operators

Let us precise that the properties and results developed in this section are well-known and already proved in many references, see [42, 55] for example. However, these results are not always exactly presented as we need. In this case, we give a detailed proof for the reader's convenience.

First, we remind two basic results concerning the fractional integrals. They are proved in [42, 55] both using the Fubini's theorem. The first one yields the *semi-group property* of the fractional integral operators:

Property 1. *Let* $g \in \mathscr{C}^0([a,b],\mathbb{R}^d)$ *and* $\alpha_1, \alpha_2 \geq 0$. *Then, we have:*

$$\forall t \in [a,b],\ I_-^{\alpha_1} \circ I_-^{\alpha_2} g(t) = I_-^{\alpha_1+\alpha_2} g(t) \quad \textit{and} \quad I_+^{\alpha_1} \circ I_+^{\alpha_2} g(t) = I_+^{\alpha_1+\alpha_2} g(t). \tag{6}$$

The following second property is occasionally called *fractional integration by parts*. It is very useful for calculus of variations involving fractional derivatives:

Property 2. *For any* $g_1, g_2 \in \mathscr{C}^0([a,b],\mathbb{R}^d)$ *and any* $\alpha \geq 0$, *we have:*

$$\int_a^b I_-^\alpha g_1 \cdot g_2 \, dt = \int_a^b g_1 \cdot I_+^\alpha g_2 \, dt. \tag{7}$$

Let us introduce the following convention: a function defined on $]a,b]$ (resp. $[a,b[$) is said to be an element of $\mathscr{C}^0([a,b],\mathbb{R}^d)$ if and only if it is continuous on $]a,b]$ (resp. $[a,b[$) and if it can be continuously extended in $t = a$ (resp. $t = b$). From Section 1.1.1, we can give the following examples:

- for $g \in \mathscr{C}^0([a,b],\mathbb{R}^d)$ and $\alpha \geq 0$, we have $I_-^\alpha g, I_+^\alpha g \in \mathscr{C}^0([a,b],\mathbb{R}^d)$;
- for $g \in \mathscr{C}^1([a,b],\mathbb{R}^d)$ and $0 < \alpha \leq 1$, we have ${}_cD_-^\alpha g, {}_cD_+^\alpha g \in \mathscr{C}^0([a,b],\mathbb{R}^d)$.

Now, we prove some results of composition between the left fractional operators. One can easily derive the analogous versions for the right ones.

Property 3. *Let* $g \in \mathscr{C}^0([a,b],\mathbb{R}^d)$ *and* $0 < \alpha \leq 1$. *Then,* ${}_cD_-^\alpha \circ I_-^\alpha g$ *is an element of* $\mathscr{C}^0([a,b],\mathbb{R}^d)$ *and for any* $t \in [a,b]$, *we have:*

$${}_cD_-^\alpha \circ I_-^\alpha g(t) = g(t). \tag{8}$$

Let us assume additionally that ${}_cD_-^\alpha g \in \mathscr{C}^0([a,b],\mathbb{R}^d)$. *Then,* $I_-^\alpha \circ {}_cD_-^\alpha g$ *is an element of* $\mathscr{C}^0([a,b],\mathbb{R}^d)$ *and for any* $t \in [a,b]$, *we have:*

$$I_-^\alpha \circ {}_cD_-^\alpha g(t) = g(t) - g(a). \tag{9}$$

Proof. Let us prove the first result. Since $I_-^\alpha g(a) = 0$, we have for any $t \in]a,b]$:

$${}_cD_-^\alpha \circ I_-^\alpha g(t) = D_-^\alpha \circ I_-^\alpha g(t) = \frac{d}{dt} \circ I_-^{1-\alpha} \circ I_-^\alpha g(t) = \frac{d}{dt}(I_-^1 g)(t) = g(t). \tag{10}$$

Hence, ${}_cD_-^\alpha \circ I_-^\alpha g$ is continuous on $]a,b]$ and can be continuously extended by $g(a)$ in $t = a$. Now, let us prove the second result. It is obvious for $\alpha = 1$. Now, let us consider $0 < \alpha < 1$. Since ${}_cD_-^\alpha g \in \mathscr{C}^0([a,b],\mathbb{R}^d)$, we have $I_-^{1-\alpha}(g - g(a)) \in \mathscr{C}^1([a,b],\mathbb{R}^d)$. Combining the first result and Equality (5), we have for any $t \in]a,b]$:

$$\begin{aligned} g(t) - g(a) &= {}_cD_-^{1-\alpha} \circ I_-^{1-\alpha}(g - g(a))(t) = I_-^\alpha \circ \frac{d}{dt} \circ I_-^{1-\alpha}(g - g(a))(t) \\ &= I_-^\alpha \circ D_-^\alpha (g - g(a))(t) = I_-^\alpha \circ {}_cD_-^\alpha g(t). \end{aligned} \tag{11}$$

Hence $I_-^\alpha \circ {}_cD_-^\alpha g$ is continuous on $]a,b]$ and can be continuously extended by 0 in $t = a$. The proof is completed. □

Finally, we prove the following fractional Cauchy-Lipschitz type theorem. Let us precise that a version of this theorem is proved in [42, Part 3.5.1, Corollary 3.26, p.205]. It gives the existence and the uniqueness of a global solution of a fractional differential equation.

Theorem 4 (Fractional Cauchy-Lipschitz theorem). *Let* $0 < \alpha \leq 1$, $F \in \mathscr{C}^0(\mathbb{R}^d \times [a,b], \mathbb{R}^d)$ *and* $A \in \mathbb{R}^d$. *Let us assume that* F *satisfies the following Lipschitz type condition:*

$$\exists K \in \mathbb{R},\ \forall (x_1, x_2, t) \in (\mathbb{R}^d)^2 \times [a,b],\ \|F(x_1,t) - F(x_2,t)\| \leq K\|x_1 - x_2\|. \tag{12}$$

Then, the following fractional Cauchy problem:

$$\begin{cases} {}_{c}D^{\alpha}_{-} g = F(g,t) \\ g(a) = A, \end{cases} \tag{13}$$

has an unique solution in $\mathscr{C}^0([a,b], \mathbb{R}^d)$. *The solution* g *is an element of* $\mathscr{C}^{[\alpha]}([a,b], \mathbb{R}^d)$ *(where* $[\alpha]$ *denotes the floor of* α*) and* g *satisfies:*

$$\forall t \in [a,b],\ g(t) = A + I^{\alpha}_{-}\big(F(g,t)\big)(t). \tag{14}$$

Proof. Firstly, let $g \in \mathscr{C}^0([a,b], \mathbb{R}^d)$. From Property 3, we conclude that g is solution of the fractional Cauchy problem (13) if and only if g satisfies Equality (14). Now, let us define:

$$\begin{array}{rcccccl} \varphi: & \mathscr{C}^0([a,b],\mathbb{R}^d) & \longrightarrow & \mathscr{C}^0([a,b],\mathbb{R}^d) & & & \\ & g & \longmapsto & \varphi(g): & [a,b] & \longrightarrow & \mathbb{R}^d \\ & & & & t & \longmapsto & A + I^{\alpha}_{-}\big(F(g,t)\big)(t). \end{array} \tag{15}$$

By induction, we prove that for any $n \in \mathbb{N}^*$ and any $g_1, g_2 \in \mathscr{C}^0([a,b], \mathbb{R}^d)$, we have:

$$\forall t \in [a,b],\ \|\varphi^n(g_1)(t) - \varphi^n(g_2)(t)\| \leq \frac{K^n (t-a)^{n\alpha}}{\Gamma(1+n\alpha)} \|g_1 - g_2\|_\infty. \tag{16}$$

Consequently, we have for any $n \in \mathbb{N}^*$ and any $g_1, g_2 \in \mathscr{C}^0([a,b], \mathbb{R}^d)$:

$$\|\varphi^n(g_1) - \varphi^n(g_2)\|_\infty \leq \frac{K^n (b-a)^{n\alpha}}{\Gamma(1+n\alpha)} \|g_1 - g_2\|_\infty. \tag{17}$$

Since $K^n(b-a)^{n\alpha}/\Gamma(1+n\alpha) \longrightarrow 0$, there exists $n \in \mathbb{N}^*$ such that φ^n is a contraction. Consequently, φ admits an unique fix point in $\mathscr{C}^0([a,b], \mathbb{R}^d)$. The proof is completed. □

1.2. Presentation of the Problem and Usual Strategy

In this section, we first give a brief presentation of the class of fractional optimal control problems interesting us in this chapter. Finally, we briefly remind the usual strategy (using the Lagrange multiplier technique) leading to a necessary condition for the existence of an optimal control, see [2, 3, 28, 29, 41] and references therein. From now and for all the rest of the chapter, we consider $0 < \alpha \leq 1$ and $A \in \mathbb{R}^d$.

An optimal control problem concerns the optimization of a quantity depending on parameters given by a controlled system. Precisely, the aim is to find a control optimizing a certain cost. Such a control is called *optimal control.* Many of these studies lead to the use of the Lagrange multiplier technique. For example, we refer to [10, 25, 34, 35] for controlled systems coded by ordinary differential equations.

In this chapter, we work in the following framework. We are interested in systems controlled by the following fractional Cauchy problem:

$$\begin{cases} {}_{\mathrm{c}}D^{\alpha}_{-}q = f(q,u,t) \\ q(a) = A, \end{cases} \tag{18}$$

where u denotes the control. The aim is to find a control u optimizing a quantity of the form:

$$\int_a^b L(q,u,t)\,dt, \tag{19}$$

where q is solution of (18).

The common strategy is first to rewrite this problem as a problem of optimization under constraint:

$$\underset{(q,u)\text{ satisfying (18)}}{\arg\min} \int_a^b L(q,u,t)\,dt. \tag{20}$$

Then, the Lagrange multiplier technique consists in the study of the critical points of the following *augmented functional*:

$$(q,u,p) \longmapsto \int_a^b L(q,u,t) - p\cdot\big({}_{\mathrm{c}}D^{\alpha}_{-}q - f(q,u,t)\big)\,dt, \tag{21}$$

where p is commonly called *Lagrange multiplier*. Finally, with a calculus of variations, such a strategy leads to the following result: a necessary condition for (q,u) to be a solution of (20) is that there exists a function p such that the following *fractional Pontryagin's system* holds:

$$\begin{cases} {}_{\mathrm{c}}D^{\alpha}_{-}q = \dfrac{\partial H}{\partial w}(q,u,p,t) \\ D^{\alpha}_{+}p = \dfrac{\partial H}{\partial x}(q,u,p,t) \\ \dfrac{\partial H}{\partial v}(q,u,p,t) = 0 \\ \big(q(a),p(b)\big) = (A,0), \end{cases} \tag{22}$$

where $H(x,v,w,t) = L(x,v,t) + w\cdot f(x,v,t)$. We refer to [2, 3, 28, 29, 41] and references therein for more details.

In this chapter, we are going to assume that f satisfies a Lipschitz type condition, see Condition (f_x lip). This assumption will allow us to write *directly* the initial problem as a simpler problem of optimization without constraint of the initial functional which is then *only dependent of the control u.* Finally, a simple calculus of variations leads us to the same result but with a new presentation. Let us remind that the method here developed is widely inspired from the Lagrange multiplier technique. Nevertheless, it allows us to give a complete proof only using classical mathematical tools adapted to the fractional case: *calculus of variations, Gronwall's Lemma, Cauchy-Lipschitz Theorem* and *stability under perturbations of differential equations*.

1.3. New Presentation of the Result

In this section, we first give rigorously the definitions concerning the class of fractional optimal control problems described in Section 1.2:

- The elements denoted $u \in \mathscr{C}^0([a,b],\mathbb{R}^m)$ are called *controls*;
- Let f be a $\mathscr{C}^2$ function of the form:

$$\begin{array}{rrcl} f: & \mathbb{R}^d \times \mathbb{R}^m \times [a,b] & \longrightarrow & \mathbb{R}^d \\ & (x,v,t) & \longmapsto & f(x,v,t). \end{array} \tag{23}$$

  It is commonly called the *constraint function*. We assume that f satisfies the following Lipschitz type condition. There exists $M \geq 0$ such that:

$$\forall (x_1,x_2,v,t) \in (\mathbb{R}^d)^2 \times \mathbb{R}^m \times [a,b],\ \|f(x_1,v,t) - f(x_2,v,t)\| \leq M\|x_1 - x_2\|; \tag{f_x lip}$$

- For any control u, let $q^{u,\alpha} \in \mathscr{C}^{[\alpha]}([a,b],\mathbb{R}^d)$ denote the unique global solution of the following fractional Cauchy problem:

$$\left\{ \begin{array}{l} {}_{\mathrm{c}}D^{\alpha}_{-} q = f(q,u,t) \\ q(a) = A. \end{array} \right. \tag{CP$^{\alpha}_{q}$}$$

  $q^{u,\alpha}$ is commonly called the *state variable* associated to u. Its existence and its uniqueness are provided by Theorem 4 and Condition (f_x lip);
- Finally, with this condition on f, the fractional optimal control problem described in Section 1.2 can be rewritten as the simpler problem of optimization of the following *cost functional*:

$$\begin{array}{rrcl} \mathcal{L}^{\alpha}: & \mathscr{C}^0([a,b],\mathbb{R}^m) & \longrightarrow & \mathbb{R} \\ & u & \longmapsto & \displaystyle\int_a^b L(q^{u,\alpha},u,t)\,dt, \end{array} \tag{24}$$

  where L is a *Lagrangian, i.e.* a $\mathscr{C}^2$ application of the form:

$$\begin{array}{rrcl} L: & \mathbb{R}^d \times \mathbb{R}^m \times [a,b] & \longrightarrow & \mathbb{R} \\ & (x,v,t) & \longmapsto & L(x,v,t). \end{array} \tag{25}$$

Hence, the existence and the uniqueness of $q^{u,\alpha}$ for any control u allow us to rewrite *directly* the initial problem as a simpler problem of optimization without constraint of the cost functional $\mathcal{L}^{\alpha}$: we do not need to introduce an augmented functional with a Lagrange multiplier. Moreover, let us note that $\mathcal{L}^{\alpha}$ is only dependent of the control.

A control optimizing $\mathcal{L}^\alpha$ is called *optimal control*. A necessary condition for a control u to be optimal is to be a *critical point* of $\mathcal{L}^\alpha$, *i.e.* to satisfy:

$$\forall \bar{u} \in \mathscr{C}^0([a,b],\mathbb{R}^m),\ D\mathcal{L}^\alpha(u)(\bar{u}) := \lim_{\varepsilon\to 0} \frac{\mathcal{L}^\alpha(u+\varepsilon\bar{u}) - \mathcal{L}^\alpha(u)}{\varepsilon} = 0. \tag{26}$$

In the following, we then focus on the characterization of the critical points of $\mathcal{L}^\alpha$. With an usual calculus of variations, we obtain the following Lemma 5 giving explicitly the value of the Gâteaux derivative of $\mathcal{L}^\alpha$:

Lemma 5. *Let u, $\bar{u} \in \mathscr{C}^0([a,b],\mathbb{R}^m)$. Then, the following equality holds:*

$$D\mathcal{L}^\alpha(u)(\bar{u}) = \int_a^b \frac{\partial L}{\partial x}(q^{u,\alpha},u,t)\cdot\bar{q} + \frac{\partial L}{\partial v}(q^{u,\alpha},u,t)\cdot\bar{u}\,dt, \tag{27}$$

where $\bar{q} \in \mathscr{C}^{[\alpha]}([a,b],\mathbb{R}^d)$ is the unique global solution of the following linearised Cauchy problem:

$$\begin{cases} {}_{c}D^\alpha_- \bar{q} = \dfrac{\partial f}{\partial x}(q^{u,\alpha},u,t)\times\bar{q} + \dfrac{\partial f}{\partial v}(q^{u,\alpha},u,t)\times\bar{u} \\ \bar{q}(a) = 0. \end{cases} \tag{LCP$^\alpha_{\bar{q}}$}$$

Proof. See Appendix A.4. □

This last result does not lead to a characterization of the critical points of $\mathcal{L}^\alpha$ yet. Then, let us introduce the following elements stemming from the Lagrange multiplier technique:

- Let H be the following application

$$\begin{array}{rccl} H: & \mathbb{R}^d\times\mathbb{R}^m\times\mathbb{R}^d\times[a,b] & \longrightarrow & \mathbb{R} \\ & (x,v,w,t) & \longmapsto & L(x,v,t) + w\cdot f(x,v,t). \end{array} \tag{28}$$

  H is commonly called the *Hamiltonian* associated to the Lagrangian L and the constraint function f;

- For any control u, let $p^{u,\alpha} \in \mathscr{C}^{[\alpha]}([a,b],\mathbb{R}^d)$ denote the unique global solution of the following fractional Cauchy problem:

$$\begin{cases} {}_{c}D^\alpha_+ p = \dfrac{\partial H}{\partial x}(q^{u,\alpha},u,p,t) = \dfrac{\partial L}{\partial x}(q^{u,\alpha},u,t) + \left(\dfrac{\partial f}{\partial x}(q^{u,\alpha},u,t)\right)^T \times p \\ p(b) = 0. \end{cases} \tag{CP$^\alpha_p$}$$

  $p^{u,\alpha}$ is commonly called the *adjoint variable* associated to u. Its existence and its uniqueness are provided by the analogous version of Theorem 4 for right fractional derivative. Since $p^{u,\alpha}(b) = 0$, we can write ${}_{c}D^\alpha_+ p^{u,\alpha} = D^\alpha_+ p^{u,\alpha}$.

Let us note that for any control u, the couple $(q^{u,\alpha},p^{u,\alpha})$ is solution of the following *fractional Hamiltonian system*:

$$\begin{cases} {}_{c}D^\alpha_- q = \dfrac{\partial H}{\partial w}(q,u,p,t) \\ D^\alpha_+ p = \dfrac{\partial H}{\partial x}(q,u,p,t). \end{cases} \tag{HS$^\alpha$}$$

Finally, the introduction of these last elements allows us to prove the following theorem:

Theorem 6. *Let $u \in \mathscr{C}^0([a,b],\mathbb{R}^m)$. Then, u is a critical point of $\mathcal{L}^\alpha$ if and only if $(q^{u,\alpha},u,p^{u,\alpha})$ is solution of the following fractional stationary equation:*

$$\frac{\partial H}{\partial v}(q,u,p,t) = 0. \tag{SE$^\alpha$}$$

Proof. Let $u, \bar{u} \in \mathscr{C}^0([a,b],\mathbb{R}^m)$. From Lemma 5, we have:

$$D\mathcal{L}^\alpha(u)(\bar{u}) = \int_a^b \frac{\partial L}{\partial x}(q^{u,\alpha},u,t)\cdot\bar{q} + \frac{\partial L}{\partial v}(q^{u,\alpha},u,t)\cdot\bar{u}\,dt. \tag{29}$$

Then:

$$D\mathcal{L}^\alpha(u)(\bar{u}) = \int_a^b \left(\frac{\partial L}{\partial x}(q^{u,\alpha},u,t) + \left(\frac{\partial f}{\partial x}(q^{u,\alpha},u,t)\right)^T \times p^{u,\alpha}\right)\cdot\bar{q}$$
$$- \left(\frac{\partial f}{\partial x}(q^{u,\alpha},u,t)\times\bar{q}\right)\cdot p^{u,\alpha} + \frac{\partial L}{\partial v}(q^{u,\alpha},u,t)\cdot\bar{u}\,dt. \tag{30}$$

From Theorem 4, since $\bar{q}$ is solution of (LCP$^\alpha_{\bar{q}}$) and $p^{u,\alpha}$ is solution of (CP$^\alpha_p$), we have:

$$\bar{q} = I^\alpha_-\left(\frac{\partial f}{\partial x}(q^{u,\alpha},u,t)\times\bar{q} + \frac{\partial f}{\partial v}(q^{u,\alpha},u,t)\times\bar{u}\right) \tag{31}$$

and

$$p^{u,\alpha} = I^\alpha_+\left(\frac{\partial L}{\partial x}(q^{u,\alpha},u,t) + \left(\frac{\partial f}{\partial x}(q^{u,\alpha},u,t)\right)^T \times p^{u,\alpha}\right). \tag{32}$$

Then, using the fractional integration by parts given in Property 2, we obtain:

$$D\mathcal{L}^\alpha(u)(\bar{u}) = \int_a^b p^{u,\alpha}\cdot\left(\frac{\partial f}{\partial x}(q^{u,\alpha},u,t)\times\bar{q} + \frac{\partial f}{\partial v}(q^{u,\alpha},u,t)\times\bar{u}\right)$$
$$- \left(\frac{\partial f}{\partial x}(q^{u,\alpha},u,t)\times\bar{q}\right)\cdot p^{u,\alpha} + \frac{\partial L}{\partial v}(q^{u,\alpha},u,t)\cdot\bar{u}\,dt. \tag{33}$$

Then:

$$\begin{aligned} D\mathcal{L}^\alpha(u)(\bar{u}) &= \int_a^b \left(\left(\frac{\partial f}{\partial v}(q^{u,\alpha},u,t)\right)^T \times p^{u,\alpha} + \frac{\partial L}{\partial v}(q^{u,\alpha},u,t)\right)\cdot\bar{u}\,dt \\ &= \int_a^b \frac{\partial H}{\partial v}(q^{u,\alpha},u,p^{u,\alpha},t)\cdot\bar{u}\,dt. \end{aligned}$$

The proof is completed by the Dubois-Raymond's lemma. □

Let us note that Lemma 5 is proved in Appendix A.4 using Lemmas 28, 29 and 30. These last three Lemmas are respectively a fractional Gronwall's Lemma, a result of stability of order 1 and a result of stability of order 2 for the fractional Cauchy problem (CP$^\alpha_q$). Hence, the proof of Theorem 6 is only based on classical mathematical tools adapted to the fractional case.

From Theorem 6, we retrieve the following result provided in [2, 3, 28, 29, 41] and references therein:

Corollary 7. *$\mathcal{L}^\alpha$ has a critical point in $\mathscr{C}^0([a,b],\mathbb{R}^m)$ if and only if there exists $(q,u,p) \in \mathscr{C}^{[\alpha]}([a,b],\mathbb{R}^d) \times \mathscr{C}^0([a,b],\mathbb{R}^m) \times \mathscr{C}^{[\alpha]}([a,b],\mathbb{R}^d)$ solution of the following fractional Pontryagin's system:*

$$\begin{cases} {}_{\mathrm{c}}D^\alpha_- q = \dfrac{\partial H}{\partial w}(q,u,p,t) \\ D^\alpha_+ p = \dfrac{\partial H}{\partial x}(q,u,p,t) \\ \dfrac{\partial H}{\partial v}(q,u,p,t) = 0 \\ \big(q(a),p(b)\big) = (A,0). \end{cases} \tag{PS$^\alpha$}$$

In the affirmative case, u is a critical point of $\mathcal{L}^\alpha$ and we have $(q,p) = (q^{u,\alpha}, p^{u,\alpha})$.

Let us note that the fractional Pontryagin's system (PS$^\alpha$) is made up of the fractional Hamiltonian system (HS$^\alpha$), the fractional stationary equation (SE$^\alpha$) and initial and final conditions. In practice, see Examples in Section 1.6, we use more Corollary 7 than Theorem 6. Let us remind that Corollary 7 was already provided in [2, 3, 28, 29, 41] and references therein without Condition (f_x lip). However, this result is proved, in each of these papers, using a Lagrange multiplier technique requiring the introduction of an augmented functional.

As we have seen in this section, fractional Pontryagin's systems emerge from the study of a class of fractional optimal control problems. They have a variational structure in the sense that they are obtained with a calculus of variations on functionals and there resolutions give explicitly the critical points of these functionals. In Section 2, our aim will be to provide them numerical schemes preserving this strong characteristic at the discrete level.

Moreover, let us make the following important remark: since a fractional Pontryagin's system emerges from a fractional optimal control problem, the main unknown is then the control u. Consequently, the convergence of the numerical scheme constructed in Section 2 is going to be considered only with respect to u.

1.4. Remarks

1.4.1. The Classical Case

In the case $\alpha = 1$, the fractional derivatives coincide with the classical one. Consequently, the fractional optimal control problem studied coincides with the classical one. Then, in this case, Corollary 7 is nothing else but the classical theorem obtained in [10, 25, 34, 35]:

Theorem 8. *$\mathcal{L}^1$ has a critical point in $\mathscr{C}^0([a,b],\mathbb{R}^m)$ if and only if there exists $(q,u,p) \in \mathscr{C}^1([a,b],\mathbb{R}^d) \times \mathscr{C}^0([a,b],\mathbb{R}^m) \times \mathscr{C}^1([a,b],\mathbb{R}^d)$ solution of the following Pontryagin's sys-*

tem:

$$\begin{cases} \dot{q} = \dfrac{\partial H}{\partial w}(q,u,p,t) \\ \dot{p} = -\dfrac{\partial H}{\partial x}(q,u,p,t) \\ \dfrac{\partial H}{\partial v}(q,u,p,t) = 0 \\ \big(q(a),p(b)\big) = (A,0). \end{cases} \tag{PS^1}$$

In the affirmative case, u is a critical point of $\mathcal{L}^1$ and we have $(q,p) = (q^{u,1}, p^{u,1})$.

1.4.2. Link with the Fractional Euler-Lagrange Equation

Let us take the constraint function $f(x,v,t) = v$ satisfying (f_x lip). In this case, the fractional optimal control problem studied is:

$$\text{optimize} \quad \int_a^b L(q,u,t)\,dt \quad \text{under the constraint} \quad \begin{cases} {}_{\mathrm{c}}D^{\alpha}_{-} q = u \\ q(a) = A. \end{cases} \tag{34}$$

Using Corollary 7, if this problem has a solution, then there exists a solution $(q,u,p) \in \mathscr{C}^{[\alpha]}([a,b],\mathbb{R}^d) \times \mathscr{C}^0([a,b],\mathbb{R}^m) \times \mathscr{C}^{[\alpha]}([a,b],\mathbb{R}^d)$ of the fractional Pontryagin's system (PS$^\alpha$) here given by:

$$\begin{cases} {}_{\mathrm{c}}D^{\alpha}_{-} q = u \\ D^{\alpha}_{+} p = \dfrac{\partial L}{\partial x}(q,u,t) \\ \dfrac{\partial L}{\partial v}(q,u,t) + p = 0 \\ \big(q(a),p(b)\big) = (A,0). \end{cases} \tag{35}$$

In the case of the existence of an optimal control u for problem (34), we obtain that the state variable associated $q^{u,\alpha}$ is solution of the following *fractional Euler-Lagrange equation*:

$$\frac{\partial L}{\partial x}(q, {}_{\mathrm{c}}D^{\alpha}_{-} q, t) + D^{\alpha}_{+}\left(\frac{\partial L}{\partial v}(q, {}_{\mathrm{c}}D^{\alpha}_{-} q, t)\right) = 0. \tag{EL$^\alpha$}$$

According to the works of Agrawal in [1], we then obtain that $q^{u,\alpha}$ is a critical point of the following *fractional Lagrangian functional*:

$$q \longrightarrow \int_a^b L(q, {}_{\mathrm{c}}D^{\alpha}_{-} q, t)\,dt. \tag{36}$$

We refer to [1] for more details concerning fractional Euler-Lagrange equations.

1.5. A Transition from the Classical to the Fractional Problem

As seen in Introduction, there is a large development in fractional calculus and consequently concerning optimal control problems with controlled systems coded by fractional differential equations. In this section, we present a link between a classical optimal control problem and its fractional version via the *Stanislavsky's formalism*, see [40, 56]. Only for this section, we assume that $[a,b] = [0,\tau]$ with $\tau > 0$ and $0 < \alpha < 1$.

1.5.1. The Stanislavsky's Formalism

The governing equation for a fluid in a homogeneous porous medium is usually the classical Richards equation and it is derived from physical principles, see [53]. Nevertheless, the fractional Richards equation plays an important role in the study of the behaviour of a fluid in a heterogeneous medium, [33, 59, 60]. A natural question then arises: *how physically understand the emergence of the fractional order when the medium changes?* Let us sketch an example of answer. A heterogeneous porous medium presents a lot of stochastic heterogeneities implying complex geometric structure and then inducing stochastic retention zones, [59]. Hence, we can assume that a change of medium induces a *modification of the time variable*. Therefore, following the work of Stanislavsky consisting in the introduction of a stochastic internal time (a "slow" time), we remind that the emergence of a fractional derivative can be the effect of a change of time. Let us give more details. However, we refer to [40, 56] for a complete study.

Let us consider T_t a stochastic process of probability density function $\rho(y,t)$. T_t represents the stochastic internal time (the "slow" time). Stanislavsky assumes that the Laplace transform $\mathrm{Lap}_{(y)}$ of $\rho(y,t)$ with respect to its first variable satisfies:

$$\forall t \geq 0,\ \forall s \in \mathbb{R},\ \mathrm{Lap}_{(y)}[\rho(y,t)](s) = \mathrm{E}_{\alpha,1}(-st^{\alpha}), \tag{37}$$

where $\mathrm{E}_{\alpha,1}$ is the Mittag-Leffler function defined in Appendix A.1. A possible construction of T_t is given in [56].

Stanislavsky studies the dynamical effects of this change of time. Precisely, for any $g \in \mathscr{C}^0([0,\tau],\mathbb{R}^d)$ and any $t \in [0,\tau]$, he studies the function $\mathcal{F}_\alpha(g)$ defined by:

$$\mathcal{F}_\alpha(g)(t) = \mathbb{E}\big(g(T_t)\big), \tag{38}$$

where $\mathbb{E}$ designates the mean value. The main property is the following result proved in [40, 56]:

Lemma 9. *Let $g \in \mathscr{C}^1([0,\tau],\mathbb{R}^d)$. Then, we have:*

$$\forall t \in [0,\tau],\ \mathbb{E}\big(\dot{g}(T_t)\big) = {}_{\mathrm{c}}D^{\alpha}_{-}\big(\mathcal{F}_\alpha(g)\big)(t). \tag{39}$$

Hence, a classical derivative is transformed into a fractional one under the introduction of the stochastic internal time of Stanislavsky.

1.5.2. Application on Fractional Optimal Control Problems

Now, let us talk about the consequences of the works of Stanislavsky on the class of fractional optimal control problems studied in this chapter. We use the notations and definitions of Section 1.5.1.

Let us assume that L and f are autonomous, *i.e.* $L(x,v,t) = L(x,v)$ and $f(x,v,t) = f(x,v)$. Then, let us assume that f satisfies the following condition:

$$\forall (q,u) \in \mathscr{C}^1([0,\tau],\mathbb{R}^d) \times \mathscr{C}^0([0,\tau],\mathbb{R}^m),\ \forall t \in [0,\tau],$$
$$\mathbb{E}\Big(f\big(q(T_t),u(T_t)\big)\Big) = f\Big(\mathbb{E}\big(q(T_t)\big),\mathbb{E}\big(u(T_t)\big)\Big). \tag{40}$$

Such a condition is satisfied by linear autonomous constraint functions, as the examples in Section 1.6. According to Lemma 9, one can prove that for any control u, $\mathcal{F}_\alpha(q^{u,1})$ satisfies:

$$\begin{cases} {}_{c}D^{\alpha}_{-}\mathcal{F}_\alpha(q^{u,1}) = f\big(\mathcal{F}_\alpha(q^{u,1}), \mathcal{F}_\alpha(u)\big) \\ \mathcal{F}_\alpha(q^{u,1})(a) = A. \end{cases} \tag{41}$$

Precisely, we have for any control u, $\mathcal{F}_\alpha(q^{u,1}) = q^{\mathcal{F}_\alpha(u),\alpha}$.

Now, let us consider the optimal control problem studied in Section 1.2 in the classical case:

$$\text{optimize the cost functional} \quad u \in \mathscr{C}^0([a,b],\mathbb{R}^d) \longmapsto \int_0^T L(q^{u,1}, u, t)\, dt. \tag{42}$$

Let us assume that the controlled system can be prone to changes inducing a modification of the time variable as it is the case with porous media and the Richards equation. We are then interested in the optimization of the following "slow" cost functional:

$$u \in \mathscr{C}^0([a,b],\mathbb{R}^d) \longmapsto \int_0^T L\big(\mathcal{F}_\alpha(q^{u,1}), \mathcal{F}_\alpha(u)\big)\, dt \tag{43}$$

which is relied to the following *fractional* cost functional:

$$u \in \mathscr{C}^0([a,b],\mathbb{R}^d) \longmapsto \int_0^T L(q^{u,\alpha}, u)\, dt \tag{44}$$

as its restriction to the image of $\mathcal{F}_\alpha$. Hence, once an optimal control of the functional (44) found, one can be interested in its projection on the image of $\mathcal{F}_\alpha$ in order to approach an optimal control of the functional (43).

Hence, the Stanislavsky's formalism is an example relying a classical optimal control problem to its fractional version via the introduction of a stochastic internal time.

1.6. Examples

In this section, we are going to study some examples of optimal control problems studied in Section 1.3 both in classical and fractional cases.

1.6.1. An Applied Classical Linear-Quadratic Example

Classical linear-quadratic examples are often studied in the literature because they are used for tracking problems. The aim of these problems is to determine a control allowing to approach as much as possible reference trajectories, [57, Part 1.4, p.49]. In this section, we study such an example, [24, Part 4.4.3, example 3, p.53]. More generally, a quadratic Lagrangian is often natural (for example in order to minimize distances) and even if the differential equations are frequently non linear, we are often leaded to study linearised versions.

In this section, we take $\alpha = d = m = 1$. We consider an evolution problem of two nondescript populations z and ξ interacting during a time interval $[a,b]$. We control the injection or the discharge of the population ξ in the system: we then control the value of the population ξ in real time. The interaction between z and ξ is governed by the following linear differential equation:

$$\dot{z} = z + \xi. \tag{45}$$

For any control ξ and any real z_0, we denote by $z_{z_0,\xi}$ the unique solution of (45) associated to ξ and satisfying the initial condition $z(a) = z_0 \in \mathbb{R}$. Let us assume that, for an initial condition $z(a) = X \in \mathbb{R}$, we know experimentally a satisfactory control ξ_X such that the evolution of the population z_{X,ξ_X} is healthy with respect to a nondescript constraint.

The problem is finally the following: *assume that the initial condition is modified (i.e. $z(a) = Y \neq X$), what is the control ξ minimizing the difference between ξ and ξ_X plus the difference between $z_{Y,\xi}$ and z_{X,ξ_X}?* More precisely, we are looking for a minimizer of the following functional:

$$\xi \in \mathscr{C}^0([a,b],\mathbb{R}) \longmapsto \frac{1}{2}\int_a^b (\xi - \xi_X)^2 + (z_{Y,\xi} - z_{X,\xi_X})^2 \, dt. \tag{46}$$

With a change of variable $A = Y - X$, $u = \xi - \xi_X$, $q = z_{Y,\xi} - z_{X,\xi_X}$ and giving the following quadratic Lagrangian and the following linear constraint function:

$$\begin{array}{rccc} L: & \mathbb{R}^2 \times [a,b] & \longrightarrow & \mathbb{R} \\ & (x,v,t) & \longmapsto & (x^2+v^2)/2 \end{array} \quad \text{and} \quad \begin{array}{rccc} f: & \mathbb{R}^2 \times [a,b] & \longrightarrow & \mathbb{R} \\ & (x,v,t) & \longmapsto & x+v, \end{array} \tag{47}$$

the problem is reduced to find an optimal control u for the cost functional $\mathcal{L}^1$ associated to L, f and A. See Section 1.3 for notations and definitions.

According to Theorem 8, we are interested in solving the Pontryagin's system (PS^1) here given by:

$$\begin{cases} \dot{q} = q + u \\ \dot{p} = -q - p \\ p + u = 0 \\ \big(q(a), p(b)\big) = (A, 0). \end{cases} \tag{48}$$

Hence, with $p + u = 0$, it is sufficient to solve the following linear Cauchy problem:

$$\begin{cases} \dot{q} = q + u \\ \dot{u} = q - u \\ \big(q(a), u(b)\big) = (A, 0). \end{cases} \tag{49}$$

Using matrix exponentials, one can prove that it exists an unique solution (q, u) to Cauchy problem (49) given by:

$$\forall t \in [a,b], \quad \begin{cases} q(t) = A\Big[\cosh\big(\sqrt{2}(t-a)\big) + \dfrac{1-R}{\sqrt{2}}\sinh\big(\sqrt{2}(t-a)\big)\Big] \\ u(t) = A\Big[\dfrac{1+R}{\sqrt{2}}\sinh\big(\sqrt{2}(t-a)\big) - R\cosh\big(\sqrt{2}(t-a)\big)\Big] \end{cases} \tag{50}$$

where

$$R = \frac{\sinh\big(\sqrt{2}(b-a)\big)}{\sqrt{2}\cosh\big(\sqrt{2}(b-a)\big) - \sinh\big(\sqrt{2}(b-a)\big)}. \tag{51}$$

Finally, we conclude that u is the unique critical point of $\mathcal{L}^1$. Hence, if there exists ξ minimizing (46), then:

$$\forall t \in [a,b], \; \xi(t) = \xi_X(t) + u(t). \tag{52}$$

1.6.2. The Fractional Linear Quadratic Example

In this section, we consider the previous example of Section 1.6.1 in the strict fractional case $0 < \alpha < 1$. Let us consider $d = m = 1$ and the couple (L, f) given in Equation (47). We want to express a critical point u of the cost functional $\mathcal{L}^\alpha$ associated to L and f. See Section 1.3 for notations and definitions.

Therefore, according to Corollary 7, we look for (q, u, p) solution of the fractional Pontryagin's system (PS$^\alpha$) here given by:

$$\begin{cases} {}_{\mathrm{c}}D^\alpha_- q = q + u \\ D^\alpha_+ p = q + p \\ p + u = 0 \\ \big(q(a), p(b)\big) = (A, 0). \end{cases} \tag{53}$$

Hence, from $u + p = 0$, it is sufficient to solve the following linear fractional Cauchy problem:

$$\begin{cases} {}_{\mathrm{c}}D^\alpha_- q = q + u \\ D^\alpha_+ u = u - q \\ \big(q(a), u(b)\big) = (A, 0). \end{cases} \tag{54}$$

Although the fractional Cauchy problem (54) is linear, the presence of the left and the right fractional derivatives is a main drawback to the explicit computation of a solution. More generally, this characteristic implies very big difficulties in order to solve the most of fractional Hamiltonian systems. Finally, the resolution of (PS$^\alpha$) is the most of time still an opened problem.

Nevertheless, in this example, we can hope getting informations on the solutions of (PS$^\alpha$) in dimension $d = m = 2$ with the help of the fractional Noether's Theorem 14 proved in Section 1.7. Indeed, in this example, the fractional Pontryagin's system (PS$^\alpha$) admits a symmetry and a constant of motion can be obtained. We refer to Section 1.7 for more details.

Let us also remind that we are going to make some numerical tests with the variational integrator constructed in Section 2. This example will be treated and graphic representations are going to be provided, see Section 2.4.2. Moreover, in dimension $d = m = 2$, we also can apply the discrete fractional Noether's Theorem 26 proved in Section 2.5. We refer to Section 2.5 for more details.

1.6.3. A Solved Fractional Example

In this section, we give a solved fractional example in the sense that we give explicitly the unique critical point of the cost functional $\mathcal{L}^\alpha$. Nevertheless, the fractional Pontryagin's system (PS$^\alpha$) is still not completely resolved.

Let us take $d = m = 1$, $[a, b] = [0, 1]$ and $0 < \alpha \leq 1$. Let us consider the following Lagrangian L and the following linear constraint function f:

$$\begin{array}{rccl} L: & \mathbb{R}^2 \times [0,1] & \longrightarrow & \mathbb{R} \\ & (x, v, t) & \longmapsto & (v^2/2) + \gamma(1-t)^\beta x \end{array} \quad \text{and} \quad \begin{array}{rccl} f: & \mathbb{R}^2 \times [0,1] & \longrightarrow & \mathbb{R} \\ & (x, v, t) & \longmapsto & \lambda x + \mu v, \end{array} \tag{55}$$

where β, γ, μ, λ are parameters in $\mathbb{R}^*$. We are looking for a critical point u of $\mathcal{L}^\alpha$ associated to L and f. See Section 1.3 for notations and definitions.

According to Corollary 7, we have to solve the fractional Pontryagin's system (PS$^\alpha$) here given by:

$$\begin{cases} {}_{c}D^{\alpha}_{-}q = \lambda q + \mu u \\ D^{\alpha}_{+}p = \lambda p + \gamma(1-t)^{\beta} \\ u + \mu p = 0 \\ \big(q(0), p(1)\big) = (A, 0). \end{cases} \tag{56}$$

From Theorem 4, the Cauchy problem ${}_{c}D^{\alpha}_{-}q = \lambda q + \mu u$ with $q(0) = A$ admits an unique solution for any $u \in \mathscr{C}^0([0,1], \mathbb{R})$. Then, from $u + \mu p = 0$, it is sufficient to solve the following fractional Cauchy problem:

$$\begin{cases} D^{\alpha}_{+}p = \lambda p + \gamma(1-t)^{\beta} \\ p(1) = 0 \end{cases} \quad \text{equivalent to} \quad \begin{cases} D^{\alpha}_{-}p_0 = \lambda p_0 + \gamma t^{\beta} \\ p_0(0) = 0, \end{cases} \tag{57}$$

with the change of unknown $p_0(t) = p(1-t)$ for any $t \in [0,1]$. The unique solution of this last fractional Cauchy problem is given in [42, Chap.3, p.137] by:

$$\forall t \in [0,1],\ p_0(t) = \int_0^t (t-y)^{\alpha-1} \mathrm{E}_{\alpha,\alpha}\big(\lambda(t-y)^{\alpha}\big)\gamma y^{\beta}\, dy, \tag{58}$$

where $\mathrm{E}_{\alpha,\alpha}$ is the Mittag-Leffler function defined in Appendix A.1. In order to get a better formulation, we make a change of variable which gives us:

$$\begin{aligned} \forall t \in [0,1],\ p_0(t) &= \gamma \int_0^t (t-y)^{\beta} y^{\alpha-1} \mathrm{E}_{\alpha,\alpha}(\lambda y^{\alpha}) dy & (59) \\ &= \gamma\Gamma(\beta+1) I^{\beta+1}_{-}\big(y^{\alpha-1}\mathrm{E}_{\alpha,\alpha}(\lambda y^{\alpha})\big)(t) & (60) \\ &= \gamma\Gamma(\beta+1) t^{\alpha+\beta} \mathrm{E}_{\alpha,\alpha+\beta+1}(\lambda t^{\alpha}). & (61) \end{aligned}$$

We refer to [42] for more details concerning the calculations. Finally, we obtain the unique critical point u of $\mathcal{L}^\alpha$ given by:

$$\forall t \in [0,1],\ u(t) = -\mu\gamma\Gamma(\beta+1)(1-t)^{\alpha+\beta}\mathrm{E}_{\alpha,\alpha+\beta+1}\big(\lambda(1-t)^{\alpha}\big). \tag{62}$$

Although we obtain the unique critical point u of $\mathcal{L}^\alpha$, let us note that this example does not provide a completely solved fractional Pontryagin's system: the state variable $q^{u,\alpha}$ is still unknown. However, this example allows to test the quality of numerical schemes giving approximations of critical points of cost functionals $\mathcal{L}^\alpha$. This will be done for the variational integrator constructed in Section 2, see Section 2.4.

1.7. A Fractional Noether's Theorem

As we have seen with concrete examples in Sections 1.6.2 and 1.6.3, fractional Pontryagin's systems (PS$^\alpha$) are the most of time not resolvable explicitly. This is a strong obstruction in order to express explicitly a critical point of a cost functional. In this section, we suggest a deviously way in order to get informations on the solutions of (PS$^\alpha$) and consequently on the critical points of the cost functional $\mathcal{L}^\alpha$ associated. Precisely, we are interested in the existence of conservation laws for fractional Pontryagin's systems admitting a symmetry (see Definition 11).

In 1918, Noether proved the existence of an explicit conservation law for any classical Euler-Lagrange equation admitting a symmetry. We refer to [6, 43, 50] for more details. Adapting her strategy to the fractional case, Cresson, Torres and Frederico proved in 2007 a preliminary result giving a conservation law for any fractional Euler-Lagrange equation admitting a symmetry, see [18, 27, 30]. Nevertheless, the conservation law obtained was not explicit but only given implicitly via a functional relation. From this first result, we have formulated in [12] a fractional Noether's theorem providing an *explicit* formulation of this conservation law via a *transfer formula*.

At the same time, in [28, 29], Torres and Frederico applied a similar strategy for fractional Pontryagin's system admitting a symmetry. Nevertheless, this conservation law is also given implicitly via a functional relation. In this section, we are going to formulate a fractional Noether's theorem providing an *explicit* formulation of this conservation law via an other *transfer formula*.

We first review the definition of a one parameter group of diffeomorphisms:

Definition 10. *Let $n \in \mathbb{N}^*$. For any real s, let $\phi(s,\cdot) : \mathbb{R}^n \longrightarrow \mathbb{R}^n$ be a diffeomorphism. Then, $\Phi = \{\phi(s,\cdot)\}_{s\in\mathbb{R}}$ is a one parameter group of diffeomorphisms of $\mathbb{R}^n$ if it satisfies:*

1. $\phi(0,\cdot) = Id_{\mathbb{R}^n}$;
2. $\forall s, s' \in \mathbb{R},\ \phi(s,\cdot) \circ \phi(s',\cdot) = \phi(s+s',\cdot)$;
3. *ϕ is of class $\mathscr{C}^2$.*

Usual examples of one parameter groups of diffeomorphisms are given by translations and rotations. The action of three one parameter groups of diffeomorphisms on an Hamiltonian allows to define the notion of a symmetry for a fractional Pontryagin's system:

Definition 11. *For $i = 1,2,3$, let $\Phi_i = \{\phi_i(s,\cdot)\}_{s\in\mathbb{R}}$ be a one parameter group of diffeomorphisms of $\mathbb{R}^d$, $\mathbb{R}^m$ and $\mathbb{R}^d$ respectively. Let L be a Lagrangian, f be a constraint function and H be the associated Hamiltonian. H is said to be ${}_{\mathrm{c}}D^{\alpha}_{-}$-invariant under the action of $(\Phi_i)_{i=1,2,3}$ if it satisfies for any (q,u,p) solution of (PS^α) and any $s \in \mathbb{R}$:*

$$H\Big(\phi_1(s,q),\phi_2(s,u),\phi_3(s,p),t\Big) - \phi_3(s,p)\cdot {}_{\mathrm{c}}D^{\alpha}_{-}\big(\phi_1(s,q)\big) = H(q,u,p,t) - p\cdot {}_{\mathrm{c}}D^{\alpha}_{-}q. \tag{63}$$

From this notion, Torres and Frederico proved in [28, 29] the following result:

Lemma 12. *Let L be a Lagrangian, f be a constraint function and H be the associated Hamiltonian. Let us assume that H is ${}_{\mathrm{c}}D^{\alpha}_{-}$-invariant under the action of three one parameter groups of diffeomorphisms $(\Phi_i)_{i=1,2,3}$. Then, the following equality holds for any solution (q,u,p) of (PS^α):*

$${}_{\mathrm{c}}D^{\alpha}_{-}\left(\frac{\partial\phi_1}{\partial s}(0,q)\right)\cdot p - \frac{\partial\phi_1}{\partial s}(0,q)\cdot D^{\alpha}_{+}p = 0. \tag{64}$$

Then, from this result, Torres and Frederico defined a notion of *fractional-conserved quantity*. Nevertheless, this result did not provide exactly a constant of motion. In this section, using a transfer formula, we are going to write the left term of Equation (64) as an explicit classical derivative and then we obtain a real constant of motion. Let us provide this transfer formula:

Lemma 13 (Transfer formula). *Let $g_1, g_2 \in \mathscr{C}^\infty([a,b],\mathbb{R}^d)$ satisfying the following condition* (C)*:*

$$\text{the sequences of functions } \left(I_-^{p-\alpha}\big(g_1 - g_1(a)\big)\cdot g_2^{(p)}\right)_{p\in\mathbb{N}^*} \text{ and } (g_1^{(p)}\cdot I_+^{p-\alpha} g_2)_{p\in\mathbb{N}^*} \text{ converge uniformly to } 0 \text{ on } [a,b].$$

Then, the following equality holds:

$$ {}_{\mathrm{c}}D_-^\alpha g_1 \cdot g_2 - g_1 \cdot D_+^\alpha g_2 = \frac{d}{dt}\left[\sum_{r=0}^{\infty}(-1)^r I_-^{r+1-\alpha}\big(g_1 - g_1(a)\big)\cdot g_2^{(r)} + g_1^{(r)}\cdot I_+^{r+1-\alpha} g_2\right]. \tag{65}$$

Proof. In [12], under a similar condition than Condition (C), we have proved:

$$D_-^\alpha g_1 \cdot g_2 = \frac{d}{dt}\left[\sum_{r=0}^{\infty}(-1)^r I_-^{r+1-\alpha} g_1 \cdot g_2^{(r)}\right] \text{ and } -g_1 \cdot D_+^\alpha g_2 = \frac{d}{dt}\left[\sum_{r=0}^{\infty} g_1^{(r)}\cdot I_+^{r+1-\alpha} g_2\right]. \tag{66}$$

Consequently, writing ${}_{\mathrm{c}}D_-^\alpha g_1 = D_-^\alpha\big(g_1 - g_1(a)\big)$, the proof is completed. □

A discussion is provided in [12] concerning the condition (C): one can prove that this condition is satisfied for any couple of analytic functions for example. Then, combining Lemmas 12 and 13, we prove:

Theorem 14 (A fractional Noether's theorem). *Let L be a Lagrangian, f be a constraint function and H be the associated Hamiltonian. Let us assume that H is ${}_{\mathrm{c}}D_-^\alpha$-invariant under the action of three one parameter groups of diffeomorphisms $(\Phi_i)_{i=1,2,3}$. Let (q,u,p) be a solution of* (PS$^\alpha$) *and let g denote $\partial\phi_1/\partial s(0,q)$. If g and p satisfy Condition* (C)*, then the following equality holds:*

$$\frac{d}{dt}\left[\sum_{r=0}^{\infty}(-1)^r I_-^{r+1-\alpha}\big(g - g(a)\big)\cdot p^{(r)} + g^{(r)}\cdot I_+^{r+1-\alpha} p\right] = 0. \tag{67}$$

This theorem then provides an explicit algorithmic way to compute a constant of motion for any fractional Pontryagin's systems admitting a symmetry. Nevertheless, the conservation law is only given as a series of functions: in most cases, it is not easily computable. However, an arbitrary closed approximation of this quantity can be obtained with a truncation.

Let us note that the fractional linear-quadratic example developed in Section 1.6.2 gives a concrete example of fractional Pontryagin's system admitting a symmetry:

Example 15. *Let us consider $d = m = 2$, the following quadratic Lagrangian and the following linear constraint function*

$$\begin{array}{rccc} L: & \mathbb{R}^2\times\mathbb{R}^2\times[0,1] & \longrightarrow & \mathbb{R} \\ & (x,v,t) & \longmapsto & (\|x\|^2+\|v\|^2)/2 \end{array} \quad \text{and} \quad \begin{array}{rccc} f: & \mathbb{R}^2\times\mathbb{R}^2\times[0,1] & \longrightarrow & \mathbb{R}^2 \\ & (x,v,t) & \longmapsto & x+v. \end{array} \tag{68}$$

Then, let us consider the three one parameter groups of diffeomorphisms given by the following rotations:

$$\begin{array}{rccc} \phi_i: & \mathbb{R}\times\mathbb{R}^2 & \longrightarrow & \mathbb{R}^2 \\ & (s,x_1,x_2) & \longmapsto & \begin{pmatrix} \cos(s\theta_i) & -\sin(s\theta_i) \\ \sin(s\theta_i) & \cos(s\theta_i)\end{pmatrix}\begin{pmatrix} x_1 \\ x_2 \end{pmatrix}, \end{array} \tag{69}$$

for $i = 1, 2, 3$ and where θ_1, $\theta_2 \in \mathbb{R}$ and $\theta_3 = -\theta_1$. With these parameters, one can prove that the Hamiltonian H associated to L and f is ${}_{c}D^{\alpha}_{-}$-invariant under the action of $(\Phi_i)_{i=1,2,3}$. Consequently, the fractional Pontryagin's system given in (53) *(in dimension* 2*) is not resolvable but admits a symmetry and then admits an explicit conservation law given by the fractional Noether's Theorem 14. As said previously, this constant of motion is not explicitly computable. However, truncating the series, one can provide an approximation of this quantity.*

2. Variational Integrator for Fractional Pontryagin's Systems

In general, fractional differential equations are very difficult to solve. One can find some solved examples in [42, 52, 55] using Mittag-Leffler functions, Fourier and Laplace transforms. Additionally, fractional Pontryagin's systems, as fractional Euler-Lagrange equations provided in [1], present an asymmetry in the sense that left and right fractional derivatives are involved. It is an additional drawback in order to solve explicitly the most of fractional Pontryagin's systems. In this section, we then develop a numerical approach treating them.

Let us remind that there exist many works concerning the statement of discrete operators approaching the fractional derivatives (see [23, 31, 51]) and then concerning numerical schemes for fractional differential equations (see [21, 45, 49, 48]). In particular, one can find studies concerning the discretization of fractional Euler-Lagrange equations [4, 13] and fractional Pontryagin's systems [2, 3, 8, 19, 41].

Nevertheless, as we have seen in Section 1.3, a fractional Pontryagin's system admits an intrinsic variational structure: its solutions correspond to the critical points of a functional. In this chapter, we want to construct a numerical scheme for fractional Pontryagin's systems preserving at the discrete level this strong property.

A *variational integrator* is a numerical scheme preserving the variational structure of a system at the discrete level. Precisely, let us consider a differential system coming from a variational principle (*i.e. its solutions correspond to the critical points of a functional*). Then, a variational integrator is the numerical scheme constructed as follows:

- firstly, one have to define a discrete version of the functional;
- secondly one have to form a discrete variational principle on it.

Hence, a numerical scheme is obtained and it is called variational integrator. It preserves the variational structure at the discrete level in the sense that *its discrete solutions correspond to the discrete critical points of the discrete functional*. Let us remind that variational integrators are well-developed for classical Euler-Lagrange equations in [36, 46] and let us remind that we have developed a variational integrator for fractional Euler-Lagrange equations in [13]. In this section, we are going to construct a variational integrator for fractional Pontryagin's systems.

Let us introduce the following notations available in the whole chapter. Let $N \in \mathbb{N}^*$, $h = (b-a)/N$ denote the step size of discretization and $\boldsymbol{T} = (t_k)_{k=0,\dots,N} = (a +$

$kh)_{k=0,\dots,N}$ be the classical partition of the interval $[a,b]$. Let us assume that N is sufficiently large in order to satisfy the following condition:

$$2h^\alpha M < 1, \qquad \text{(cond } h\text{)}$$

where M is the Lipschitz coefficient of the constraint function f, see Condition (f_x lip).

2.1. Definitions and Results about Discrete Fractional Calculus

2.1.1. Reminder about Discrete Fractional Derivatives of Grünwald-Letnikov

For the sequel, we need the introduction of discrete operators approximating the fractional derivatives of Riemann-Liouville and Caputo. As in [19, 23], let us define Δ_-^α and Δ_+^α the following discrete analogous of D_-^α and D_+^α respectively:

$$\begin{array}{rccc} \Delta_-^\alpha : & (\mathbb{R}^d)^{N+1} & \longrightarrow & (\mathbb{R}^d)^N \\ & \boldsymbol{G} & \longmapsto & \left(\dfrac{1}{h^\alpha} \displaystyle\sum_{r=0}^{k} \alpha_r G_{k-r} \right)_{k=1,\dots,N} \end{array} \tag{70}$$

and

$$\begin{array}{rccc} \Delta_+^\alpha : & (\mathbb{R}^d)^{N+1} & \longrightarrow & (\mathbb{R}^d)^N \\ & \boldsymbol{G} & \longmapsto & \left(\dfrac{1}{h^\alpha} \displaystyle\sum_{r=0}^{N-k} \alpha_r G_{k+r} \right)_{k=0,\dots,N-1} \end{array}, \tag{71}$$

where the elements $(\alpha_r)_{r\in\mathbb{N}}$ are defined by $\alpha_0 := 1$ and

$$\forall r \in \mathbb{N}^*, \ \alpha_r := \frac{(-\alpha)(1-\alpha)\dots(r-1-\alpha)}{r!}. \tag{72}$$

These discrete fractional operators are approximations of the continuous ones. Indeed, passing to the limit $h \to 0$, these discrete operators correspond to the definition of the fractional derivatives of Grünwald-Letnikov (1867) coinciding with the Riemann-Liouville's ones. We refer to [52] for more details.

Finally, according to Equation (4), we define ${}_{\mathrm{c}}\Delta_-^\alpha$ and ${}_{\mathrm{c}}\Delta_+^\alpha$ the following discrete analogous of ${}_{\mathrm{c}}D_-^\alpha$ and ${}_{\mathrm{c}}D_+^\alpha$ respectively:

$$\begin{array}{rccc} {}_{\mathrm{c}}\Delta_-^\alpha : & (\mathbb{R}^d)^{N+1} & \longrightarrow & (\mathbb{R}^d)^N \\ & \boldsymbol{G} & \longmapsto & \left((\Delta_-^\alpha(\boldsymbol{G} - G_0))_k \right)_{k=1,\dots,N} \end{array} \tag{73}$$

and

$$\begin{array}{rccc} {}_{\mathrm{c}}\Delta_+^\alpha : & (\mathbb{R}^d)^{N+1} & \longrightarrow & (\mathbb{R}^d)^N \\ & \boldsymbol{G} & \longmapsto & \left((\Delta_+^\alpha(\boldsymbol{G} - G_N))_k \right)_{k=0,\dots,N-1} \end{array}. \tag{74}$$

Let us note that we preserve some continuous properties at the discrete level. In particular, $G_0 = 0$ (resp. $G_N = 0$) implies ${}_{\mathrm{c}}\Delta_-^\alpha \boldsymbol{G} = \Delta_-^\alpha \boldsymbol{G}$ (resp. ${}_{\mathrm{c}}\Delta_+^\alpha \boldsymbol{G} = \Delta_+^\alpha \boldsymbol{G}$). Additionally, in the classical case $\alpha = 1$, these discrete fractional derivatives coincide with the usual backward and forward Euler's approximations of d/dt with a (-1) term in the right case:

$$\forall k = 1, \dots, N,\ (\Delta_-^1 \boldsymbol{G})_k = ({}_{\mathrm{c}}\Delta_-^1 \boldsymbol{G})_k = \frac{G_k - G_{k-1}}{h} \tag{75}$$

and

$$\forall k = 0, \dots, N-1,\ (\Delta_+^1 \boldsymbol{G})_k = ({}_{\mathrm{c}}\Delta_+^1 \boldsymbol{G})_k = \frac{G_k - G_{k+1}}{h}. \tag{76}$$

2.1.2. Results Concerning the Discrete Fractional Derivatives

In this section, we prove two important properties preserved from the continuous level to the discrete one. For the sequel, we first need the introduction of the following *shift operators*:

$$\begin{array}{rccl} \sigma: & (\mathbb{R}^n)^{N+1} & \longrightarrow & (\mathbb{R}^n)^N \\ & \boldsymbol{G} & \longmapsto & (G_{k+1})_{k=0,\dots,N-1} \end{array} \quad \text{and} \quad \begin{array}{rccl} \sigma^{-1}: & (\mathbb{R}^n)^{N+1} & \longrightarrow & (\mathbb{R}^n)^N \\ & \boldsymbol{G} & \longmapsto & (G_{k-1})_{k=1,\dots,N}, \end{array} \tag{77}$$

where the integer n is d or m.

The first property is the following: considering the quadrature formula of Gauss as approximation of the integral, we can prove the following discrete fractional integration by parts:

Property 16 (Discrete fractional integration by parts). *Let $\boldsymbol{G}^1$, $\boldsymbol{G}^2 \in (\mathbb{R}^d)^{N+1}$ satisfying $G_0^1 = G_N^2 = 0$, then we have:*

$$h \sum_{k=1}^{N} ({}_{\mathrm{c}}\Delta_-^\alpha \boldsymbol{G}^1)_k \cdot \sigma^{-1}(\boldsymbol{G}^2)_k = h \sum_{k=0}^{N-1} \sigma(\boldsymbol{G}^1)_k \cdot ({}_{\mathrm{c}}\Delta_+^\alpha \boldsymbol{G}^2)_k. \tag{DFIBP}$$

Proof. Since $G_0^1 = G_N^2 = 0$, we have ${}_{\mathrm{c}}\Delta_-^\alpha \boldsymbol{G}^1 = \Delta_-^\alpha \boldsymbol{G}^1$ and ${}_{\mathrm{c}}\Delta_+^\alpha \boldsymbol{G}^2 = \Delta_+^\alpha \boldsymbol{G}^2$. Then, we have:

$$h \sum_{k=1}^{N} (\Delta_-^\alpha \boldsymbol{G}^1)_k \cdot \sigma^{-1}(\boldsymbol{G}^2)_k = h \sum_{k=0}^{N-1} (\Delta_-^\alpha \boldsymbol{G}^1)_{k+1} \cdot G_k^2 = h^{1-\alpha} \sum_{k=0}^{N-1} \sum_{r=0}^{k+1} \alpha_r G_{k+1-r}^1 \cdot G_k^2. \tag{78}$$

Finally, since $G_0^1 = G_N^2 = 0$, the following equalities hold:

$$\begin{aligned} h \sum_{k=1}^{N} (\Delta_-^\alpha \boldsymbol{G}^1)_k \cdot \sigma^{-1}(\boldsymbol{G}^2)_k &= h^{1-\alpha} \sum_{k=0}^{N-1} \sum_{r=0}^{k} \alpha_r G_{k+1-r}^1 \cdot G_k^2 \\ = h^{1-\alpha} \sum_{r=0}^{N-1} \sum_{k=r}^{N-1} \alpha_r G_{k+1-r}^1 \cdot G_k^2 &= h^{1-\alpha} \sum_{r=0}^{N-1} \sum_{k=0}^{N-r-1} \alpha_r G_{k+1}^1 \cdot G_{k+r}^2 \\ = h^{1-\alpha} \sum_{k=0}^{N-1} G_{k+1}^1 \cdot \left(\sum_{r=0}^{N-k-1} \alpha_r G_{k+r}^2 \right) &= h^{1-\alpha} \sum_{k=0}^{N-1} G_{k+1}^1 \cdot \left(\sum_{r=0}^{N-k} \alpha_r G_{k+r}^2 \right), \end{aligned} \tag{79}$$

which concludes the proof. □

This last result is very useful for discrete calculus of variations involving discrete fractional derivatives, see proof of Theorem 19. Secondly let us prove the following discrete version of the fractional Cauchy-Lipschitz Theorem 14 proved in Section 1.1.2:

Theorem 17 (Discrete fractional Cauchy-Lipschitz theorem). *Let $F \in \mathscr{C}^0(\mathbb{R}^d \times [a,b], \mathbb{R}^d)$ satisfying the following Lipschitz type condition:*

$$\exists K \in \mathbb{R},\ \forall (x_1, x_2, t) \in (\mathbb{R}^d)^2 \times [a,b],\ \|F(x_1,t) - F(x_2,t)\| \leq K\|x_1 - x_2\|, \quad (80)$$

with $h^\alpha K < 1$. Then, the following discrete fractional Cauchy problem:

$$\begin{cases} {}_{c}\Delta_{-}^{\alpha} \boldsymbol{Q} = F(\boldsymbol{Q}, \boldsymbol{T}) \\ Q_0 = A \end{cases} \quad (81)$$

has an unique solution $\boldsymbol{Q} \in (\mathbb{R}^d)^{N+1}$.

Proof. We are going to construct by induction the solution $\boldsymbol{Q}$ of (81). Our method uses the classical fix point theorem concerning the contraction mappings. Indeed, let us choose $Q_0 = A$. Then, for any $k = 1, \ldots, N$, Q_k has to satisfy:

$$Q_k = h^\alpha F(Q_k, t_k) + Q_0 - \sum_{r=1}^{k-1} \alpha_r (Q_{k-r} - Q_0). \quad (82)$$

However, for any $k = 1, \ldots, N$, the application $h^\alpha F(\cdot, t_k) + Q_0 - \sum_{r=1}^{k-1} \alpha_r (Q_{k-r} - Q_0)$ is a contraction and consequently admits an unique fix point. Hence, we first construct Q_1, then Q_2, etc. By induction, we construct a solution $\boldsymbol{Q}$ of (81) and such a construction assures its uniqueness. □

2.2. Construction of the Variational Integrator

2.2.1. First Step of Construction

As said in introduction of this section, in order to complete the first step of construction of a variational integrator, we have to provide a discrete version of $\mathcal{L}^\alpha$. In this way, let us give the following definition:

- The elements $\boldsymbol{U} \in (\mathbb{R}^m)^{N+1}$ are called the *discrete controls*;
- For any discrete control $\boldsymbol{U}$, let $\boldsymbol{Q}^{\boldsymbol{U},\alpha} \in (\mathbb{R}^d)^{N+1}$ denote the unique solution of the following discrete Cauchy problem:

$$\begin{cases} {}_{c}\Delta_{-}^{\alpha} \boldsymbol{Q} = f(\boldsymbol{Q}, \boldsymbol{U}, \boldsymbol{T}) \\ Q_0 = A \in \mathbb{R}^d. \end{cases} \quad (\mathrm{CP}_{\boldsymbol{Q}}^{\alpha})$$

  $\boldsymbol{Q}^{\boldsymbol{U},\alpha}$ is called the *discrete state variable* associated to $\boldsymbol{U}$. Its existence and its uniqueness are provided by Theorem 17 and Conditions (f_x lip) and (cond h);
- Finally, we define the following *discrete cost functional*:

$$\begin{array}{rccl} \mathcal{L}_h^\alpha : & (\mathbb{R}^m)^{N+1} & \longrightarrow & \mathbb{R} \\ & \boldsymbol{U} & \longmapsto & h \displaystyle\sum_{k=1}^{N} L(Q_k^{\boldsymbol{U},\alpha}, U_k, t_k). \end{array} \quad (83)$$

Hence, we have provided a discrete version $\mathcal{L}_h^\alpha$ to the cost functional $\mathcal{L}^\alpha$. Now, the second step of the construction of the variational integrator is to characterize the *discrete critical points* of the discrete cost functional $\mathcal{L}_h^\alpha$ with the help of a discrete calculus of variations.

Let us make the following remark: such a characterization implies to be a necessary condition for the existence of an optimizer of the discrete cost functional $\mathcal{L}_h^\alpha$. In fact, in this section, we have defined an actual discrete fractional optimal control problem.

2.2.2. Second Step of Construction

The second step of construction of a variational integrator consists in forming a discrete variational principle on $\mathcal{L}_h^\alpha$. Precisely, we focus on the characterization of its *discrete critical points*, *i.e.* the elements $\boldsymbol{U} \in (\mathbb{R}^m)^{N+1}$ satisfying:

$$\forall \bar{\boldsymbol{U}} \in (\mathbb{R}^m)^{N+1}, \; D\mathcal{L}_h^\alpha(\boldsymbol{U})(\bar{\boldsymbol{U}}) := \lim_{\varepsilon \to 0} \frac{\mathcal{L}_h^\alpha(\boldsymbol{U}+\varepsilon\bar{\boldsymbol{U}}) - \mathcal{L}_h^\alpha(\boldsymbol{U})}{\varepsilon} = 0. \tag{84}$$

With a discrete calculus of variations, we obtain the following discrete version of Lemma 5 giving explicitly the value of the Gâteaux derivative of $\mathcal{L}_h^\alpha$.

Lemma 18. *Let $\boldsymbol{U}, \bar{\boldsymbol{U}} \in (\mathbb{R}^m)^{N+1}$. Then, the following equality holds:*

$$D\mathcal{L}_h^\alpha(\boldsymbol{U})(\bar{\boldsymbol{U}}) = h \sum_{k=1}^{N} \left[\frac{\partial L}{\partial x}(Q_k^{\boldsymbol{U},\alpha}, U_k, t_k) \cdot \bar{Q}_k + \frac{\partial L}{\partial v}(Q_k^{\boldsymbol{U},\alpha}, U_k, t_k) \cdot \bar{U}_k \right], \tag{85}$$

where $\bar{\boldsymbol{Q}} \in (\mathbb{R}^d)^{N+1}$ is the unique solution of the following linearised discrete fractional Cauchy problem:

$$\begin{cases} {}_{\mathrm{c}}\Delta_-^\alpha \bar{\boldsymbol{Q}} = \dfrac{\partial f}{\partial x}(\boldsymbol{Q}^{\boldsymbol{U},\alpha}, \boldsymbol{U}, \boldsymbol{T}) \times \bar{\boldsymbol{Q}} + \dfrac{\partial f}{\partial v}(\boldsymbol{Q}^{\boldsymbol{U},\alpha}, \boldsymbol{U}, \boldsymbol{T}) \times \bar{\boldsymbol{U}} \\ \bar{Q}_0 = 0. \end{cases} \tag{LCP$^\alpha_{\bar{\boldsymbol{Q}}}$}$$

Proof. See Appendix B.3. □

This last result does not lead to a characterization of the critical points of $\mathcal{L}_h^\alpha$ yet. As in the continuous case, we then introduce the notion of discrete adjoint variable: for any discrete control $\boldsymbol{U}$, let $\boldsymbol{P}^{\boldsymbol{U},\alpha} \in (\mathbb{R}^d)^{N+1}$ denote the unique solution of the following shifted discrete Cauchy problem:

$$\begin{cases} {}_{\mathrm{c}}\Delta_+^\alpha \boldsymbol{P} &= \dfrac{\partial H}{\partial x}(\sigma(\boldsymbol{Q}^{\boldsymbol{U},\alpha}), \sigma(\boldsymbol{U}), \boldsymbol{P}, \sigma(\boldsymbol{T})) \\ &= \dfrac{\partial L}{\partial x}(\sigma(\boldsymbol{Q}^{\boldsymbol{U},\alpha}), \sigma(\boldsymbol{U}), \sigma(\boldsymbol{T})) + \left(\dfrac{\partial f}{\partial x}(\sigma(\boldsymbol{Q}^{\boldsymbol{U},\alpha}), \sigma(\boldsymbol{U}), \sigma(\boldsymbol{T}))\right)^T \times \boldsymbol{P} \\ P_N &= 0. \end{cases} \tag{σCP$^\alpha_{\boldsymbol{P}}$}$$

$\boldsymbol{P}^{\boldsymbol{U},\alpha}$ is called the *discrete adjoint variable* associated to $\boldsymbol{U}$. Its existence and its uniqueness are provided by the analogous of Theorem 17 for right discrete fractional derivative and by Conditions (f_x lip) and (cond h). Let us note that, since $\boldsymbol{P}_N^{\boldsymbol{U},\alpha} = 0$, we can write ${}_{\mathrm{c}}\Delta_+^\alpha \boldsymbol{P}^{\boldsymbol{U},\alpha} = \Delta_+^\alpha \boldsymbol{P}^{\boldsymbol{U},\alpha}$.

The presence of shift operators in the definition of the discrete adjoint variable is the consequence of the change of sums in the discrete fractional integration by parts (DFIBP) (see Property 16). We refer to the proof of Theorem 19 for more details. We also refer to Remark 21 for a discussion about the presence of the shift operators.

Finally, let us note that for any discrete control $\boldsymbol{U}$, the couple $(\boldsymbol{Q}^{\boldsymbol{U},\alpha}, \boldsymbol{P}^{\boldsymbol{U},\alpha})$ is solution of the following *shifted discrete fractional Hamiltonian system*:

$$\begin{cases} {}_{\mathrm{c}}\Delta_{-}^{\alpha} \boldsymbol{Q} = \dfrac{\partial H}{\partial w}(\boldsymbol{Q}, \boldsymbol{U}, \sigma^{-1}(\boldsymbol{P}), \boldsymbol{T}) \\ \Delta_{+}^{\alpha} \boldsymbol{P} = \dfrac{\partial H}{\partial x}(\sigma(\boldsymbol{Q}), \sigma(\boldsymbol{U}), \boldsymbol{P}, \sigma(\boldsymbol{T})). \end{cases} \tag{$\sigma\mathrm{HS}_h^\alpha$}$$

Finally, the introduction of this last discrete element allows us to prove the following theorem:

Theorem 19. *Let $\boldsymbol{U} \in (\mathbb{R}^m)^{N+1}$. Then, $\boldsymbol{U}$ is a discrete critical point of $\mathcal{L}_h^\alpha$ if and only if $(\boldsymbol{Q}^{\boldsymbol{U},\alpha}, \boldsymbol{U}, \boldsymbol{P}^{\boldsymbol{U},\alpha})$ is solution of the following shifted discrete fractional stationary equation:*

$$\frac{\partial H}{\partial v}(\boldsymbol{Q}, \boldsymbol{U}, \sigma^{-1}(\boldsymbol{P}), \boldsymbol{T}) = 0. \tag{$\sigma\mathrm{SE}_h^\alpha$}$$

Proof. Let $\boldsymbol{U}, \bar{\boldsymbol{U}} \in (\mathbb{R}^m)^{N+1}$. From Lemma 18, we have:

$$h^{-1} D\mathcal{L}_h^\alpha(\boldsymbol{U})(\bar{\boldsymbol{U}}) = \sum_{k=1}^{N} \left[\frac{\partial L}{\partial x}(Q_k^{\boldsymbol{U},\alpha}, U_k, t_k) + \left(\frac{\partial f}{\partial x}(Q_k^{\boldsymbol{U},\alpha}, U_k, t_k) \right)^T \times \sigma^{-1}(\boldsymbol{P}^{\boldsymbol{U},\alpha})_k \right] \cdot \bar{Q}_k$$
$$- \sum_{k=1}^{N} \left(\left(\frac{\partial f}{\partial x}(Q_k^{\boldsymbol{U},\alpha}, U_k, t_k) \right)^T \times \sigma^{-1}(\boldsymbol{P}^{\boldsymbol{U},\alpha})_k \right) \cdot \bar{Q}_k + \sum_{k=1}^{N} \frac{\partial L}{\partial v}(Q_k^{\boldsymbol{U},\alpha}, U_k, t_k) \cdot \bar{U}_k. \tag{86}$$

Then:

$$h^{-1} D\mathcal{L}_h^\alpha(\boldsymbol{U})(\bar{\boldsymbol{U}}) = \sum_{k=0}^{N-1} ({}_{\mathrm{c}}\Delta_{+}^{\alpha} \boldsymbol{P}^{\boldsymbol{U},\alpha})_k \cdot \sigma(\bar{\boldsymbol{Q}})_k - \sum_{k=1}^{N} \left(\frac{\partial f}{\partial x}(Q_k^{\boldsymbol{U},\alpha}, U_k, t_k) \times \bar{Q}_k \right) \cdot \sigma^{-1}(\boldsymbol{P}^{\boldsymbol{U},\alpha})_k$$
$$+ \sum_{k=1}^{N} \frac{\partial L}{\partial v}(Q_k^{\boldsymbol{U},\alpha}, U_k, t_k) \cdot \bar{U}_k. \tag{87}$$

From the discrete fractional integration by parts (DFIBP) (see Property 16), we obtain:

$$h^{-1} D\mathcal{L}_h^\alpha(\boldsymbol{U})(\bar{\boldsymbol{U}}) = \sum_{k=1}^{N} \left(({}_{\mathrm{c}}\Delta_{-}^{\alpha} \bar{\boldsymbol{Q}})_k - \frac{\partial f}{\partial x}(Q_k^{\boldsymbol{U},\alpha}, U_k, t_k) \times \bar{Q}_k \right) \cdot \sigma^{-1}(\boldsymbol{P}^{\boldsymbol{U},\alpha})_k$$
$$+ \sum_{k=1}^{N} \frac{\partial L}{\partial v}(Q_k^{\boldsymbol{U},\alpha}, U_k, t_k) \cdot \bar{U}_k. \tag{88}$$

Since $\bar{\boldsymbol{Q}}$ is solution of ($\mathrm{LCP}_{\bar{\boldsymbol{Q}}}^\alpha$), we have:

$$\begin{aligned} h^{-1} D\mathcal{L}_h^\alpha(\boldsymbol{U})(\bar{\boldsymbol{U}}) &= \sum_{k=1}^{N} \left(\frac{\partial f}{\partial v}(Q_k^{\boldsymbol{U},\alpha}, U_k, t_k) \times \bar{U}_k \right) \cdot \sigma^{-1}(\boldsymbol{P}^{\boldsymbol{U},\alpha})_k + \sum_{k=1}^{N} \frac{\partial L}{\partial v}(Q_k^{\boldsymbol{U},\alpha}, U_k, t_k) \cdot \bar{U}_k \\ &= \sum_{k=1}^{N} \left(\left(\frac{\partial f}{\partial v}(Q_k^{\boldsymbol{U},\alpha}, U_k, t_k) \right)^T \times \sigma^{-1}(\boldsymbol{P}^{\boldsymbol{U},\alpha})_k + \frac{\partial L}{\partial v}(Q_k^{\boldsymbol{U},\alpha}, U_k, t_k) \right) \cdot \bar{U}_k. \end{aligned}$$

Finally:

$$D\mathcal{L}_h^\alpha(\boldsymbol{U})(\bar{\boldsymbol{U}}) = h\sum_{k=1}^{N} \frac{\partial H}{\partial v}(Q_k^{\boldsymbol{U},\alpha}, U_k, \sigma^{-1}(\boldsymbol{P}^{\boldsymbol{U},\alpha})_k, t_k)\cdot \bar{U}_k. \tag{89}$$

The proof is completed. □

Finally, from Theorem 19, we obtain the following result leading to the variational integrator constructed:

Corollary 20. *$\mathcal{L}_h^\alpha$ has a discrete critical point if and only if there exists $(\boldsymbol{Q},\boldsymbol{U},\boldsymbol{P}) \in (\mathbb{R}^d)^{N+1}\times(\mathbb{R}^m)^{N+1}\times(\mathbb{R}^d)^{N+1}$ solution of the following shifted discrete fractional Pontryagin's system:*

$$\begin{cases} {}_{\mathrm{c}}\Delta_-^\alpha \boldsymbol{Q} = \dfrac{\partial H}{\partial w}(\boldsymbol{Q},\boldsymbol{U},\sigma^{-1}(\boldsymbol{P}),\boldsymbol{T}) \\ \Delta_+^\alpha \boldsymbol{P} = \dfrac{\partial H}{\partial x}(\sigma(\boldsymbol{Q}),\sigma(\boldsymbol{U}),\boldsymbol{P},\sigma(\boldsymbol{T})) \\ \dfrac{\partial H}{\partial v}(\boldsymbol{Q},\boldsymbol{U},\sigma^{-1}(\boldsymbol{P}),\boldsymbol{T}) = 0 \\ (Q_0,P_N) = (A,0). \end{cases} \tag{$\sigma\mathrm{PS}_h^\alpha$}$$

In this case, $\boldsymbol{U}$ is a discrete critical point of $\mathcal{L}_h^\alpha$ and we have $(\boldsymbol{Q},\boldsymbol{P}) = (\boldsymbol{Q}^{\boldsymbol{U},\alpha},\boldsymbol{P}^{\boldsymbol{U},\alpha})$.

Let us note that ($\sigma\mathrm{PS}_h^\alpha$) is made up of the shifted discrete Hamiltonian system ($\sigma\mathrm{HS}_h^\alpha$), the shifted stationary equation ($\sigma\mathrm{SE}_h^\alpha$) and initial and final conditions.

Hence, we have constructed the variational integrator ($\sigma\mathrm{PS}_h^\alpha$) for the fractional Pontryagin's system (PS^α). It is then a numerical scheme for (PS^α) preserving its variational structure in the sense that the discrete solutions $\boldsymbol{U}$ obtained correspond to the discrete critical points of the discrete version $\mathcal{L}_h^\alpha$ of $\mathcal{L}^\alpha$.

Remark 21. *Let us note that the variational integrator ($\sigma\mathrm{PS}_h^\alpha$) does not correspond with a direct discretization of (PS^α) as it is done in [19]. There is an emergence of shift operators caused by the conservation at the discrete level of the variational structure. However, it is proved that the use of shifted numerical schemes allows to obtain more stability for some fractional differential equations, see [48, 49].*

Remark 22. *Let us remind the following remark: since a fractional Pontryagin's system emerges from a fractional optimal control problem, the main unknown is then the control u. Consequently, the convergence of the variational integrator ($\sigma\mathrm{PS}_h^\alpha$) is going to be considered only with respect to u. Let us note that the value of U_0 does not take place in the variational integrator ($\sigma\mathrm{PS}_h^\alpha$): it is a free value. Nevertheless, this is totally coherent with the fact that this value does not take place neither in the definition of $\mathcal{L}_h^\alpha$. Hence, in numerical tests in Section 2.4, the error between an exact solution u of (PS^α) and a numerical solution $\boldsymbol{U}$ obtained with ($\sigma\mathrm{PS}_h^\alpha$) is going to be evaluated on $\|u(t_k) - U_k\|$ for $k \in \{1,\dots,N\}$ only.*

2.3. Link with the Discrete Fractional Euler-Lagrange Equation

Let us take the constraint function $f(x, v, t) = v$ satisfying (f_x lip). In this case, applying Corollary 20, we know that there exists a critical point of $\mathcal{L}_h^\alpha$ if and only if there exists a solution $(\boldsymbol{Q}, \boldsymbol{U}, \boldsymbol{P}) \in (\mathbb{R}^d)^{N+1} \times (\mathbb{R}^m)^{N+1} \times (\mathbb{R}^d)^{N+1}$ of the shifted discrete fractional Pontryagin's system ($\sigma\mathrm{PS}_h^\alpha$) here given by:

$$\begin{cases} {}_{\mathrm{c}}\Delta_-^\alpha \boldsymbol{Q} = \boldsymbol{U} \\ \Delta_+^\alpha \boldsymbol{P} = \dfrac{\partial L}{\partial x}(\sigma(\boldsymbol{Q}), \sigma(\boldsymbol{U}), \sigma(\boldsymbol{T})) \\ \dfrac{\partial L}{\partial v}(\boldsymbol{Q}, \boldsymbol{U}, \boldsymbol{T}) + \sigma^{-1}(\boldsymbol{P}) = 0 \\ (Q_0, P_N) = (A, 0). \end{cases} \tag{90}$$

In the affirmative case, it implies that $\boldsymbol{Q}$ is then a discrete solution of the following *discrete fractional Euler-Lagrange equation*:

$$\frac{\partial L}{\partial x}(\boldsymbol{Q}, {}_{\mathrm{c}}\Delta_-^\alpha \boldsymbol{Q}, \boldsymbol{T}) + \Delta_+^\alpha \left(\frac{\partial L}{\partial v}(\boldsymbol{Q}, {}_{\mathrm{c}}\Delta_-^\alpha \boldsymbol{Q}, \boldsymbol{T}) \right) = 0. \tag{EL$_h^\alpha$}$$

Finally, according to our works in [13], we then obtain that $\boldsymbol{Q}$ is a critical point of the following *discrete fractional Lagrangian functional*:

$$\boldsymbol{Q} \longrightarrow h \sum_{k=1}^{N} L(Q_k, ({}_{\mathrm{c}}\Delta_-^\alpha \boldsymbol{Q})_k, t_k). \tag{91}$$

We refer to [13] for more details concerning discrete fractional Euler-Lagrange equations.

2.4. Numerical Tests

In the following numerical tests, according to Remark 22, we are going to give graphic representations only of discrete solutions $\boldsymbol{U}$ and the study of the convergence of the variational integrator ($\sigma\mathrm{PS}_h^\alpha$) is only going to be evaluated on the convergence of the discrete control to the continuous one.

2.4.1. The Linear-Quadratic Example

Linear-quadratic examples are often studied in the literature because they are used for tracking problems. The aim of these problems is to determine a control allowing to approach as much as possible reference trajectories, [57, Part 1.4, p.49]. In this section, we study such an example, [24, Part 4.4.3, example 3, p.53]. More generally, a quadratic Lagrangian is often natural (for example in order to minimize distances) and even if the constraint functions are frequently non linear, we are often leaded to study linearised versions.

Let us choose $d = m = A = 1$ and $[a, b] = [0, 1]$. Then, let us take the following quadratic Lagrangian and linear constraint function:

$$\begin{array}{rccc} L: & \mathbb{R}^2 \times [0,1] & \longrightarrow & \mathbb{R} \\ & (x, v, t) & \longmapsto & (x^2 + v^2)/2 \end{array} \quad \text{and} \quad \begin{array}{rccc} f: & \mathbb{R}^2 \times [0,1] & \longrightarrow & \mathbb{R} \\ & (x, v, t) & \longmapsto & x + v. \end{array} \tag{92}$$

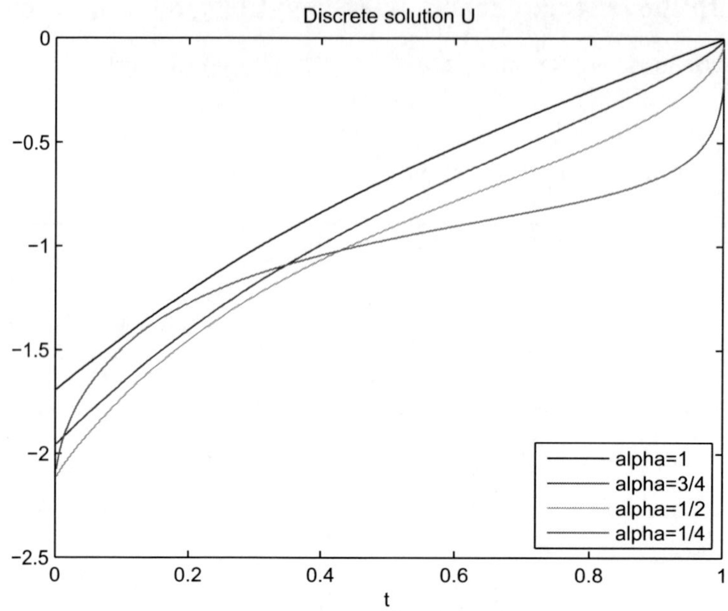

Figure 1.

Let us give the graphic representations of the numerical solutions $\boldsymbol{U}$ given by $(\sigma\mathrm{PS}_h^\alpha)$ for $N = 500$ and for $\alpha = 1, 3/4, 1/2, 1/4$: We have seen in Sections 1.6.1 and 1.6.2 that the fractional Pontryagin's system (PS^α) is explicitly solved only in the classical case $\alpha = 1$ and we obtained the following unique critical point of $\mathcal{L}^1$:

$$\forall t \in [0,1],\ u(t) = \frac{\cosh(\sqrt{2})}{R_1}\sinh(\sqrt{2}t) - \frac{\sinh(\sqrt{2})}{R_1}\cosh(\sqrt{2}t), \tag{93}$$

where $R_1 = \sqrt{2}\cosh(\sqrt{2}) - \sinh(\sqrt{2})$. Hence, we can only test the convergence of the variational integrator $(\sigma\mathrm{PS}_h^\alpha)$ for $\alpha = 1$. We give the following graphic representing the logarithm of the error $\max\big(|u(t_k) - U_k|, k = 1, \ldots, N\big)$ versus the logarithm of h and the identity function for comparison: In this example with $\alpha = 1$, the convergence seems then obtained with order 1. Nevertheless, we do not know the exact solution of (PS^α) in the strict fractional case $0 < \alpha < 1$. Consequently, we can not study the behaviour of the error in this case.

2.4.2. A Solved Fractional Example

In this section, we are going to compute $(\sigma\mathrm{PS}_h^\alpha)$ in the framework of the example solved in the strict fractional case seen in Section 1.6.3. Consequently, for this example, we can test the convergence of the variational integrator $(\sigma\mathrm{PS}_h^\alpha)$ for any $0 < \alpha \leq 1$.

Then, let us choose $d = m = A = 1$ and $[a, b] = [0, 1]$. Then, let us take the following Lagrangian and linear constraint function:

$$\begin{array}{rccc} L: & \mathbb{R}^2 \times [0,1] & \longrightarrow & \mathbb{R} \\ & (x,v,t) & \longmapsto & (1-t)x + (v^2/2) \end{array} \quad \text{and} \quad \begin{array}{rccc} f: & \mathbb{R}^2 \times [0,1] & \longrightarrow & \mathbb{R} \\ & (x,v,t) & \longmapsto & x+v. \end{array} \tag{94}$$

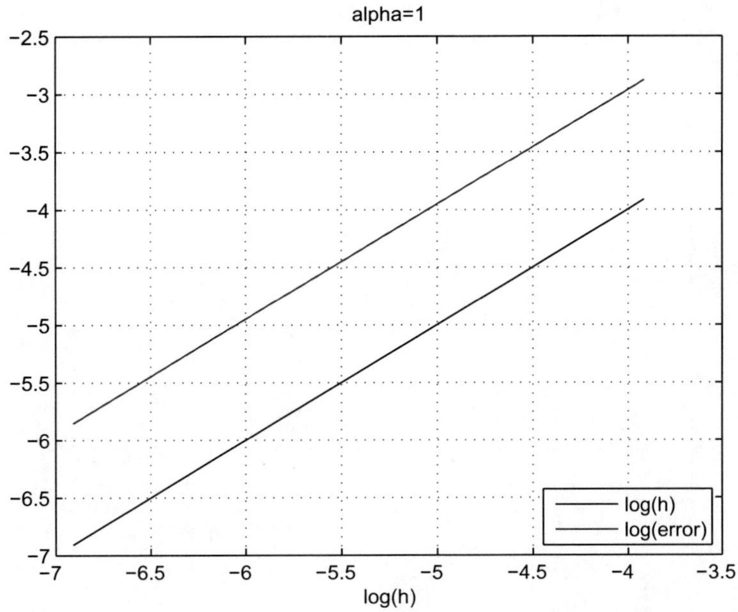

Figure 2.

Let us give the graphic representations of the numerical solutions $\boldsymbol{U}$ given by $(\sigma \mathrm{PS}_h^\alpha)$ for $N = 500$ and for $\alpha = 1, 3/4, 1/2, 1/4$: As we have seen in Section 1.6.3, the fractional Pontryagin's system (PS^α) is explicitly solved for any $0 < \alpha \leq 1$ and we obtained the following unique critical point of $\mathcal{L}^\alpha$:

$$\forall t \in [0,1],\ u(t) = -(1-t)^{\alpha+1}\mathrm{E}_{\alpha,\alpha+2}\big((1-t)^\alpha\big), \tag{95}$$

where $\mathrm{E}_{\alpha,\alpha+2}$ is the Mittag-Leffler function with parameter $(\alpha, \alpha+2)$. Let us test the convergence of the variational integrator $(\sigma \mathrm{PS}_h^\alpha)$ for any $0 < \alpha \leq 1$. We give the following graphics representing the logarithm of the error $\max\big(|u(t_k) - U_k|, k = 1, \ldots, N\big)$ versus the logarithm of h and the identity function for comparison for $\alpha = 1,\ 3/4,\ 1/2,\ 1/4$: For this example, the convergence seems then obtained for any $\alpha = 1, 3/4, 1/2, 1/4$ and still with order 1. Hence, the graphics obtained in these Sections 2.4.1 and 2.4.2 make us confident with respect to the quality of $(\sigma \mathrm{PS}_h^\alpha)$ both in the classical and strict fractional cases.

2.5. A Discrete Fractional Noether's Theorem

In this section, we study the existence of discrete conservation laws for shifted discrete fractional Pontryagin's systems $(\sigma \mathrm{PS}_h^\alpha)$. Precisely, following the same strategy as in the continuous case, we introduce the notion of *discrete symmetry* for such systems and prove a discrete fractional Noether's theorem providing an *explicit computable* discrete constant of motion. Let us note that this work is strongly inspired from our study in [12] where we have provided a discrete fractional Noether's theorem for discrete fractional Euler-Lagrange equations admitting a discrete symmetry.

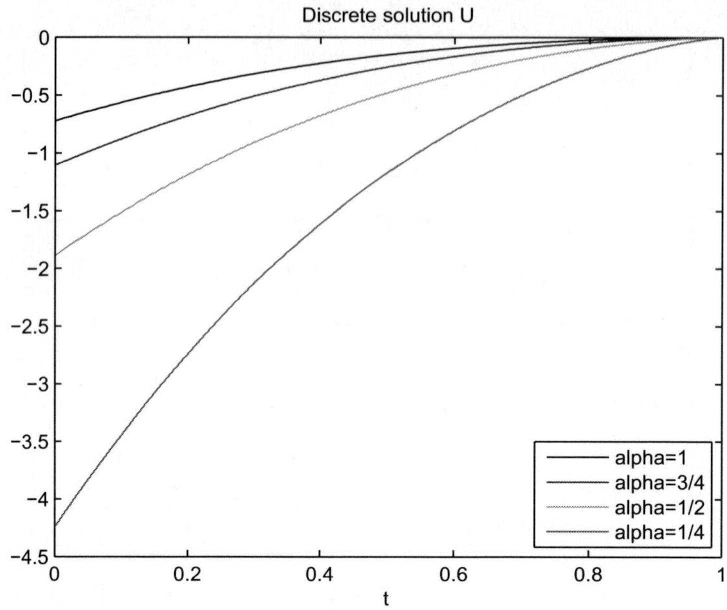

Figure 3.

As in the continuous case, a discrete symmetry for a shifted discrete fractional Pontryagin's system ($\sigma\mathrm{PS}_h^\alpha$) is based on the action of three one parameter groups of diffeomorphisms on the Hamiltonian associated:

Definition 23. *Let $\Phi_i = \{\phi_i(s,\cdot)\}_{s\in\mathbb{R}}$, for $i=1,2,3$, be three one parameter groups of diffeomorphisms of $\mathbb{R}^d$, $\mathbb{R}^m$ and $\mathbb{R}^d$ respectively. Let L be a Lagrangian, f be a constraint function and H be the associated Hamiltonian. H is said to be ${}_{\mathrm{c}}\Delta_-^\alpha$-invariant under the action of $(\Phi_i)_{i=1,2,3}$ if it satisfies: for any $(\boldsymbol{Q},\boldsymbol{U},\boldsymbol{P})$ solution of ($\sigma\mathrm{PS}_h^\alpha$) and any $s\in\mathbb{R}$*

$$H\Big(\phi_1(s,\boldsymbol{Q}),\phi_2(s,\boldsymbol{U}),\phi_3(s,\sigma^{-1}(\boldsymbol{P})),\boldsymbol{T}\Big) - \phi_3(s,\sigma^{-1}(\boldsymbol{P}))\cdot {}_{\mathrm{c}}\Delta_-^\alpha\Big(\phi_1(s,\boldsymbol{Q})\Big) = H\big(\boldsymbol{Q},\boldsymbol{U},\sigma^{-1}(\boldsymbol{P}),\boldsymbol{T}\big) - \sigma^{-1}(\boldsymbol{P})\cdot {}_{\mathrm{c}}\Delta_-^\alpha \boldsymbol{Q}. \tag{96}$$

From this notion, we prove the following discrete version of Lemma 12:

Lemma 24. *Let L be a Lagrangian, f be a constraint function and H be the associated Hamiltonian. Let us assume that H is ${}_{\mathrm{c}}\Delta_-^\alpha$-invariant under the action of three one parameter groups of diffeomorphisms $(\Phi_i)_{i=1,2,3}$. Then, the following equality holds for any solution $(\boldsymbol{Q},\boldsymbol{U},\boldsymbol{P})$ solution of ($\sigma\mathrm{PS}_h^\alpha$):*

$$\frac{\partial\phi_1}{\partial s}(0,\boldsymbol{Q})\cdot\sigma^{-1}(\Delta_+^\alpha\boldsymbol{P}) - {}_{\mathrm{c}}\Delta_-^\alpha\left(\frac{\partial\phi_1}{\partial s}(0,\boldsymbol{Q})\right)\cdot\sigma^{-1}(\boldsymbol{P}) = 0. \tag{97}$$

Proof. Let us differentiate (96) with respect to s and let us invert the operator ${}_{\mathrm{c}}\Delta_-^\alpha$ and

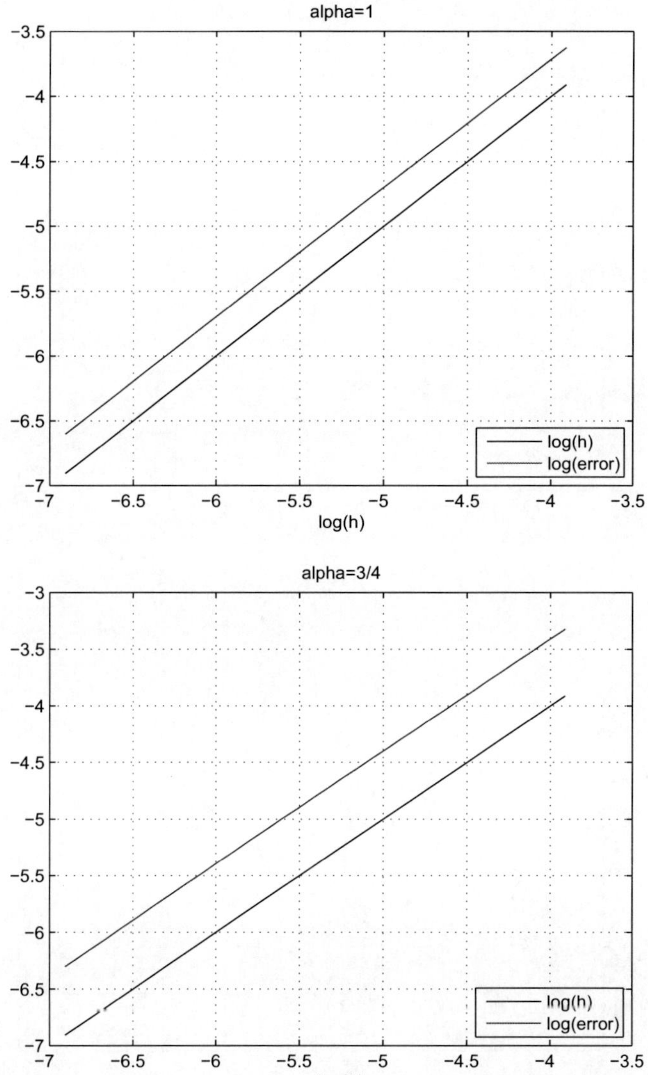

Figure 4.

$\partial/\partial s$. Taking $s=0$, we finally obtain:

$$\frac{\partial H}{\partial x}(\star)\cdot\frac{\partial\phi_1}{\partial s}(0,\boldsymbol{Q})+\frac{\partial H}{\partial v}(\star)\cdot\frac{\partial\phi_2}{\partial s}(0,\boldsymbol{U})+\frac{\partial H}{\partial w}(\star)\cdot\frac{\partial\phi_3}{\partial s}(0,\sigma^{-1}(\boldsymbol{P}))$$
$$-\frac{\partial\phi_3}{\partial s}(0,\sigma^{-1}(\boldsymbol{P}))\cdot{}_{\mathrm{c}}\Delta^{\alpha}_{-}\boldsymbol{Q}-\sigma^{-1}(\boldsymbol{P})\cdot{}_{\mathrm{c}}\Delta^{\alpha}_{-}\left(\frac{\partial\phi_1}{\partial s}(0,\boldsymbol{Q})\right)=0, \quad (98)$$

where $\star=\left(\boldsymbol{Q},\boldsymbol{U},\sigma^{-1}(\boldsymbol{P}),\boldsymbol{T}\right)$. Since $(\boldsymbol{Q},\boldsymbol{U},\boldsymbol{P})$ is solution of $(\sigma\mathrm{PS}^{\alpha}_{h})$, we obtain (97). □

Let us remind that our aim is to provide an explicit discrete constant of motion for shifted discrete fractional Pontryagin's systems $(\sigma\mathrm{PS}^{\alpha}_{h})$ exhibiting a discrete symmetry. Our

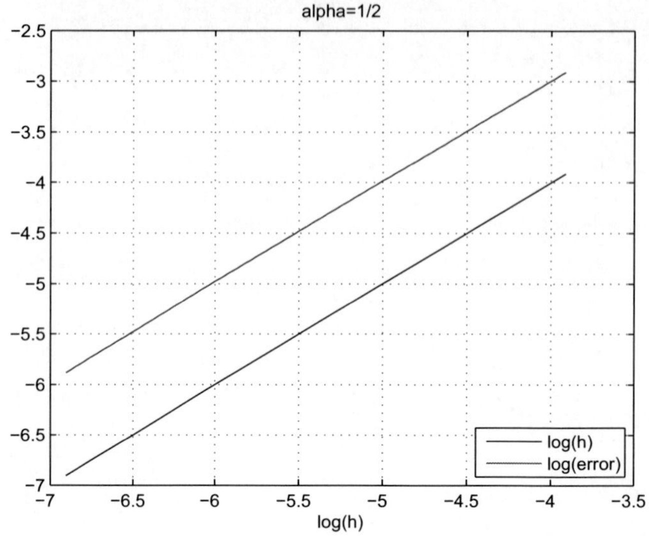

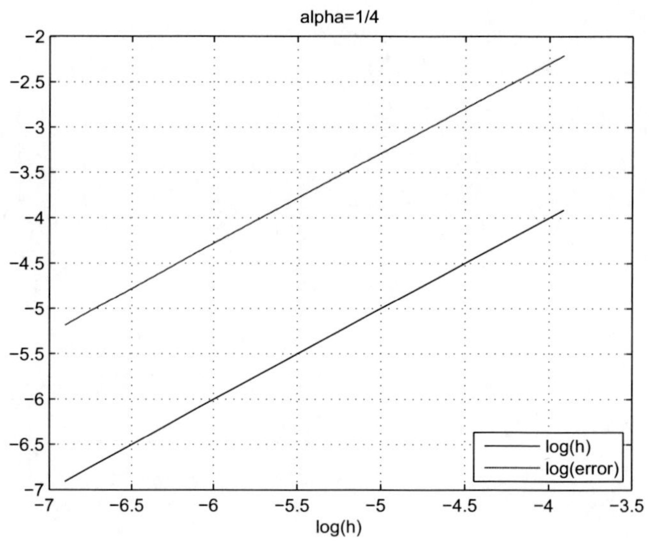

Figure 5.

result is based on Lemma 24 and on the following implication:

$$\forall \boldsymbol{G} \in \mathbb{R}^{N+1},\ \Delta^1_- \boldsymbol{G} = 0 \Longrightarrow \exists c \in \mathbb{R},\ \forall k = 0, \ldots, N,\ G_k = c. \tag{99}$$

Namely, if the discrete derivative of $\boldsymbol{G}$ vanishes, then $\boldsymbol{G}$ is constant. Consequently, our aim is to write the left term of (97) as an explicit discrete derivative (*i.e.* as Δ^1_- of an explicit quantity). In this way, we are going to use a *discrete transfer formula* as it is done in [12] for discrete fractional Euler-Lagrange equations admitting a discrete symmetry.

Nevertheless, we have first to introduce some square matrices of length $(N+1)$. First, $B_1 := \mathrm{Id}_{N+1}$ and then, for any $r \in \{2, \ldots, N\}$, the square matrices $B_r \in \mathcal{M}_{N+1}$ defined by:

$$\forall i,j = 0,\dots,N,\ (B_r)_{i,j} := \delta_{\{1\leq i\leq N-1\}}\delta_{\{1\leq j\leq N-r\}}\delta_{\{0\leq i-j\leq r-1\}} - \delta_{\{j=0\}}\delta_{\{r\leq i\}}, \tag{100}$$

where δ is the Kronecker symbol. Secondly, we define the square matrices $C_r \in \mathcal{M}_{N+1}$ by:

$$\forall r = 1,\dots,N,\ \forall i,j = 0,\dots,N,\ (C_r)_{i,j} := \delta_{\{r\leq i\}}\delta_{\{j=0\}}. \tag{101}$$

Finally, we define the square matrices $A_r \in \mathcal{M}_{N+1}$ by:

$$\forall r = 1,\dots,N,\ A_r := \alpha_r B_r + \beta_r^\alpha C_r, \tag{102}$$

where $\beta_r^\alpha = \sum_{k=0}^r \alpha_k$. Examples of matrices $A_r \in \mathcal{M}_{N+1}$ for $N = 5$ are given in Appendix B.4.

Lemma 25 (Discrete transfer formula)**.** *Let $\boldsymbol{G}^1, \boldsymbol{G}^2 \in (\mathbb{R}^d)^{N+1}$ satisfying $G_N^2 = 0$. Then, the following equality holds:*

$$\boldsymbol{G}^1 \cdot \sigma^{-1}(\Delta_+^\alpha \boldsymbol{G}^2) - ({}_{\mathrm{c}}\Delta_-^\alpha \boldsymbol{G}^1)\cdot\sigma^{-1}(\boldsymbol{G}^2) = h^{1-\alpha}\Delta_-^1\Big[\sum_{r=1}^N A_r \times \big(\boldsymbol{G}^1\cdot\sigma^{r-1}(\boldsymbol{G}^2)\big)\Big]. \tag{103}$$

Proof. See Appendix B.4. □

Consequently, combining Lemmas 24 and 25, we prove:

Theorem 26 (Discrete fractional Noether's theorem)**.** *Let L be a Lagrangian, f be a constraint function and H be the associated Hamiltonian. Let us assume that H is ${}_{\mathrm{c}}\Delta_-^\alpha$-invariant under the action of three one parameter groups of diffeomorphisms $(\Phi_i)_{i=1,2,3}$. Then, the following equality holds for any solution $(\boldsymbol{Q},\boldsymbol{U},\boldsymbol{P})$ of $(\sigma\mathrm{PS}_h^\alpha)$:*

$$\Delta_-^1\left[\sum_{r=1}^N A_r \times \left(\frac{\partial\phi_1}{\partial s}(0,\boldsymbol{Q})\cdot\sigma^{r-1}(\boldsymbol{P})\right)\right] = 0. \tag{104}$$

According to Equation (99), this theorem provides a discrete constant of motion for any shifted discrete fractional Pontryagin's systems $(\sigma\mathrm{PS}_h^\alpha)$ exhibiting a discrete symmetry. Moreover, this discrete conservation law is not only *explicit* but also *computable* in a finite number of steps. Let us see a concrete example:

Example 27. *Let us consider $d = m = 2$, the following quadratic Lagrangian and the following linear constraint function:*

$$\begin{array}{rrcl} L: & \mathbb{R}^2\times\mathbb{R}^2\times[a,b] & \longrightarrow & \mathbb{R} \\ & (x,v,t) & \longmapsto & (\|x\|^2+\|v\|^2)/2 \end{array} \quad \text{and} \quad \begin{array}{rrcl} f: & \mathbb{R}^2\times\mathbb{R}^2\times[a,b] & \longrightarrow & \mathbb{R}^2 \\ & (x,v,t) & \longmapsto & x+v. \end{array} \tag{105}$$

Then, we consider the three one parameter groups of diffeomorphisms given by the following rotations:

$$\begin{array}{rrcl} \phi_i: & \mathbb{R}\times\mathbb{R}^2 & \longrightarrow & \mathbb{R}^2 \\ & (s,x_1,x_2) & \longmapsto & \begin{pmatrix} \cos(s\theta_i) & -\sin(s\theta_i) \\ \sin(s\theta_i) & \cos(s\theta_i) \end{pmatrix}\begin{pmatrix} x_1 \\ x_2 \end{pmatrix}, \end{array} \tag{106}$$

for $i = 1, 2, 3$ *and where* θ_1, $\theta_2 \in \mathbb{R}$ *and* $\theta_3 = -\theta_1$. *With these parameters, one can prove that the Hamiltonian* H *associated to* L *and* f *is* ${}_{\mathrm{c}}\Delta_{-}^{\alpha}$*-invariant under the action of* $(\Phi_i)_{i=1,2,3}$. *Consequently, the fractional Pontryagin's system* $(\sigma\mathrm{PS}_h^\alpha)$ *admits a symmetry and then admits an explicit discrete conservation law given by the discrete fractional Noether's Theorem 26.*

We choose $A = (1, 2)$, $N = 100$ *and* $\theta_1 = \theta_2 = -\theta_3 = 1$. *Let us compute* $(\sigma\mathrm{PS}_h^\alpha)$ *for* $\alpha = 1$, $3/4$, $1/2$, $1/4$. *Then, we denote* $\boldsymbol{Q} = (\boldsymbol{Q}^1, \boldsymbol{Q}^2)$ *and* $\boldsymbol{P} = (\boldsymbol{P}^1, \boldsymbol{P}^2)$ *the discrete solutions obtained and we denote* $\boldsymbol{G} = \partial\phi_1/\partial s(0, \boldsymbol{Q}) = (-\boldsymbol{Q}^2, \boldsymbol{Q}^1)$. *We are then interested in the value of:*

$$\sum_{r=1}^{N} A_r \times \left(\boldsymbol{G} \cdot \sigma^{r-1}(\boldsymbol{P})\right). \tag{107}$$

Let us see the graphics obtained by the computation of $(\sigma\mathrm{PS}_h^\alpha)$ *and by the computation of the quantity given in Equation* (107) *for* $\alpha = 1$, $3/4$, $1/2$, $1/4$:

As expected from Theorem 26, we obtain discrete constants of motion for this discrete fractional Pontryagin's system $(\sigma\mathrm{PS}_h^\alpha)$ *admitting a discrete symmetry and for any* $\alpha = 1$, $3/4$, $1/2$, $1/4$. *In this specific example, the constant obtained is zero.*

A. Appendix

A.1. A Fractional Gronwall's Lemma

Let us recall the definition extracted from [42, 52, 55] of the Mittag-Leffler function $\mathrm{E}_{\alpha_1,\alpha_2}$ of parameters α_1, $\alpha_2 \geq 0$:

$$\forall t \in \mathbb{R},\ \mathrm{E}_{\alpha_1,\alpha_2}(t) := \sum_{k=0}^{\infty} \frac{t^k}{\Gamma(\alpha_1 k + \alpha_2)}. \tag{108}$$

Now, let us give the following Lemma proved in [21, 58]. For the reader's convenience, we recall the proof:

Lemma 28 (A fractional Gronwall's lemma)**.** *Let* $g \in \mathscr{C}^0([a,b], \mathbb{R})$ *and* $\alpha > 0$. *Let us assume that there exist* K_1, $K_2 \geq 0$ *such that:*

$$\forall t \in [a,b],\ 0 \leq g(t) \leq K_1 I_{-}^{\alpha} g(t) + K_2. \tag{109}$$

Then, g *satisfies:*

$$\forall t \in [a,b],\ 0 \leq g(t) \leq K_2 \mathrm{E}_{\alpha,1}\big(K_1 (t-a)^\alpha\big). \tag{110}$$

Proof. Using Property 1, Assumption (109) implies by induction that for any $n \in \mathbb{N}^*$ and any $t \in [a,b]$:

$$0 \leq g(t) \leq K_1^n I_{-}^{n\alpha} g(t) + K_2 \sum_{k=0}^{n-1} K_1^k I_{-}^{k\alpha}(1)(t) = K_1^n I_{-}^{n\alpha} g(t) + K_2 \sum_{k=0}^{n-1} \frac{\big(K_1 (t-a)^\alpha\big)^k}{\Gamma(\alpha k + 1)}. \tag{111}$$

Since g is continuous on $[a,b]$ and non-negative, g is bounded by 0 and by a constant $K_3 \geq 0$. Hence, using the Stirling asymptotic formula, we have for any $t \in [a,b]$:

$$0 \leq K_1^n I_{-}^{n\alpha} g(t) = \frac{K_1^n}{\Gamma(\alpha n)} \int_a^t (t-y)^{n\alpha - 1} g(y)\,dy \leq K_3 \frac{\big(K_1 (t-a)^\alpha\big)^n}{\Gamma(\alpha n + 1)} \underset{n \to +\infty}{\longrightarrow} 0. \tag{112}$$

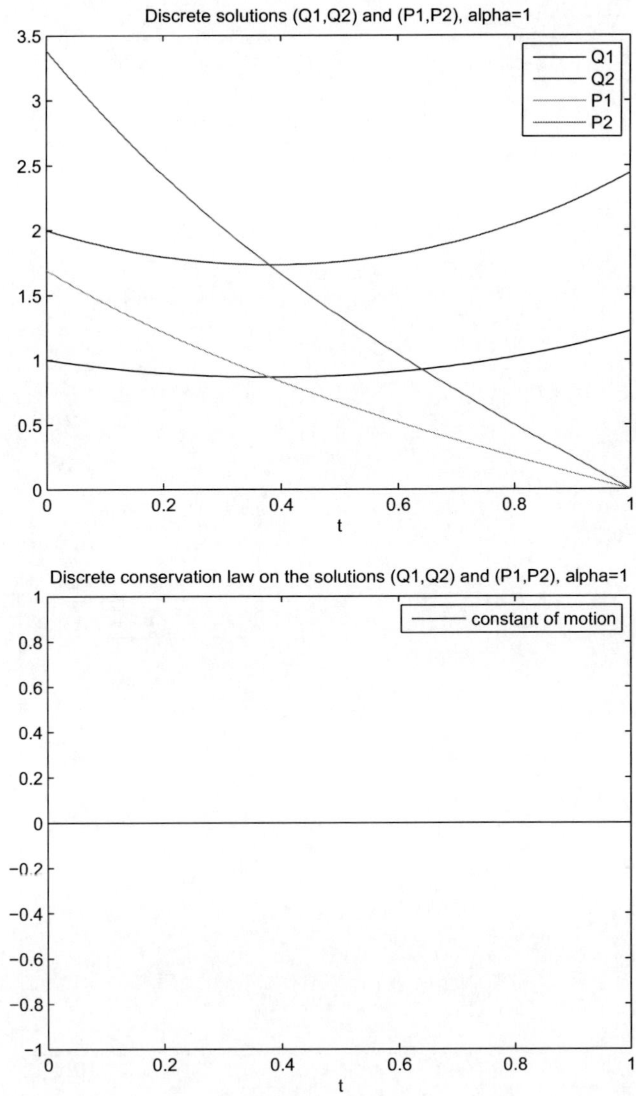

Figure 6.

Finally, making n tend to ∞ in Inequality (111), the proof is completed. □

Let us note that $\mathrm{E}_{1,1}$ is the exponential function. Consequently, in the classical case $\alpha = 1$, the fractional Gronwall's lemma is nothing else but the classical Gronwall's lemma.

A.2. Result of Stability of Order 1

In this section, we use the notations and definitions given in Section 1.3. We prove the following Lemma with the help of Lemma 28.

Lemma 29. *Let $u, \bar{u} \in \mathscr{C}^0([a,b], \mathbb{R}^m)$. Then, there exists a constant $C_1 \geq 0$ such that:*

$$\forall |\varepsilon| < 1,\ \forall t \in [a,b],\ \|q^{u+\varepsilon\bar{u},\alpha}(t) - q^{u,\alpha}(t)\| \leq C_1|\varepsilon|. \tag{113}$$

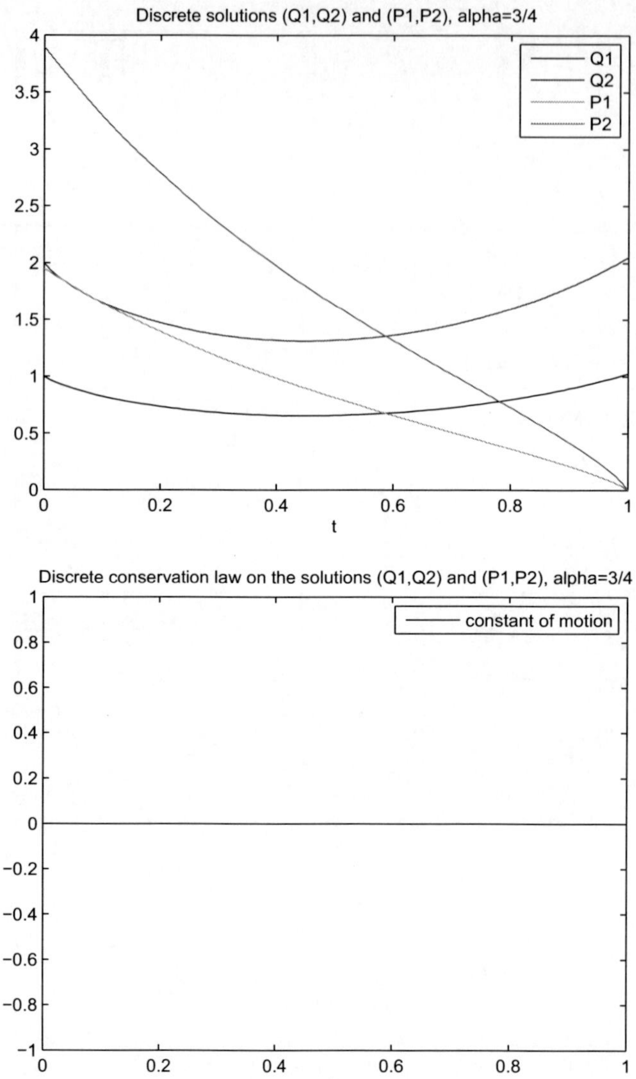

Figure 7.

Proof. From Theorem 4, we have for any $|\varepsilon| < 1$ and any $t \in [a,b]$:

$$q^{u,\alpha}(t) = A + I^{\alpha}_{-}\big(f(q^{u,\alpha},u,t)\big)(t) \quad \text{and} \quad q^{u+\varepsilon\bar{u},\alpha}(t) = A + I^{\alpha}_{-}\big(f(q^{u+\varepsilon\bar{u},\alpha},u+\varepsilon\bar{u},t)\big)(t). \tag{114}$$

Hence, for any $|\varepsilon| < 1$ and any $t \in [a,b]$, we have:

$$\begin{aligned} \|q^{u+\varepsilon\bar{u},\alpha}(t) - q^{u,\alpha}(t)\| &= \|I^{\alpha}_{-}\big(f(q^{u+\varepsilon\bar{u},\alpha},u+\varepsilon\bar{u},t) - f(q^{u,\alpha},u,t)\big)(t)\| & (115)\\ &\leq I^{\alpha}_{-}\Big(\|f(q^{u+\varepsilon\bar{u},\alpha},u+\varepsilon\bar{u},t) - f(q^{u,\alpha},u,t)\|\Big)(t). & (116) \end{aligned}$$

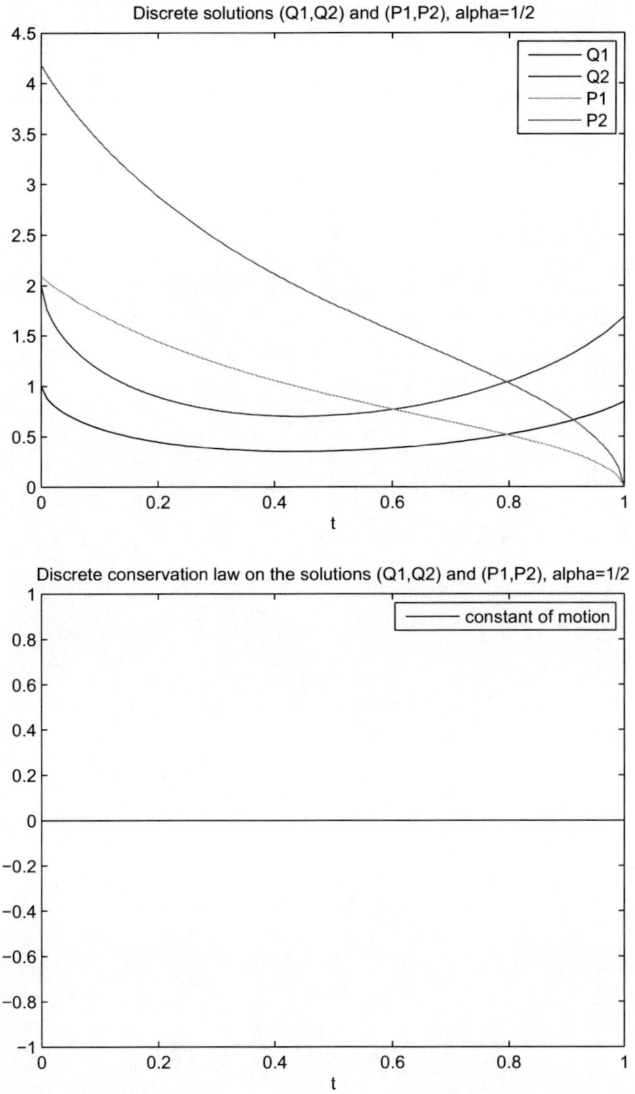

Figure 8.

We have for any $|\varepsilon| < 1$ and any $y \in [a, b]$:

$$\begin{aligned} &\|f\left(q^{u+\varepsilon\bar{u},\alpha}(y), u(y)+\varepsilon\bar{u}(y), y\right) - f\left(q^{u,\alpha}(y), u(y), y\right)\| \\ &\quad \leq \|f\left(q^{u+\varepsilon\bar{u},\alpha}(y), u(y)+\varepsilon\bar{u}(y), y\right) - f\left(q^{u,\alpha}(y), u(y)+\varepsilon\bar{u}(y), y\right)\| \\ &\qquad + \|f\left(q^{u,\alpha}(y), u(y)+\varepsilon\bar{u}(y), y\right) - f\left(q^{u,\alpha}(y), u(y), y\right)\|. \end{aligned} \quad (117)$$

From Condition (f_x lip) and with a Taylor's expansion of order 1 with explicit remainder,

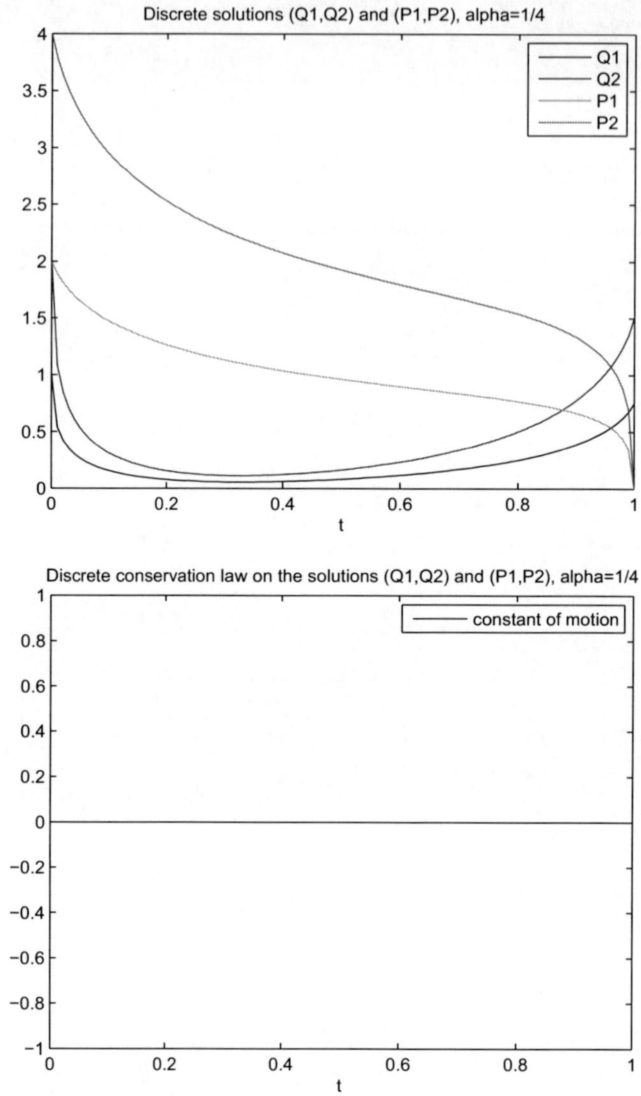

Figure 9.

the following inequality holds for any $|\varepsilon| < 1$ and any $y \in [a, b]$:

$$\begin{aligned} &\|f\left(q^{u+\varepsilon\bar{u},\alpha}(y), u(y)+\varepsilon\bar{u}(y), y\right) - f\left(q^{u,\alpha}(y), u(y), y\right)\| \\ &\qquad \leq M\|q^{u+\varepsilon\bar{u},\alpha}(y) - q^{u,\alpha}(y)\| + |\varepsilon| \left\| \frac{\partial f}{\partial v}\left(q^{u,\alpha}(y), \xi^{\varepsilon}(y), y\right) \times \bar{u}(y) \right\|. \quad (118) \end{aligned}$$

where $\xi^{\varepsilon}(y) \in [u(y), u(y) + \varepsilon\bar{u}(y)] \subset [-M_1, M_1]^m$ with $M_1 \geq 0$ independent of $|\varepsilon| < 1$ since u and $\bar{u}$ are continuous on $[a, b]$.

Thus, since $\partial f/\partial v$, $q^{u,\alpha}$ and $\bar{u}$ are continuous, there exists $M_2 \geq 0$ such that for any

$|\varepsilon| < 1$, we have:

$$\forall y \in [a,b], \left\| \frac{\partial f}{\partial v}(q^{u,\alpha}(y), \xi^{\varepsilon}(y), y) \times \bar{u}(y) \right\| \leq M_2$$

$$\text{and then, } \forall t \in [a,b],\ I^{\alpha}_{-}\left(\left\| \frac{\partial f}{\partial v}(q^{u,\alpha}, \xi^{\varepsilon}, t) \times \bar{u} \right\| \right)(t) \leq M_3, \quad (119)$$

where $M_3 := M_2(b-a)^\alpha/\Gamma(\alpha+1)$ is independent of $|\varepsilon| < 1$.

Finally, from Inequalities (116), (118) and (119), we have:

$$\forall |\varepsilon| < 1, \forall t \in [a,b],\ \|q^{u+\varepsilon\bar{u},\alpha}(t) - q^{u,\alpha}(t)\| \leq M I^{\alpha}_{-}\left(\|q^{u+\varepsilon\bar{u},\alpha} - q^{u,\alpha}\| \right)(t) + M_3|\varepsilon|. \quad (120)$$

Using the fractional Gronwall's Lemma 28, we conclude that:

$$\begin{aligned} \forall |\varepsilon| < 1,\ \forall t \in [a,b],\ \|q^{u+\varepsilon\bar{u},\alpha}(t) - q^{u,\alpha}(t)\| &\leq M_3|\varepsilon| \mathrm{E}_{\alpha,1}\big(M(t-a)^\alpha\big) \quad (121) \\ &\leq M_3|\varepsilon| \mathrm{E}_{\alpha,1}\big(M(b-a)^\alpha\big). \quad (122) \end{aligned}$$

Defining $C_1 := M_3 \mathrm{E}_{\alpha,1}\big(M(b-a)^\alpha\big)$, the proof is completed. □

A.3. Result of Stability of Order 2

In this section, we use the notations and definitions given in Section 1.3. We prove the following Lemma with the help of Lemmas 28 and 29.

Lemma 30. *Let $u, \bar{u} \in \mathscr{C}^0([a,b], \mathbb{R}^m)$. There exists a constant $C \geq 0$ such that:*

$$\forall |\varepsilon| < 1,\ \forall t \in [a,b],\ \|q^{u+\varepsilon\bar{u},\alpha}(t) - q^{u,\alpha}(t) - \varepsilon\bar{q}(t)\| \leq C\varepsilon^2. \quad (123)$$

where $\bar{q}$ is the unique global solution of the following linearised Cauchy's problem:

$$\begin{cases} {}_{\mathrm{c}}D^{\alpha}_{-}\bar{q} = \dfrac{\partial f}{\partial x}(q^{u,\alpha}, u, t) \times \bar{q} + \dfrac{\partial f}{\partial v}(q^{u,\alpha}, u, t) \times \bar{u} \\ \bar{q}(a) = 0. \end{cases} \quad (\mathrm{LCP}^{\alpha}_{\bar{q}})$$

The existence and the uniqueness of $\bar{q}$ are given by Theorem 4 and Condition (f_x lip).

Proof. We proceed in the same manner than for Lemma 29. We have for any $|\varepsilon| < 1$ and any $t \in [a,b]$:

$$\begin{aligned} &\|q^{u+\varepsilon\bar{u},\alpha}(t) - q^{u,\alpha}(t) - \varepsilon\bar{q}(t)\| \\ &\leq I^{\alpha}_{-}\left(\left\| f(q^{u+\varepsilon\bar{u},\alpha}, u+\varepsilon\bar{u}, t) - f(q^{u,\alpha}, u, t) - \varepsilon\frac{\partial f}{\partial x}(q^{u,\alpha}, u, t) \times \bar{q} - \varepsilon\frac{\partial f}{\partial v}(q^{u,\alpha}, u, t) \times \bar{u} \right\| \right)(t). \end{aligned} \quad (124)$$

With a Taylor's expansion's of order 2 with explicit remainder, we have for any $|\varepsilon| < 1$ and any $y \in [a,b]$:

$$\begin{aligned} \Big\| f\big(q^{u+\varepsilon\bar{u},\alpha}(y), u(y)+\varepsilon\bar{u}(y), y\big) - f\big(q^{u,\alpha}(y), u(y), y\big) \\ -\varepsilon \frac{\partial f}{\partial x}\big(q^{u,\alpha}(y), u(y), y\big) \times \bar{q}(y) - \varepsilon \frac{\partial f}{\partial v}\big(q^{u,\alpha}(y), u(y), y\big) \times \bar{u}(y) \Big\| \\ \leq \Big\| \frac{\partial f}{\partial x}\big(q^{u,\alpha}(y), u(y), y\big) \times \big(q^{u+\varepsilon\bar{u},\alpha}(y) - q^{u,\alpha}(y) - \varepsilon\bar{q}(y)\big) \Big\| \\ + \Big\| \frac{1}{2}\nabla^2 f\big(\xi_1^\varepsilon(y), \xi_2^\varepsilon(y), y\big)\big(q^{u+\varepsilon\bar{u},\alpha}(y) - q^{u,\alpha}(y), \varepsilon\bar{u}(y), 0\big)^2 \Big\|, \end{aligned} \tag{125}$$

where for any $y \in [a,b]$:

- $\xi_1^\varepsilon(y) \in [q^{u,\alpha}(y), q^{u+\varepsilon\bar{u},\alpha}(y)] \subset [\|q^{u,\alpha}\|_\infty - C_1, \|q^{u,\alpha}\|_\infty + C_1]^d$ according to Lemma 29. Then, there exists $M_1 \geq 0$ such that $\xi_1^\varepsilon(y) \in [-M_1, M_1]^d$ for any $|\varepsilon| < 1$ and any $y \in [a,b]$;
- $\xi_2^\varepsilon(y) \in [u(y), u(y)+\varepsilon\bar{u}(y)] \subset [-M_2, M_2]^m$ with $M_2 \geq 0$ independent of $|\varepsilon| < 1$ and $y \in [a,b]$ since u and $\bar{u}$ are continuous on $[a,b]$.

Since $\nabla^2 f$ is continuous on the compact $[-M_1, M_1]^d \times [-M_2, M_2]^m \times [a,b]$, there exists $M_3 \geq 0$ such that for any $|\varepsilon| < 1$ and any $y \in [a,b]$, we have:

$$\begin{aligned} \Big\| \frac{1}{2}\nabla^2 f\big(\xi_1^\varepsilon(y), \xi_2^\varepsilon(y), y\big)\big(q^{u+\varepsilon\bar{u},\alpha}(y) - q^{u,\alpha}(y), \varepsilon\bar{u}(y), 0\big)^2 \Big\| \\ \leq M_3\big(\|q^{u+\varepsilon\bar{u},\alpha}(y) - q^{u,\alpha}(y)\|^2 + 2\|q^{u+\varepsilon\bar{u},\alpha}(y) - q^{u,\alpha}(y)\|\|\varepsilon\bar{u}(y)\| + \|\varepsilon\bar{u}(y)\|^2\big). \end{aligned} \tag{126}$$

Then, from Lemma 29 and since $\bar{u}$ is continuous, there exists $M_4 \geq 0$ such that:

$$\forall |\varepsilon| < 1,\ \forall y \in [a,b],\ \Big\| \frac{1}{2}\nabla^2 f\big(\xi_1^\varepsilon(y), \xi_2^\varepsilon(y), y\big)\big(q^{u+\varepsilon\bar{u},\alpha}(y) - q^{u,\alpha}(y), \varepsilon\bar{u}(y), 0\big)^2 \Big\| \leq M_4\varepsilon^2. \tag{127}$$

Consequently, from Inequality (125) and since f satisfies Condition (f_x lip) (implying $\|\partial f/\partial x\|_\infty \leq M$), we have for any $|\varepsilon| < 1$ and any $y \in [a,b]$:

$$\begin{aligned} \Big\| f\big(q^{u+\varepsilon\bar{u},\alpha}(y), u(y)+\varepsilon\bar{u}(y), y\big) - f\big(q^{u,\alpha}(y), u(y), y\big) \\ -\varepsilon \frac{\partial f}{\partial x}\big(q^{u,\alpha}(y), u(y), y\big) \times \bar{q}(y) - \varepsilon \frac{\partial f}{\partial v}\big(q^{u,\alpha}(y), u(y), y\big) \times \bar{u}(y) \Big\| \\ \leq 2M\|q^{u+\varepsilon\bar{u},\alpha}(y) - q^{u,\alpha}(y) - \varepsilon\bar{q}(y)\| + M_4\varepsilon^2. \end{aligned} \tag{128}$$

Finally, using the previous inequality and Inequality (124), we obtain for any $|\varepsilon| < 1$ and any $t \in [a,b]$:

$$\|q^{u+\varepsilon\bar{u},\alpha}(t) - q^{u,\alpha}(t) - \varepsilon\bar{q}(t)\| \leq 2MI_-^\alpha\big(\|q^{u+\varepsilon\bar{u},\alpha} - q^{u,\alpha} - \varepsilon\bar{q}\|\big)(t) + M_5\varepsilon^2, \tag{129}$$

where $M_5 := M_4(b-a)^\alpha/\Gamma(1+\alpha)$ is independent of $|\varepsilon|<1$. Finally, from the fractional Gronwall's Lemma 28, we conclude that:

$$\forall|\varepsilon|<1,\ \forall t\in[a,b],\ \|q^{u+\varepsilon\bar{u},\alpha}(t)-q^{u,\alpha}(t)-\varepsilon\bar{q}(t)\| \le M_5\varepsilon^2 \mathrm{E}_{\alpha,1}\big(2M(b-a)^\alpha\big). \quad (130)$$

Defining $C := M_5\mathrm{E}_{\alpha,1}\big(2M(b-a)^\alpha\big)$, the proof is completed. □

A.4. Proof of Lemma 5

In this section, we prove Lemma 5 and consequently, we use the notations and definitions given in Section 1.3.

Let $u,\ \bar{u}\in\mathscr{C}^0([a,b],\mathbb{R}^m)$ and $\bar{q}\in\mathscr{C}^{[\alpha]}([a,b],\mathbb{R}^d)$ the unique global solution of (LCP$^\alpha_{\bar{q}}$). From Lemma 30, we have:

$$\forall|\varepsilon|<1,\ q^{u+\varepsilon\bar{u},\alpha} = q^{u,\alpha}+\varepsilon\bar{q}+h^\varepsilon, \quad (131)$$

where $\|h^\varepsilon\|_\infty \le C\varepsilon^2$. In particular, since $\bar{q}$ is continuous, there exists $M_1\ge 0$ such that:

$$\forall|\varepsilon|<1,\ \forall y\in[a,b],\ [q^{u,\alpha}(y),q^{u+\varepsilon\bar{u},\alpha}(y)]\subset[-M_1,M_1]^d. \quad (132)$$

In the same way, since u and $\bar{u}$ are continuous, there exists $M_2\ge 0$ such that:

$$\forall|\varepsilon|<1,\ \forall y\in[a,b],\ [u(y),u(y)+\varepsilon\bar{u}(y)]\subset[-M_2,M_2]^m. \quad (133)$$

We have:

$$\forall|\varepsilon|<1,\ \mathcal{L}^\alpha(u+\varepsilon\bar{u})-\mathcal{L}^\alpha(u) = \int_a^b L(q^{u+\varepsilon\bar{u},\alpha},u+\varepsilon\bar{u},t)-L(q^{u,\alpha},u,t)\,dt. \quad (134)$$

With a Taylor's expansion of order 2 with explicit remainder, we have for any $|\varepsilon|<1$ and any $y\in[a,b]$:

$$\begin{aligned}&\Big|L\big(q^{u+\varepsilon\bar{u},\alpha}(y),u(y)+\varepsilon\bar{u}(y),y\big)-L\big(q^{u,\alpha}(y),u(y),y\big)\\ &\qquad -\varepsilon\frac{\partial L}{\partial x}\big(q^{u,\alpha}(y),u(y),y\big)\cdot\bar{q}(y)-\varepsilon\frac{\partial L}{\partial v}\big(q^{u,\alpha}(y),u(y),y\big)\cdot\bar{u}(y)\Big|\\ &\le\Big|\frac{\partial L}{\partial x}\big(q^{u,\alpha}(y),u(y),y\big)\cdot h^\varepsilon(y)\Big|+\Big|\frac{1}{2}\nabla^2 L\big(\xi_1^\varepsilon(y),\xi_2^\varepsilon(y),y\big)\big(\varepsilon\bar{q}(y)+h^\varepsilon(y),\varepsilon\bar{u}(y),0\big)^2\Big|,\end{aligned} \quad (135)$$

where for any $|\varepsilon|<1$ and any $y\in[a,b]$, $\xi_1^\varepsilon(y)\in[q^{u,\alpha}(y),q^{u+\varepsilon\bar{u},\alpha}(y)]\subset[-M_1,M_1]^d$ and $\xi_2^\varepsilon(y)\in[u(y),u(y)+\varepsilon\bar{u}(y)]\subset[-M_2,M_2]^m$. Since L is of class $\mathscr{C}^2$, we obtain easily that there exists $M_3\ge 0$ such that for any $|\varepsilon|<1$ and any $y\in[a,b]$:

$$\begin{aligned}&\Big|L\big(q^{u+\varepsilon\bar{u},\alpha}(y),u(y)+\varepsilon\bar{u}(y),y\big)-L\big(q^{u,\alpha}(y),u(y),y\big)\\ &\qquad -\varepsilon\frac{\partial L}{\partial x}\big(q^{u,\alpha}(y),u(y),y\big)\cdot\bar{q}(y)-\varepsilon\frac{\partial L}{\partial v}\big(q^{u,\alpha}(y),u(y),y\big)\cdot\bar{u}(y)\Big|\le M_3\varepsilon^2.\end{aligned} \quad (136)$$

Consequently, we have for any $0 < |\varepsilon| < 1$:

$$\left\| \frac{L(q^{u+\varepsilon\bar{u},\alpha}, u+\varepsilon\bar{u}, t) - L(q^{u,\alpha}, u, t)}{\varepsilon} - \frac{\partial L}{\partial x}(q^{u,\alpha}, u, t) \cdot \bar{q} - \frac{\partial L}{\partial v}(q^{u,\alpha}, u, t) \cdot \bar{u} \right\|_\infty \leq M_3 \varepsilon. \tag{137}$$

Hence:

$$\lim_{\varepsilon \to 0} \frac{\mathcal{L}^\alpha(u+\varepsilon\bar{u}) - \mathcal{L}^\alpha(u)}{\varepsilon} = \int_a^b \frac{\partial L}{\partial x}(q^{u,\alpha}, u, t) \cdot \bar{q} + \frac{\partial L}{\partial v}(q^{u,\alpha}, u, t) \cdot \bar{u} \, dt. \tag{138}$$

The proof is completed.

B. Appendix B

B.1. Result of Stability of Order 1

In this section, we use the notations and definitions given in Section 2.2. Let us prove the following discrete version of Lemma 29:

Lemma 31. *Let $\boldsymbol{U}, \bar{\boldsymbol{U}} \in (\mathbb{R}^m)^{N+1}$. Then, there exists a constant $C_1 \geq 0$ such that:*

$$\forall |\varepsilon| < 1, \ \forall k = 0, \ldots, N, \ \|Q_k^{\boldsymbol{U}+\varepsilon\bar{\boldsymbol{U}},\alpha} - Q_k^{\boldsymbol{U},\alpha}\| \leq C_1 |\varepsilon|. \tag{139}$$

Proof. First, let us prove by induction that:

$$\forall k = 0, \ldots, N, \ \exists R_k \geq 0, \ \forall |\varepsilon| < 1, \ \|Q_k^{\boldsymbol{U}+\varepsilon\bar{\boldsymbol{U}},\alpha} - Q_k^{\boldsymbol{U},\alpha}\| \leq R_k |\varepsilon|. \tag{140}$$

For $k = 0$, let us take $R_0 = 0$ since $Q_0^{\boldsymbol{U}+\varepsilon\bar{\boldsymbol{U}},\alpha} = Q_0^{\boldsymbol{U},\alpha} = A$ for any $|\varepsilon| < 1$. Let $k \in \{1, \ldots, N\}$ and let us assume that the result (140) is satisfied for any $n = 0, \ldots, k-1$. Let us prove that the result (140) is then satisfied for $n = k$. Since $\boldsymbol{Q}^{\boldsymbol{U},\alpha}$ (resp. $\boldsymbol{Q}^{\boldsymbol{U}+\varepsilon\bar{\boldsymbol{U}},\alpha}$) is solution of (CP$^\alpha_{\boldsymbol{Q}}$) associated to $\boldsymbol{U}$ (resp. to $\boldsymbol{U} + \varepsilon\bar{\boldsymbol{U}}$), we have:

$$Q_k^{\boldsymbol{U},\alpha} = h^\alpha f(Q_k^{\boldsymbol{U},\alpha}, U_k, t_k) + A - \sum_{r=1}^{k} \alpha_r (Q_{k-r}^{\boldsymbol{U},\alpha} - A) \tag{141}$$

and for any $|\varepsilon| < 1$:

$$Q_k^{\boldsymbol{U}+\varepsilon\bar{\boldsymbol{U}},\alpha} = h^\alpha f(Q_k^{\boldsymbol{U}+\varepsilon\bar{\boldsymbol{U}},\alpha}, U_k + \varepsilon\bar{U}_k, t_k) + A - \sum_{r=1}^{k} \alpha_r (Q_{k-r}^{\boldsymbol{U}+\varepsilon\bar{\boldsymbol{U}},\alpha} - A). \tag{142}$$

Consequently, for any $|\varepsilon| < 1$:

$$\|Q_k^{\boldsymbol{U}+\varepsilon\bar{\boldsymbol{U}},\alpha} - Q_k^{\boldsymbol{U},\alpha}\| \leq h^\alpha \|f(Q_k^{\boldsymbol{U}+\varepsilon\bar{\boldsymbol{U}},\alpha}, U_k + \varepsilon\bar{U}_k, t_k) - f(Q_k^{\boldsymbol{U},\alpha}, U_k, t_k)\| + \sum_{r=1}^{k} |\alpha_r| \|Q_{k-r}^{\boldsymbol{U}+\varepsilon\bar{\boldsymbol{U}},\alpha} - Q_{k-r}^{\boldsymbol{U},\alpha}\|, \tag{143}$$

and then, with the induction hypothesis, we have for any $|\varepsilon| < 1$:

$$\|Q_k^{U+\varepsilon\bar{U},\alpha} - Q_k^{U,\alpha}\| \leq h^\alpha \|f(Q_k^{U+\varepsilon\bar{U},\alpha}, U_k + \varepsilon\bar{U}_k, t_k) - f(Q_k^{U,\alpha}, U_k + \varepsilon\bar{U}_k, t_k)\|$$
$$+ h^\alpha \|f(Q_k^{U,\alpha}, U_k + \varepsilon\bar{U}_k, t_k) - f(Q_k^{U,\alpha}, U_k, t_k)\| + \sum_{r=1}^{k} |\alpha_r| R_{k-r} |\varepsilon|. \quad (144)$$

Finally, using Condition (f_x lip) and a Taylor's expansion of order 1 with explicit remainder, we prove:

$$\forall\, |\varepsilon| < 1,\ \|Q_k^{U+\varepsilon\bar{U},\alpha} - Q_k^{U,\alpha}\| \leq h^\alpha M \|Q_k^{U+\varepsilon\bar{U},\alpha} - Q_k^{U,\alpha}\|$$
$$+ h^\alpha\, |\varepsilon| \left\| \frac{\partial f}{\partial v}(Q_k^{U,\alpha}, \xi^\varepsilon, t_k) \times \bar{U}_k \right\| + \sum_{r=1}^{k} |\alpha_r| R_{k-r} |\varepsilon|, \quad (145)$$

where $\xi^\varepsilon \in [U_k, U_k + \varepsilon\bar{U}_k] \subset [-M_1, M_1]^m$ with $M_1 \geq 0$ independent of $|\varepsilon| < 1$. Hence, since $\partial f / \partial v$ is continuous, we can conclude that there exists $M_2 \geq 0$ such that:

$$\forall\, |\varepsilon| < 1,\ \left\| \frac{\partial f}{\partial v}(Q_k^{U,\alpha}, \xi^\varepsilon, t_k) \times \bar{U}_k \right\| \leq M_2. \quad (146)$$

Consequently, since $h^\alpha M < 1$ from Condition (cond h), we have:

$$\forall\, |\varepsilon| < 1,\ \|Q_k^{U+\varepsilon\bar{U},\alpha} - Q_k^{U,\alpha}\| \leq \frac{|\varepsilon|}{1 - h^\alpha M} \left(h^\alpha M_2 + \sum_{r=1}^{k} R_{k-r} |\alpha_r| \right). \quad (147)$$

We then define $R_k := (h^\alpha M_2 + \sum_{r=1}^{k} R_{k-r}|\alpha_r|)/(1 - h^\alpha M)$ independent $|\varepsilon| < 1$ which concludes the induction. To complete the proof, we have just to define $C_1 = \max\{R_k,\ k = 0, \ldots, N\}$. □

B.2. Result of Stability of Order 2

In this section, we use the notations and definitions given in Section 2.2. We prove the following discrete version of Lemma 30 with the help of Lemma 31:

Lemma 32. *Let $\boldsymbol{U}, \bar{\boldsymbol{U}} \in (\mathbb{R}^m)^{N+1}$. Then, there exists a constant $C \geq 0$ such that:*

$$\forall |\varepsilon| < 1,\ \forall k = 0, \ldots, N,\ \|Q_k^{U+\varepsilon\bar{U},\alpha} - Q_k^{U,\alpha} - \varepsilon\bar{Q}_k\| \leq C\varepsilon^2, \quad (148)$$

where $\bar{\boldsymbol{Q}}$ is the unique solution of the following linearised discrete fractional Cauchy problem:

$$\begin{cases} {}_{\mathrm{c}}\Delta_-^\alpha \bar{\boldsymbol{Q}} = \dfrac{\partial f}{\partial x}(\boldsymbol{Q}^{U,\alpha}, \boldsymbol{U}, \boldsymbol{T}) \times \bar{\boldsymbol{Q}} + \dfrac{\partial f}{\partial v}(\boldsymbol{Q}^{U,\alpha}, \boldsymbol{U}, \boldsymbol{T}) \times \bar{\boldsymbol{U}} \\ \bar{Q}_0 = 0. \end{cases} \quad (\mathrm{LCP}^\alpha_{\bar{Q}})$$

Its existence and its uniqueness are provided by Theorem 17 and Conditions (f_x lip) and (cond h).

Proof. We proceed in the same manner that for Lemma 31. Let us prove by induction that:

$$\forall k = 0, \dots, N,\ \exists R_k \geq 0,\ \forall |\varepsilon| < 1,\ \|Q_k^{U+\varepsilon \bar{U},\alpha} - Q_k^{U,\alpha} - \varepsilon \bar{Q}_k\| \leq R_k \varepsilon^2. \tag{149}$$

For $k = 0$, let us take $R_0 = 0$ since $Q_0^{U+\varepsilon \bar{U},\alpha} = Q_0^{U,\alpha} = A$ for any $|\varepsilon| < 1$ and $\bar{Q}_0 = 0$. Let $k \in \{1, \dots, N\}$ and let us assume that the result (149) is satisfied for any $n = 0, \dots, k-1$. Let us prove that the result (149) is then satisfied for $n = k$. Since $\boldsymbol{Q}^{U,\alpha}$ (resp. $\boldsymbol{Q}^{U+\varepsilon \bar{U},\alpha}$) is solution of (LCP$^\alpha_{\boldsymbol{Q}}$) associated to $\boldsymbol{U}$ (resp. to $\boldsymbol{U} + \varepsilon \bar{\boldsymbol{U}}$) and since $\bar{\boldsymbol{Q}}$ is solution of (LCP$^\alpha_{\bar{\boldsymbol{Q}}}$), we have with a Taylor's expansion of order 2 with explicit remainder:

$$\forall |\varepsilon| < 1,\ Q_k^{U+\varepsilon \bar{U},\alpha} - Q_k^{U,\alpha} - \varepsilon \bar{Q}_k = h^\alpha \frac{\partial f}{\partial x}(Q_k^{U,\alpha}, U_k, t_k) \times (Q_k^{U+\varepsilon \bar{U},\alpha} - Q_k^{U,\alpha} - \varepsilon \bar{Q}_k)$$
$$+ h^\alpha \left(\frac{1}{2} \nabla^2 f(\xi_1^\varepsilon, \xi_2^\varepsilon, t_k)(Q_k^{U+\varepsilon \bar{U},\alpha} - Q_k^{U,\alpha}, \varepsilon \bar{U}_k, 0)^2 \right) - \sum_{r=1}^{k} \alpha_r (Q_{k-r}^{U+\varepsilon \bar{U},\alpha} - Q_{k-r}^{U,\alpha} - \varepsilon \bar{Q}_{k-r}), \tag{150}$$

where:

- $\xi_1^\varepsilon \in [Q_k^{U,\alpha}, Q_k^{U+\varepsilon \bar{U},\alpha}] \subset [\|Q_k^{U,\alpha}\| - C_1, \|Q_k^{U,\alpha}\| + C_1]$ from Lemma 31. Then, $\xi_1^\varepsilon \in (-M_1, M_1]^d$ with $M_1 \geq 0$ independent of $|\varepsilon| < 1$;
- $\xi_2^\varepsilon \in [U_k, U_k + \varepsilon \bar{U}_k] \subset [-M_2, M_2]^m$ with $M_2 \geq 0$ independent of $|\varepsilon| < 1$.

Since $\nabla^2 f(\cdot, \cdot, t_k)$ is continuous, we conclude that there exists $M_3 \geq 0$ such that:

$$\forall |\varepsilon| < 1,\ \left\| \frac{1}{2} \nabla^2 f(\xi_1^\varepsilon, \xi_2^\varepsilon, t_k)(Q_k^{U+\varepsilon \bar{U},\alpha} - Q_k^{U,\alpha}, \varepsilon \bar{U}_k, 0)^2 \right\|$$
$$\leq M_3 (\|Q_k^{U+\varepsilon \bar{U},\alpha} - Q_k^{U,\alpha}\|^2 + 2 \|Q_k^{U+\varepsilon \bar{U},\alpha} - Q_k^{U,\alpha}\| \|\varepsilon \bar{U}_k\| + \|\varepsilon \bar{U}_k\|^2). \tag{151}$$

Hence, from Lemma 31, there exists $M_4 \geq 0$ such that:

$$\forall |\varepsilon| < 1,\ \left\| \frac{1}{2} \nabla^2 f(\xi_1^\varepsilon, \xi_2^\varepsilon, t_k)(Q_k^{U+\varepsilon \bar{U},\alpha} - Q_k^{U,\alpha}, \varepsilon \bar{U}_k, 0)^2 \right\| \leq M_4 \varepsilon^2. \tag{152}$$

From Equality (150) and Condition (f_x lip), Inequality (152) and the induction hypothesis, we obtain:

$$\forall |\varepsilon| < 1,\ \|Q_k^{U+\varepsilon \bar{U},\alpha} - Q_k^{U,\alpha} - \varepsilon \bar{Q}_k\| \leq 2h^\alpha M \|Q_k^{U+\varepsilon \bar{U},\alpha} - Q_k^{U,\alpha} - \varepsilon \bar{Q}_k\|$$
$$+ \sum_{r=1}^{k} |\alpha_r| R_{k-r} \varepsilon^2 + h^\alpha M_4 \varepsilon^2. \tag{153}$$

Finally, since $2h^\alpha M < 1$ from Condition (cond h), we have:

$$\forall |\varepsilon| < 1,\ \|Q_k^{U+\varepsilon \bar{U},\alpha} - Q_k^{U,\alpha} - \varepsilon \bar{Q}_k\| \leq \frac{\varepsilon^2}{1 - 2h^\alpha M} \left(h^\alpha M_4 + \sum_{r=1}^{k} R_{k-r} |\alpha_r| \right). \tag{154}$$

We then define $R_k := (h^\alpha M_4 + \sum_{r=1}^{k} R_{k-r} |\alpha_r|)/(1 - 2h^\alpha M)$ which concludes the induction. In order to complete the proof, we just have to define $C := \max\{R_k,\ k = 0, \dots, N\}$. □

B.3. Proof of Lemma 18

In this section, we prove Lemma 18 and consequently, we use notations and definitions given in Section 2.2.

Let $\boldsymbol{U}, \bar{\boldsymbol{U}} \in (\mathbb{R}^m)^{N+1}$ and $\bar{\boldsymbol{Q}} \in (\mathbb{R}^d)^{N+1}$ the unique solution of $(\text{LCP}^{\alpha}_{\bar{\boldsymbol{Q}}})$. From Lemma 32, we have:

$$\forall k = 0, \dots, N,\ \forall |\varepsilon| < 1,\ Q_k^{\boldsymbol{U}+\varepsilon\bar{\boldsymbol{U}},\alpha} = Q_k^{\boldsymbol{U},\alpha} + \varepsilon \bar{Q}_k + H_k^{\varepsilon}, \tag{155}$$

where for any $k = 0, \dots, N$ and for any $|\varepsilon| < 1$, $\|H_k^{\varepsilon}\| \le C\varepsilon^2$. In particular, there exists $M_1 \ge 0$ such that:

$$\forall k = 0, \dots, N,\ \forall |\varepsilon| < 1,\ [Q_k^{\boldsymbol{U},\alpha}, Q_k^{\boldsymbol{U}+\varepsilon\bar{\boldsymbol{U}},\alpha}] \subset [-M_1, M_1]^d. \tag{156}$$

In the same way, there exists $M_2 \ge 0$ such that:

$$\forall k = 0, \dots, N,\ \forall |\varepsilon| < 1, [U_k, U_k + \varepsilon \bar{U}_k] \subset [-M_2, M_2]^m. \tag{157}$$

We have:

$$\forall |\varepsilon| < 1,\ \mathcal{L}_h^{\alpha}(\boldsymbol{U}+\varepsilon\bar{\boldsymbol{U}}) - \mathcal{L}_h^{\alpha}(\boldsymbol{U}) = h\sum_{k=1}^{N} \Big[L(Q_k^{\boldsymbol{U}+\varepsilon\bar{\boldsymbol{U}},\alpha}, U_k + \varepsilon\bar{U}_k, t_k) - L(Q_k^{\boldsymbol{U},\alpha}, U_k, t_k) \Big]. \tag{158}$$

With a Taylor's expansion of order 2 with explicit remainder, we have for any $|\varepsilon| < 1$ and any $k = 0, \dots, N$:

$$\begin{aligned} \Big| L(Q_k^{\boldsymbol{U}+\varepsilon\bar{\boldsymbol{U}},\alpha}, U_k + \varepsilon\bar{U}_k, t_k) - L(Q_k^{\boldsymbol{U},\alpha}, U_k, t_k) - \varepsilon \frac{\partial L}{\partial x}(Q_k^{\boldsymbol{U},\alpha}, U_k, t_k) \cdot \bar{Q}_k - \varepsilon \frac{\partial L}{\partial v}(Q_k^{\boldsymbol{U},\alpha}, U_k, t_k) \cdot \bar{U}_k \Big| \\ \le \Big| \frac{\partial L}{\partial x}(Q_k^{\boldsymbol{U},\alpha}, U_k, t_k) \cdot H_k^{\varepsilon} \Big| + \Big| \frac{1}{2} \nabla^2 L(\xi_1^{\varepsilon}, \xi_2^{\varepsilon}, t_k)(\varepsilon\bar{Q}_k + H_k^{\varepsilon}, \varepsilon\bar{U}_k, 0)^2 \Big|, \end{aligned} \tag{159}$$

where $\xi_1^{\varepsilon} \in [Q_k^{\boldsymbol{U},\alpha}, Q_k^{\boldsymbol{U}+\varepsilon\bar{\boldsymbol{U}},\alpha}] \subset [-M_1, M_1]^d$ and $\xi_2^{\varepsilon} \in [U_k, U_k + \varepsilon\bar{U}_k] \subset [-M_2, M_2]^m$. Since L is of class $\mathscr{C}^2$, we obtain easily that there exists $M_3 \ge 0$ such that for any $|\varepsilon| < 1$ and any $k = 0, \dots, N$:

$$\begin{aligned} \Big| L(Q_k^{\boldsymbol{U}+\varepsilon\bar{\boldsymbol{U}},\alpha}, U_k + \varepsilon\bar{U}_k, t_k) - L(Q_k^{\boldsymbol{U},\alpha}, U_k, t_k) \\ - \varepsilon \frac{\partial L}{\partial x}(Q_k^{\boldsymbol{U},\alpha}, U_k, t_k) \cdot \bar{Q}_k - \varepsilon \frac{\partial L}{\partial v}(Q_k^{\boldsymbol{U},\alpha}, U_k, t_k) \cdot \bar{U}_k \Big| \le M_3 \varepsilon^2. \end{aligned} \tag{160}$$

Consequently, we have for any $0 < |\varepsilon| < 1$ and any $k = 0, \dots, N$:

$$\begin{aligned} \Bigg| \frac{L(Q_k^{\boldsymbol{U}+\varepsilon\bar{\boldsymbol{U}},\alpha}, U_k + \varepsilon\bar{U}_k, t_k) - L(Q_k^{\boldsymbol{U},\alpha}, U_k, t_k)}{\varepsilon} \\ - \frac{\partial L}{\partial x}(Q_k^{\boldsymbol{U},\alpha}, U_k, t_k) \cdot \bar{Q}_k - \frac{\partial L}{\partial v}(Q_k^{\boldsymbol{U},\alpha}, U_k, t_k) \cdot \bar{U}_k \Bigg| \le M_3 \varepsilon. \end{aligned} \tag{161}$$

Hence:

$$\lim_{\varepsilon \to 0} \frac{\mathcal{L}_h^{\alpha}(\boldsymbol{U}+\varepsilon\bar{\boldsymbol{U}}) - \mathcal{L}_h^{\alpha}(\boldsymbol{U})}{\varepsilon} = h\sum_{k=1}^{N} \left[\frac{\partial L}{\partial x}(Q_k^{\boldsymbol{U},\alpha}, U_k, t_k) \cdot \bar{Q}_k + \frac{\partial L}{\partial v}(Q_k^{\boldsymbol{U},\alpha}, U_k, t_k) \cdot \bar{U}_k \right]. \tag{162}$$

The proof is completed.

B.4. Proof of Lemma 25

In this section, we use the notations and definitions given in Section 2.5. Let us prove Lemma 25.

Let $\boldsymbol{G}^1, \boldsymbol{G}^2 \in (\mathbb{R}^d)^{N+1}$ satisfying $G^2_N = 0$. First, let us denote for any $k = 1, \dots, N$:

$$X_k := h^\alpha \Big[G^1_k \cdot \sigma^{-1}(\Delta^\alpha_+ \boldsymbol{G}^2)_k - ({}_{\rm c}\Delta^\alpha_- \boldsymbol{G}^1)_k \cdot \sigma^{-1}(\boldsymbol{G}^2)_k \Big]. \tag{163}$$

Our aim is to write $\boldsymbol{X}$ as an explicit discrete derivative (*i.e.* as Δ^1_- of an explicit quantity). We have for any $k = 1, \dots, N$:

$$X_k = G^1_k \cdot \left(\sum_{r=0}^{N+1-k} \alpha_r G^2_{k+r-1} \right) - \left(\sum_{r=0}^{k} \alpha_r (G^1_{k-r} - G^1_0) \right) \cdot G^2_{k-1} = \alpha_1 h (\Delta^1_- \boldsymbol{G}^1 \cdot \boldsymbol{G}^2)_k + Y_k + Z_k, \tag{164}$$

where for any $k = 1, \dots, N$:

$$Y_k := \left(\sum_{r=0}^{k} \alpha_r \right) G^1_0 \cdot G^2_{k-1} = \beta^\alpha_k G^1_0 \cdot G^2_{k-1} \tag{165}$$

and

$$Z_k := \left[G^1_k \cdot \left(\sum_{r=2}^{N-k} \alpha_r G^2_{k+r-1} \right) - \left(\sum_{r=2}^{k} \alpha_r G^1_{k-r} \right) \cdot G^2_{k-1} \right]. \tag{166}$$

Our aim is then to write $\boldsymbol{Y}$ and $\boldsymbol{Z}$ as explicit discrete derivatives. We then define for any $i = 0, \dots, N$, $V_i := h \sum_{r=1}^{i} Y_r$ and $W_i := h \sum_{j=1}^{i} Z_j$. Hence, we have $\Delta^1_- \boldsymbol{V} = \boldsymbol{Y}$ and $\Delta^1_- \boldsymbol{W} = \boldsymbol{Z}$ and then, $\boldsymbol{X} = \Delta^1_-(\alpha_1 h \boldsymbol{G}^1 \cdot \boldsymbol{G}^2 + \boldsymbol{V} + \boldsymbol{W})$. Our aim is then to explicit $\boldsymbol{V}$ and $\boldsymbol{W}$. For any $i = 0, \dots, N$, we have:

$$V_i = h \sum_{r=1}^{i} \beta^\alpha_r G^1_0 \cdot G^2_{r-1} = h \sum_{r=1}^{i} \beta^\alpha_r G^1_0 \cdot \sigma^{r-1}(\boldsymbol{G}^2)_0 = h \sum_{r=1}^{N} \sum_{j=0}^{N} \beta^\alpha_r C_r(i,j) G^1_j \cdot \sigma^{r-1}(\boldsymbol{G}^2)_j. \tag{167}$$

For any $i = 0, \dots, N$, we have:

$$\begin{aligned}
W_i &= h \sum_{j=1}^{i} \left[G^1_j \cdot \left(\sum_{r=2}^{N-j} \alpha_r G^2_{j+r-1} \right) - \left(\sum_{r=2}^{j} \alpha_r G^1_{j-r} \right) \cdot G^2_{j-1} \right] && (168)\\
&= h \sum_{j=1}^{i} \sum_{r=2}^{N-j} \alpha_r G^1_j \cdot G^2_{j+r-1} - h \sum_{j=2}^{i} \sum_{r=2}^{j} \alpha_r G^1_{j-r} \cdot G^2_{j-1} && (169)\\
&= h \sum_{j=1}^{i} \sum_{r=2}^{N-j} \alpha_r G^1_j \cdot \sigma^{r-1}(\boldsymbol{G}^2)_j - h \sum_{r=2}^{i} \sum_{j=r}^{i} \alpha_r G^1_{j-r} \cdot G^2_{j-1} && (170)\\
&= h \sum_{j=1}^{i} \sum_{r=2}^{N-j} \alpha_r G^1_j \cdot \sigma^{r-1}(\boldsymbol{G}^2)_j - h \sum_{r=2}^{i} \sum_{j=0}^{i-r} \alpha_r G^1_j \cdot \sigma^{r-1}(\boldsymbol{G}^2)_j. && (171)
\end{aligned}$$

The following equality holds for any $r = 2, \dots, N$ and any $i, j = 0, \dots, N$:

$$\begin{aligned}
&\delta_{\{1 \le j \le i\}} \delta_{\{2 \le r \le N-j\}} - \delta_{\{2 \le r \le i\}} \delta_{\{0 \le j \le i-r\}} \\
&\qquad = \delta_{\{1 \le i \le N-1\}} \delta_{\{1 \le j \le N-r\}} \delta_{\{0 \le i-j \le r-1\}} - \delta_{\{j=0\}} \delta_{\{r \le i\}}. \qquad (172)
\end{aligned}$$

Consequently, we have for any $i = 0, \dots, N$:

$$W_i = h \sum_{r=2}^{N} \sum_{j=0}^{N} \alpha_r B_r(i,j) G_j^1 \cdot \sigma^{r-1}(\boldsymbol{G}^2)_j. \tag{173}$$

Finally, we have for any $i = 0, \dots, N$:

$$\alpha_1 h G_i^1 \cdot G_i^2 + V_i + W_i = h \sum_{r=1}^{N} \sum_{j=0}^{N} A_r(i,j) G_j^1 \cdot \sigma^{r-1}(\boldsymbol{G}^2)_j. \tag{174}$$

Finally, the following equality holds:

$$\boldsymbol{X} = h \Delta_-^1 \Big[\sum_{r=1}^{N} A_r \times \big(\boldsymbol{G}^1 \cdot \sigma^{r-1}(\boldsymbol{G}^2) \big) \Big], \tag{175}$$

which concludes the proof.

Now, let us see some examples of matrices $A_r \in \mathcal{M}_{N+1}$ for $N = 5$:

$$A_1 = \begin{pmatrix} \alpha_1 & 0 & 0 & 0 & 0 & 0 \\ \beta_1^\alpha & \alpha_1 & 0 & 0 & 0 & 0 \\ \beta_1^\alpha & 0 & \alpha_1 & 0 & 0 & 0 \\ \beta_1^\alpha & 0 & 0 & \alpha_1 & 0 & 0 \\ \beta_1^\alpha & 0 & 0 & 0 & \alpha_1 & 0 \\ \beta_1^\alpha & 0 & 0 & 0 & 0 & \alpha_1 \end{pmatrix}, \quad A_2 = \begin{pmatrix} 0 & 0 & 0 & 0 & 0 & 0 \\ 0 & \alpha_2 & 0 & 0 & 0 & 0 \\ \beta_2^\alpha - \alpha_2 & \alpha_2 & \alpha_2 & 0 & 0 & 0 \\ \beta_2^\alpha - \alpha_2 & 0 & \alpha_2 & \alpha_2 & 0 & 0 \\ \beta_2^\alpha - \alpha_2 & 0 & 0 & \alpha_2 & 0 & 0 \\ \beta_2^\alpha - \alpha_2 & 0 & 0 & 0 & 0 & 0 \end{pmatrix},$$

$$A_3 = \begin{pmatrix} 0 & 0 & 0 & 0 & 0 & 0 \\ 0 & \alpha_3 & 0 & 0 & 0 & 0 \\ 0 & \alpha_3 & \alpha_3 & 0 & 0 & 0 \\ \beta_3^\alpha \quad \alpha_3 & \alpha_3 & \alpha_3 & 0 & 0 & 0 \\ \beta_3^\alpha - \alpha_3 & 0 & \alpha_3 & 0 & 0 & 0 \\ \beta_3^\alpha - \alpha_3 & 0 & 0 & 0 & 0 & 0 \end{pmatrix}, \quad A_4 = \begin{pmatrix} 0 & 0 & 0 & 0 & 0 & 0 \\ 0 & \alpha_4 & 0 & 0 & 0 & 0 \\ 0 & \alpha_4 & 0 & 0 & 0 & 0 \\ 0 & \alpha_4 & 0 & 0 & 0 & 0 \\ \beta_4^\alpha - \alpha_4 & \alpha_4 & 0 & 0 & 0 & 0 \\ \beta_4^\alpha - \alpha_4 & 0 & 0 & 0 & 0 & 0 \end{pmatrix}$$

and

$$A_5 = \begin{pmatrix} 0 & 0 & 0 & 0 & 0 & 0 \\ 0 & 0 & 0 & 0 & 0 & 0 \\ 0 & 0 & 0 & 0 & 0 & 0 \\ 0 & 0 & 0 & 0 & 0 & 0 \\ 0 & 0 & 0 & 0 & 0 & 0 \\ \beta_5^\alpha - \alpha_5 & 0 & 0 & 0 & 0 & 0 \end{pmatrix}.$$

References

[1] O.P. Agrawal. Formulation of Euler-Lagrange equations for fractional variational problems. *J. Math. Anal. Appl.*, 272(1):368–379, 2002.

[2] O.P. Agrawal. A general formulation and solution scheme for fractional optimal control problems. *Nonlinear Dynam.*, 38(1-4):323–337, 2004.

[3] O.P. Agrawal. A formulation and numerical scheme for fractional optimal control problems. *J. Vib. Control*, 14(9-10):1291–1299, 2008.

[4] O.P. Agrawal. A general finite element formulation for fractional variational problems. *J. Math. Anal. Appl.*, 337(1):1–12, 2008.

[5] R. Almeida, A.B. Malinowska, and D.F.M. Torres. A fractional calculus of variations for multiple integrals with application to vibrating string. *J. Math. Phys.*, 51(3):033503, 12, 2010.

[6] V. I. Arnold. *Mathematical methods of classical mechanics*, volume 60 of *Graduate Texts in Mathematics*. Springer-Verlag, New York, 1979.

[7] R. L. Bagley and R. A. Calico. Fractional order state equations for the control of viscoelastically damped structures. *Journal of Guidance, Control, and Dynamics*, 14:304–311, 1991.

[8] D. Baleanu, O. Defterli, and O.P. Agrawal. A central difference numerical scheme for fractional optimal control problems. *J. Vib. Control*, 15(4):583–597, 2009.

[9] D. Baleanu and S.I. Muslih. Lagrangian formulation of classical fields within Riemann-Liouville fractional derivatives. *Phys. Scripta*, 72(2-3):119–121, 2005.

[10] M. Bergmann. *Optimisation aérodynamique par réduction de modèle POD et contrôle optimal. Application au sillage laminaire d'un cylindre circulaire*. PhD thesis, Institut National Polytechnique de Lorraine, 2004.

[11] B. Bonnard, L. Faubourg, and E. Trélat. *Mécanique céleste et contrôle des véhicules spatiaux*, volume 51 of *Mathématiques & Applications*. Springer-Verlag, Berlin, 2006.

[12] L. Bourdin, J. Cresson, and I. Greff. A continuous/discrete fractional Noether's theorem. *Commun. Nonlinear Sci. Numer. Simul.*, 71:14–23, 2013.

[13] L. Bourdin, J. Cresson, I. Greff, and P. Inizan. Variational integrator for fractional Euler-Lagrange equations. *Appl. Numer. Math.*, 18(4):878–887, 2013.

[14] J.A.E. Bryson and Y.C. Ho. *Applied optimal control*. Hemisphere Publishing Corp. Washington, D. C., 1975. *Optimization, estimation, and control*, Revised printing.

[15] R. Burlacu and A. Cavache. On a class of optimal control problems in mathematical biology. *Sci. Ann. Univ. Agric. Sci. Vet. Med.*, 45:41–54 (2003), 2002. *Annual Symposium on Mathematics Applied in Biology & Biophysics*.

[16] L. Cantaluppi. Optimal control of a dam under seasonal electricity prices. *European J. Oper. Res.*, 17(1):57–70, 1984.

[17] F. Comte. Opérateurs fractionnaires en économétrie et en finance. *Prépublication MAP5*, 2001.

[18] J. Cresson. Fractional embedding of differential operators and Lagrangian systems. *J. Math. Phys.*, 48(3):033504, 34, 2007.

[19] O. Defterli. A numerical scheme for two-dimensional optimal control problems with memory effect. *Comput. Math. Appl.*, 59(5):1630–1636, 2010.

[20] G. Demange and J.-C. Rochet. Méthodes Mathématiques de la Finance. *Economica*, 1992.

[21] K. Diethelm. *The analysis of fractional differential equations*, volume 2004 of *Lecture Notes in Mathematics*. Springer-Verlag, Berlin, 2010. An application-oriented exposition using differential operators of Caputo type.

[22] N.M. Dmitruk. Optimization of the von Neumann model of an expanding economy by optimal control methods. *Vestsī Nats. Akad. Navuk Belarusī Ser. Fīz.-Mat. Navuk*, (4):48–53, 141, 2000.

[23] F. Dubois, A-C. Galucio, and N. Point. Introduction à la dérivation fractionnaire. Théorie et applications. *Série des Techniques de l'ingénieur*, 2009.

[24] L.C. Evans. *An introduction to mathematical optimal control theory.*

[25] J. Favier. *Contrôle d'écoulements : approche expérimentale et modélisation de dimension réduite*. PhD thesis, Institut National Polytechnique de Toulouse, 2007.

[26] A.V. Fedoseev. Optimal control in a three-sector model of the economy of an underdeveloped country. *Ž. Vyčisl. Mat. i Mat. Fiz.*, 12:925–942, 1085, 1972.

[27] G.S.F. Frederico and D.F.M. Torres. A formulation of Noether's theorem for fractional problems of the calculus of variations. *J. Math. Anal. Appl.*, 334(2):834–846, 2007.

[28] G.S.F. Frederico and D.F.M. Torres. Fractional conservation laws in optimal control theory. *Nonlinear Dynam.*, 53(3):215–222, 2008.

[29] G.S.F. Frederico and D.F.M. Torres. Fractional optimal control in the sense of Caputo and the fractional Noether's theorem. *Int. Math. Forum*, 3(9-12):479–493, 2008.

[30] G.S.F. Frederico and D.F.M. Torres. Fractional Noether's theorem in the Riesz-Caputo sense. *Appl. Math. Comput.*, 217(3):1023–1033, 2010.

[31] A. C. Galucio, J.-F. Deü, and F. Dubois. The G^α-scheme for approximation of fractional derivatives: application to the dynamics of dissipative systems. *J. Vib. Control*, 14(9-10):1597–1605, 2008.

[32] R.V. Gamkrelidze. Discovery of the maximum principle. In *Mathematical events of the twentieth century*, pages 85–99. Springer, Berlin, 2006.

[33] E. Gerolymatou, I. Vardoulakis, and R. Hilfer. Modelling infiltration by means of a nonlinear fractional diffusion model. *J. Phys. D: Appl. Phys.*, 39:4104, 2006.

[34] M.D. Gunzburger. Introduction into mathematical aspects of flow control and optimization. *Lecture series - Von KÃ¡rmÃ¡n Institute for Fluid Dynamics*, 5:A1–A20, 1997.

[35] M.D. Gunzburger. Lagrange multiplier techniques. *Lecture series - Von KÃ¡rmÃ¡n Institute for Fluid Dynamics*, 5:C1–C18, 1997.

[36] E. Hairer, C. Lubich, and G. Wanner. *Geometric numerical integration*, volume 31 of *Springer Series in Computational Mathematics*. Springer-Verlag, Berlin, second edition, 2006. Structure-preserving algorithms for ordinary differential equations.

[37] M.R. Hestenes. *Calculus of variations and optimal control theory*. Robert E. Krieger Publishing Co. Inc., Huntington, N.Y., 1980. Corrected reprint of the 1966 original.

[38] R. Hilfer. Applications of fractional calculus in physics. *World Scientific*, River Edge, New Jersey, 2000.

[39] R. Hilfer. Fractional calculus and regular variation in thermodynamics. In *Applications of fractional calculus in physics*, pages 429–463. *World Sci. Publ.*, River Edge, NJ, 2000.

[40] P. Inizan. *Dynamique fractionnaire pour le chaos hamiltonien*. PhD thesis, Université de Pau et des Pays de l'Adour, 2010.

[41] Z.D. Jelicic and N. Petrovacki. Optimality conditions and a solution scheme for fractional optimal control problems. *Struct. Multidiscip. Optim.*, 38(6):571–581, 2009.

[42] A.A. Kilbas, H.M. Srivastava, and J.J. Trujillo. *Theory and applications of fractional differential equations*, volume 204 of *North-Holland Mathematics Studies*. Elsevier Science B.V., Amsterdam, 2006.

[43] Y. Kosmann-Schwarzbach. *Les théorèmes de Noether*. Histoire des Mathématiques (Palaiseau). Éditions de l'École Polytechnique, Palaiseau, second edition, 2006. Invariance et lois de conservation au XXe siècle.

[44] P. Lévy. L'addition des variables aléatoires définies sur une circonférence. *Bull. Soc. Math. France*, 67:1–41, 1939.

[45] Yanqin Liu and Baogui Xin. Numerical solutions of a fractional predator-prey system. *Adv. Difference Equ.*, pages Art. ID 190475, 11, 2011.

[46] J.E. Marsden and M. West. Discrete mechanics and variational integrators. *Acta Numer.*, 10:357–514, 2001.

[47] F. Martin-Robine. Histoire du principe de moindre action: trois siècles de principes variationnels de Fermat à Feynman. *Histoire des sciences*. Vuibert, 2006.

[48] M.M. Meerschaert, H.-P. Scheffler, and C. Tadjeran. Finite difference methods for two-dimensional fractional dispersion equation. *J. Comput. Phys.*, 211(1):249–261, 2006.

[49] M.M. Meerschaert and C. Tadjeran. Finite difference approximations for fractional advection-dispersion flow equations. *J. Comput. Appl. Math.*, 172(1):65–77, 2004.

[50] E. Noether. Invariant variation problems. *Transport Theory Statist. Phys.*, 1(3):186–207, 1971. Translated from the German (Nachr. Akad. Wiss. Göttingen Math.-Phys. Kl. II 1918, 235–257).

[51] K.B. Oldham and J. Spanier. *The fractional calculus*. Academic Press [A subsidiary of Harcourt Brace Jovanovich, Publishers], New York-London, 1974. Theory and applications of differentiation and integration to arbitrary order, With an annotated chronological bibliography by Bertram Ross, *Mathematics in Science and Engineering*, Vol. 111.

[52] I. Podlubny. *Fractional differential equations*, volume 198 of *Mathematics in Science and Engineering*. Academic Press Inc., San Diego, CA, 1999. An introduction to fractional derivatives, fractional differential equations, to methods of their solution and some of their applications.

[53] L.A. Richards. Capillary conduction of liquids through porous mediums. *Phys.*, 1:318, 1931.

[54] F. Riewe. Mechanics with fractional derivatives. *Phys. Rev. E (3)*, 55(3, part B):3581–3592, 1997.

[55] S.G. Samko, A.A. Kilbas, and O.I. Marichev. *Fractional integrals and derivatives*. Gordon and Breach Science Publishers, Yverdon, 1993. Theory and applications, Translated from the 1987 Russian original.

[56] A. A. Stanislavsky. Hamiltonian formalism of fractional systems. *Eur. Phys. J. B Condens. Matter Phys.*, 49(1):93–101, 2006.

[57] E. Trélat. *Contrôle optimal*. Mathématiques Concrètes. Vuibert, Paris, 2005. Théorie & applications.

[58] H. Ye, J. Gao, and Y. Ding. A generalized Gronwall inequality and its application to a fractional differential equation. *J. Math. Anal. Appl.*, 328(2):1075–1081, 2007.

[59] A. Zoia, M.-C. Néel, and A. Cortis. Continuous-time random-walk model of transport in variably saturated heterogeneous porous media. *Phys. Rev. E*, 81(3):031104, Mar 2010.

[60] A. Zoia, M.-C. Néel, and M. Joelson. Mass transport subject to time-dependent flow with nonuniform sorption in porous media. *Phys. Rev. E*, 80:056301, 2009.

In: Fractional Calculus in Analysis, Dynamics ... ISBN: 978-1-62948-635-2
Editor: Jacky Cresson, pp. 179-200

Chapter 4

FRACTAL TRAPS AND FRACTIONAL DYNAMICS

Pierre Inizan*
Institut de Mécanique Céleste et de Calcul des Éphémérides,
Observatoire de Paris, Paris, France

Abstract

Anomalous diffusion may arise in typical chaotic Hamiltonian systems. According to G.M. Zaslavsky's analysis, this behavior is induced by sticky zones within the phase space, which may trap trajectories for a long time. A description can be done by means of fractional kinetics equations. However, the dynamical origin of those fractional derivatives is still unclear. We provide in this article an attempt for a possible explanation. Starting from R. Hilfer's work, an expression for the average infinitesimal evolution of trajectories sets is given by using Poincaré recurrence times. The fractal structures of the traps, described by G.M. Zaslavsky, are then taken into account, and it is shown that in this case, the derivative associated to this evolution may become fractional, with order equal to the transport exponent of the diffusion process.

PACS: 05.45-a, 52.35.Mw, 96.50.Fm

Keywords: Fractional dynamics, Hamiltonian systems, chaotic dynamics

AMS Subject Classification: 53D, 37C, 65P

1. Introduction

Fractional calculus [26, 21, 22] is efficiently used in several fields of physics [12, 24]. For example, it may be used to take into account memory effects and anomalous transport. Several equations of physics have hence been generalized to non-integer orders so as to provide new models. Among them, one find the Euler-Lagrange equation [23, 1, 2, 5, 7] and the diffusion equation [20, 4, 9, 18, 27, 29, 6].

However, reasons for emergence of such operators are still unclear and the use of this formalism often remains heuristic. R. Hilfer [10, 14, 11] and G.M. Zaslavsky [31, 34,

*E-mail address: pierre.inizan@imcce.fr

25, 33] have tried through different ways to understand the physical origin of fractional derivatives. Both of their models rely on the *recurrence time* notion.

The first of those authors studies the evolution operator of a subsystem and shows that after a temporal renormalization, the associated infinitesimal generator is a fractional derivative. However, the interpretation of this operator may seem difficult and the renormalization procedure is ambivalent.

Zaslavsky is interested in chaotic Hamiltonian systems. He makes fractional derivatives appear in the diffusion equation related to the kinetic (i.e. probabilistic) description of the system. Without a true justification for the introduction of those derivatives, he nevertheless connects the transport exponent μ with the fractional orders of derivation and the coefficients of the self-similar structures which appear in the phase space around resonance areas.

In the present article, we study the dynamics of an Hamiltonian system, presented in section 2. With ideas taken from Hilfer, we focus in section 3 on the evolution of a phase space subset under the Hamiltonian flow. More precisely, the associated infinitesimal generator is considered. In several examples we show that it is proportional to the usual derivative d/dt. Then we precise our model by taking into account the phase space structure described by Zaslavsky and sumed up in section 4. In that case, we prove in section 5 that the infinitesimal generator may turn into a fractional derivative of order μ. We discuss the relevance of this exponent in section 6, before concluding in section 7.

2. Studied System

Let H be an Hamiltonian defined on a compact manifold M. The induced flow is denoted ϕ^t. Let m be a measure defined on M. Let G be a measurable subset of M and m' a measure adapted to G (such that $m'(G) > 0$). Two cases may happen: if $m(G) > 0$, then we can choose $m' = m$. Conversely if $m(G) = 0$ (important case in this article), m cannot measure subsets of G, so m' must differ from m.

Let us suppose that we only have access to G. Thus we are interested in the dynamics restricted to G, and we consider the measurable space $(G, \mathcal{T}', m')$, where $\mathcal{T}'$ is a σ-algebra of G over which m' is defined. We introduce G_{att}, the "attractive" subset of G:

$$G_{att} = \left\{x \in G \,|\, \exists t_0 > 0,\, \forall\, t > t_0,\, \phi^t x \in G\right\}.$$

If $x \in G_{att}$, after some time it becomes possible to completely follow the trajectory starting from x. The assumption that we only have access to G is thus invisible concerning the dynamics on G_{att}. Conversely, trajectories starting from $G \backslash G_{att}$ leave G and then cannot be tracked. Fortunately, from Poincaré recurrence theorem, if $m(G) > 0$, then almost all trajectories come back into G. More precisely, we may define the Poincaré recurrence time as

$$\forall x \in G \backslash G_{att},\; \tau_G(x) = \inf\left\{t > 0 \,|\, \phi^t x \in G,\, \exists t_0 \in (0,t),\, \phi^{t_0} x \notin G\right\}.$$

We remark that if $\tau_G(x) < \infty$, then $\phi^{\tau_G(x)} x \in \partial G \cap (G \backslash G_{att})$, where ∂G is the boundary of G. Let G_{ext} be the set of the starting points of trajectories which never come back into G, i.e. points x such that $\tau_G(x) = \infty$:

$$G_{ext} = \left\{x \in G \,|\, \exists t_0 > 0,\, \forall\, t > t_0,\, \phi^t x \notin G\right\}.$$

Theorem 1 (Poincaré recurrence theorem). *The set G_{ext} is negligible:* $m(G_{ext}) = 0$.

In that case, G is said m-recurrent. Until the end, if $m(G) = 0$, we suppose that G is m'-recurrent. We note G_{rec} the set of trajectories which alternatively wander inside and outside G:

$$G_{rec} = G \backslash (G_{att} \cup G_{ext}) .$$

We may remark that if $x \in G_{rec}$, then $\phi^{\tau_G(x)} x \in G_{rec}$. Although it is impossible to have a continuous description of the dynamics within G_{rec}, we may then track by intermittence the trajectories stemming from this set, thanks to recurrence times. Following Hilfer [11], we introduce the mapping

$$\begin{array}{rccc} S : & G_{rec} & \longrightarrow & G_{rec} \\ & x & \longmapsto & \phi^{\tau_G(x)} x. \end{array} \qquad (1)$$

Iterations of S permit to follow the temporal evolution of a point of G_{rec}. Let us remark that for all $k \geq 1$, $S^k x \in \partial G$. Since $m'(G) = m'(G_{att}) + m'(G_{rec})$, it is now possible to track almost all trajectories starting from G, at least by intermittence.

The following characteristic times will also be useful:

$$\forall x \in G_{rec}, \quad \tau_r(x) = \inf \left\{ t > 0 \,|\, \phi^t x \notin G \right\}, \quad \tau_e(x) = \inf \left\{ t > 0 \,|\, \phi^{\tau_r(x)+t} x \in G \right\}.$$

The time $\tau_r(x)$ is the time that the trajectory starting from x stays in G before leaving, while $\tau_e(x)$ is the time this trajectory then spends outside G (see figure 1). Those times verify

$$\tau_G(x) = \tau_r(x) + \tau_e(x).$$

We may also note that by continuity of the Hamiltonian flow, if $\tau_r(x) = 0$, then $x \in \partial G$.

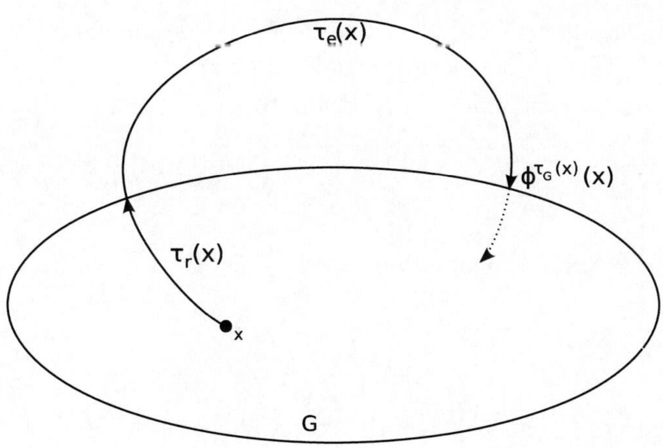

Figure 1. Characteristic times.

In order to obtain global informations on the dynamics inside G, sets of trajectories - i.e. evolution of subsets of G - should be studied, for instance through the evolution of their measures.

This problem is studied in details in [11, 10]. The operator S is redefined at precision Δt on the set of measures on G and appears as a convolution product. For a measure ρ on G and a subset $A \subset G$, Hilfer obtains

$$S(\Delta t)\rho(A)(t) = p_{\Delta t} * \rho(A)(t).$$

Then he focuses on the induced dynamics after a renormalization in time scale (*continuous time-limit* in [11] and *ultra-long time limit* in [10]) and obtains a new operator associated to a new time step, $\tilde{S}(\widetilde{\Delta t})$. In that case, he shows that the characteristic derivative of this operator, more precisely the infinitesimal generator [8, p.356] $\mathcal{G}$ associated to the semi-group $\left(\tilde{S}(\widetilde{\Delta t})\right)_{\widetilde{\Delta t} \geq 0}$ and defined by

$$\mathcal{G}\rho(A)(t) = \lim_{\widetilde{\Delta t} \to 0} \frac{\tilde{S}(\widetilde{\Delta t})\rho(A)(t) - \rho(A)(t)}{\widetilde{\Delta t}},$$

may be equal (up to the sign) to the Marchaud fractional derivative of order α [26, p.109], with $\alpha \in (0,1)$. Actually, this approach is part of a deeper questioning on time structure and irreversibility [14, 13].

However some points may still seem unclear, such as the signification of $S(\Delta t)$ and the renormalization procedures. Furthermore, the exponent α remains unspecified.

While keeping a similar approach, we propose here a simple dynamical model for which we study the infinitesimal generator. In several examples, it is proportional to the ordinary derivative. Then we use Zaslavsky analysis on Hamiltonian chaotic systems: it that case, the generator may become a fractional derivative.

3. Construction of a Simple Model

Let us recall our objective: we would like to describe the dynamics of our system restricted to G, in a global way, i.e. by considering sets of trajectories.

To do so, we introduce the mapping N given by

$$\begin{array}{rccc} N: & \mathcal{T}' & \longrightarrow & \mathcal{F}(\mathbb{R},\mathbb{R}) \\ & A & \longmapsto & N_A, \end{array}$$

where N_A is a real-valued function defined by

$$\begin{array}{rccc} N_A: & \mathbb{R} & \longrightarrow & \mathbb{R}^+ \\ & t & \longmapsto & m'((\phi^t A) \cap G). \end{array}$$

Let $A \in \mathcal{T}'$ and $t_0 \in \mathbb{R}$. We want to know the infinitesimal evolution of $N_A(t_0)$. Several successive steps will lead us to a general formula.

3.1. Model with One Trap

Let $\Delta t > 0$. We consider the following binary dynamics: all of the trajectories which leave G are "trapped" within $P \subset \Gamma$, with $P \cap G = \emptyset$. Then they come back after $2\Delta t$, and stay in G during a multiple of Δt, until possibly leaving again.

We may then split G with the two following sets:

$$\begin{aligned} G_0(\Delta t) &= \{x \in G \,|\, \tau_r(x) \geq \Delta t\}\,, & (2)\\ G_1(\Delta t) &= \{x \in G \,|\, \tau_r(x) < \Delta t,\, \tau_e(x) = 2\Delta t\}\,. & (3) \end{aligned}$$

We remark that $G_1(\Delta t)$ may also be written as

$$\begin{aligned} G_1(\Delta t) &= \{x \in G \,|\, \tau_r(x) = 0,\, \tau_e(x) = 2\Delta t\}\,,\\ &= \{x \in G \,|\, \tau_r(x) < \Delta t\}\,,\\ &= \{x \in G \,|\, \tau_r(x) = 0\}\,. \end{aligned}$$

This set is directly linked to trap P.
Hence we have $G_0(\Delta t) \cap G_1(\Delta t) = \emptyset$ and $G = G_0(\Delta t) \cup G_1(\Delta t)$.
As in [10], we define the numbers

$$p_k(\Delta t) = \frac{m'(G_k(\Delta t))}{m'(G)}, \quad k \in \{0,1\}.$$

Those two quantities provide a probability density associated to recurrence times ($p_0(\Delta t) + p_1(\Delta t) = 1$).
We suppose that these sets are "well mixed":

$$\forall B \in \mathcal{T}',\ m'(B \cap G_0(\Delta t)) = p_0(\Delta t)m'(B),\ m'(B \cap G_1(\Delta t)) = p_1(\Delta t)m'(B).$$

Starting from $N_A(t_0) = m'(A)$, we determine the following states. The shifts will occur every Δt, so we may just consider $N_A(t_0 + n\Delta t)$, with $n \in \mathbb{N}$.

1. At t_0^+, trajectories starting from A are splitting: some of them stay in G while the others leave G during $2\Delta t$. We note A_0 the set of initial conditions of the first ones and A_1 the set of the second ones. Consequently, $m'(A_0) = p_0 m'(A)$ and $m'(A_1) = p_1 m'(A)$ (we omit the dependance of p_0 and p_1 in Δt).

2. At $t_0 + \Delta t$, only A_0 is in G:

   $$N_A(t_0 + \Delta t) = m'(A_0) = p_0 N_A(t_0).$$

   Within the trap, A_1 becomes A_{11}.
   At $t_0^+ + \Delta t$, it is now A_0 which splits similarly to A, and gives birth to A_{00} and A_{01}: $m'(A_{00}) = p_0 m'(A_0)$ and $m'(A_{01}) = p_1 m'(A_0)$.

3. Trajectories which escaped from G at t_0^+ come back at $t_0^+ + 2\Delta t$. Consequently, at $t_0 + 2\Delta t$, only A_{00} is present in G:

   $$N_A(t_0 + 2\Delta t) = m'(A_{00}) = p_0 N_A(t_0 + \Delta t).$$

   At $t_0^+ + 2\Delta t$, A_{00} splits into A_{000} and A_{001}, A_{11} comes back (it turns into A_{110}), and A_{01} stays outside G while becoming A_{011}.

4. At $t_0 + 3\Delta t$, G contains A_{000} and A_{110}. Hence we have

$$\begin{aligned} N_A(t_0 + 3\Delta t) &= m'(A_{000}) + m'(A_{110}), \\ &= p_0 m'(A_{00}) + m'(A_1), \\ &= p_0 N_A(t_0 + 2\Delta t) + p_1 N_A(t_0). \end{aligned}$$

At $t_0^+ + 3\Delta t$, A_{000} splits into A_{0000} and A_{0001}, A_{110} into A_{1100} and A_{1101}, A_{011} comes back and becomes A_{0110}. Concerning A_{001}, it stays outside G and turns into A_{0011}.

5. At $t_0 + 4\Delta t$, we find in G the sets A_{0000}, A_{1100} and A_{0110}:

$$\begin{aligned} N_A(t_0 + 4\Delta t) &= m'(A_{0000}) + m'(A_{1100}) + m'(A_{0110}), \\ &= p_0 \left(m'(A_{000}) + m'(A_{110})\right) + m'(A_{01}), \\ &= p_0 N_A(t_0 + 3\Delta t) + p_1 N_A(t_0 + \Delta t). \end{aligned}$$

A sketch of the dynamics is given in figure 2.

An immediate generalization leads to

$$\forall n \in \mathbb{Z},\ N_A(t_0 + n\Delta t) = p_0 N_A(t_0 + (n-1)\Delta t) + p_1 N_A(t_0 + (n-3)\Delta t).$$

In particular,

$$N_A(t_0 + \Delta t) = p_0 N_A(t_0) + p_1 N_A(t_0 - 2\Delta t). \tag{4}$$

Keeping in mind definition (1), we note $S(\Delta t)$ the operator of "infinitesimal" temporal evolution, which leads to the next temporal step. In this example, time is discrete and takes its values in $t_0 + \Delta t\mathbb{Z} + 2\Delta t\mathbb{Z} = t_0 + \Delta t\mathbb{Z}$.

So we have

$$S(\Delta t) N_A(t_0) = N_A(t_0 + \Delta t). \tag{5}$$

Given that $\lim_{\Delta t \to 0+} p_0(\Delta t) N_A(t_0) + p_1(\Delta t) N_A(t_0 - 2\Delta t) = N_A(t_0)$, $S(\Delta t)$ verifies

$$S(0) = \text{ id}. \tag{6}$$

Moreover, from (5), $S(\Delta t)$ also verifies

$$\forall \Delta t_1,\ \Delta t_2 > 0,\ S(\Delta t_1)\, S(\Delta t_2) = S(\Delta t_1 + \Delta t_2). \tag{7}$$

Let us remark that (4) may not be used to check property (7). By verifying (6) and (7), $(S(\Delta t))_{\Delta t \geq 0}$ defines a one-parameter semi-group. If N_A possesses a left derivative at t_0, the associated infinitesimal generator $\mathcal{G}$ is given by

$$\mathcal{G} N_A(t_0) = \lim_{\Delta t \to 0+} \frac{S(\Delta t) N_A(t_0) - N_A(t_0)}{\Delta t}, \tag{8}$$

$$= \lim_{\Delta t \to 0+} \frac{p_1(\Delta t)\left(N_A(t_0 - 2\Delta t) - N_A(t_0)\right)}{\Delta t}, \tag{9}$$

$$= -2p_1(0^+) \frac{d}{dt^-} N_A(t_0), \tag{10}$$

where $\frac{d}{dt^-} N_A(t_0)$ is the left derivative of N_A at t_0 and $p_1(0^+)$ is the right limit of p_1 at 0.

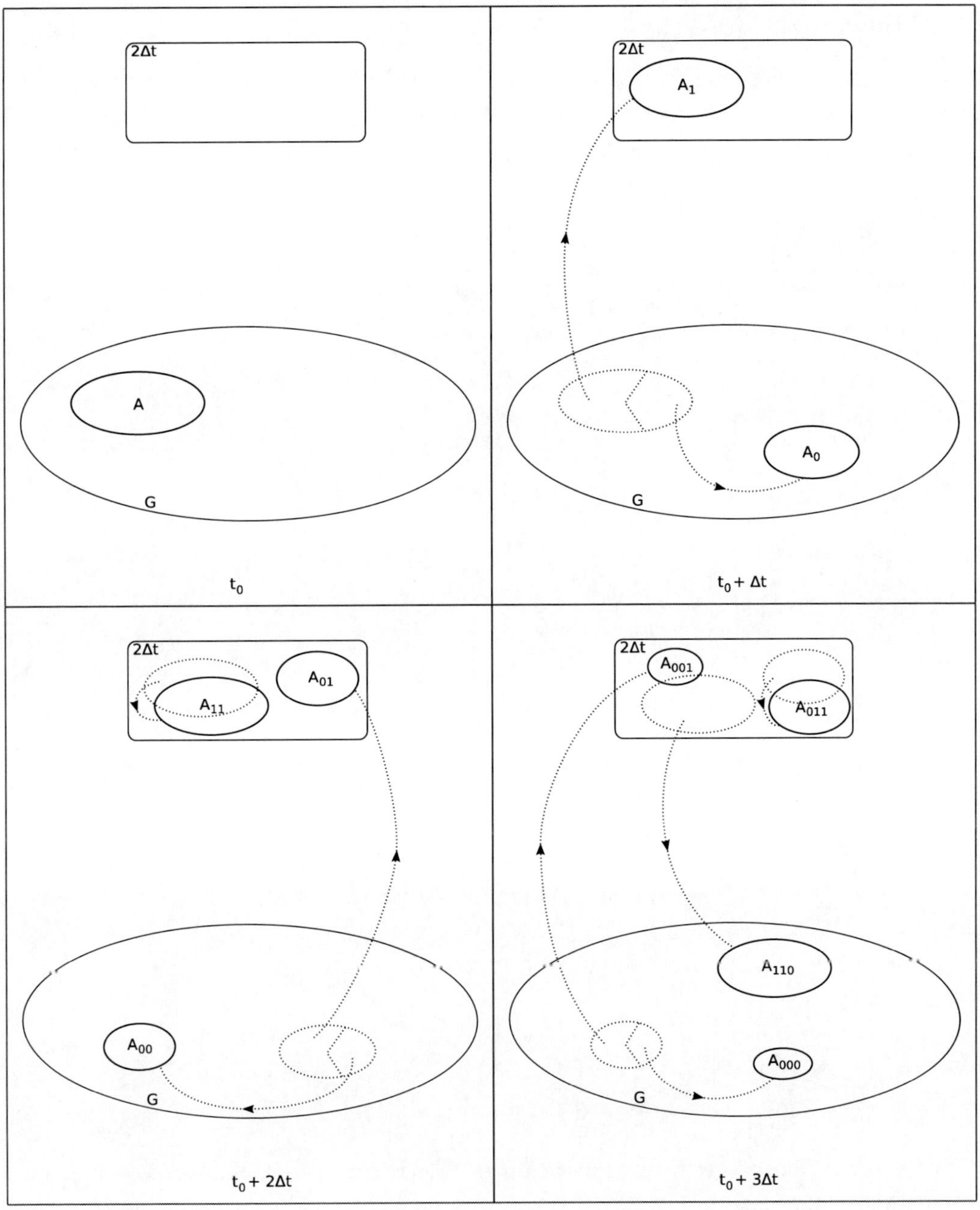

Figure 2. Model with one trap.

Remark 1. *In this example, the function* N_A *cannot be differentiable at* t_0*, unless* $N'_A(t_0) = 0$. *Indeed,* $S(\Delta t)N_A(t_0) = N_A(t_0 + \Delta t)$ *in that case, so we also have* $\mathcal{G}N_A(t_0) = \frac{d}{dt^+}N_A(t_0)$.

3.2. Model with Two Traps

We generalize the previous example by supposing that there are now two sets P_1 and P_2 outside G, which "trap" trajectories during $2\Delta t$ and $3\Delta t$ respectively. During a cycle, a trajectory wanders only through one trap. Trapped trajectories then stay in G during a multiple of Δt. Figure 3 illustrates this model.

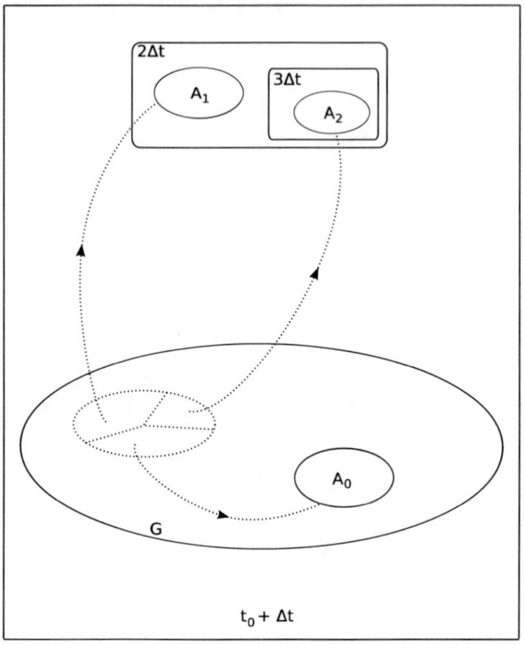

Figure 3. Model with two traps.

As previously, we introduce the following sets:

$$\begin{aligned} G_0(\Delta t) &= \{x \in G \,|\, \tau_r(x) \geq \Delta t\}\,, \\ G_1(\Delta t) &= \{x \in G \,|\, \tau_r(x) < \Delta t,\ \tau_e(x) = 2\Delta t\}\,, \\ G_2(\Delta t) &= \{x \in G \,|\, \tau_r(x) < \Delta t,\ \tau_e(x) = 3\Delta t\}\,. \end{aligned}$$

Once again, those sets form a partition of G. For $k \in \{0, 2\}$, we note $p_k(\Delta t) = \dfrac{m'(G_k(\Delta t))}{m'(G)}$. We still have $p_0(\Delta t) + p_1(\Delta t) + p_2(\Delta t) = 1$.

By proceeding similaraly to the previous model, we find:

$$N_A(t_0 + \Delta t) = p_0 N_A(t_0) + p_1 N_A(t_0 - 2\Delta t) + p_2 N_A(t_0 - 3\Delta t).$$

Time evolves here in $t_0 + \Delta t\mathbb{Z} + 2\Delta t\mathbb{Z} + 3\Delta t\mathbb{Z} = t_0 + \Delta t\mathbb{Z}$. The infinitesimal evolution operator $S(\Delta t)$ once again verifies

$$S(\Delta t) N_A(t_0) = N_A(t_0 + \Delta t),$$

thus semi-group properties (6) and (7) are still fulfilled.

Concerning the infinitesimal generator, we have

$$\begin{aligned}\mathcal{G}N_A(t_0) &= \lim_{\Delta t\to 0^+} \frac{p_1\left(N_A(t_0-2\Delta t)-N_A(t_0)\right)}{\Delta t}+\frac{p_2\left(N_A(t_0-3\Delta t)-N_A(t_0)\right)}{\Delta t},\\ &= -(2p_1(0^+)+3p_2(0^+))\frac{d}{dt^-}N_A(t_0).\end{aligned}$$

3.3. Generalizations

Let $\{P_k\}_{k\in\mathbb{N}^*}$ be a set of traps with respective trapping times $n_k\Delta t$, $n_k \in \mathbb{N}^*$. We still assume that each time a trajectory leaves G, it is trapped by exactly one trap. Hence, if we note

$$G_0(\Delta t) = \{x \in G \,|\, \tau_r(x) \geq \Delta t\}$$

and for all $k \in \mathbb{N}^*$,

$$G_k(\Delta t) = \{x \in G \,|\, \tau_r(x) < \Delta t,\ \tau_e(x) = n_k\Delta t\},$$

we still obtain a partition of G.

For all $k \in \mathbb{N}$, if we set

$$p_k(\Delta t) = \frac{m'(G_k(\Delta t))}{m'(G)}, \tag{11}$$

the evolution of N_A verifies

$$N_A(t_0+\Delta t) = \sum_{k\geq 0} p_k(\Delta t)N_A(t_0-n_k\Delta t). \tag{12}$$

Once again, the successive instants belong to $t_0+\Delta t\mathbb{Z}$. Then $S(\Delta t)N_A(t_0) = N_A(t_0+\Delta t)$, and

$$S(\Delta t)N_A(t_0) = \sum_{k\geq 0} p_k(\Delta t)N_A(t_0-n_k\Delta t). \tag{13}$$

Now we consider any trapping times, denoted $T_k(\Delta t)$ with $k \in \mathbb{N}^*$, and we suppose they are well-ordered:

$$0 < T_1(\Delta t) < \cdots < T_k(\Delta t) < \cdots.$$

We set $\mathcal{T}(\Delta t) = \left\{\sum_{k\in I} T_k(\Delta t)x_k \,|\, I \subset \mathbb{N},\ \mathrm{Card}(I) < \infty,\ \forall k \in I,\ x_k \in \mathbb{Z}\right\}$.

In that case, as soon there exists $(k_1, k_2) \in (\mathbb{N}^*)^2$ such that $\dfrac{T_{k_1}(\Delta t)}{T_{k_2}(\Delta t)} \notin \mathbb{Q}$, the group $\mathcal{T}(\Delta t)$ cannot anymore be written as $\tau_0\mathbb{Z}$, but is dense in $\mathbb{R}$. In particular, it is no more possible to define a minimal time step. But $\mathcal{T}(\Delta t)$ remains countable, so it is still possible to move to the next step: the operator $S(\Delta t)$ still makes sense, but is no longer equal to $N_A(t_0+\Delta t)$. Consequently, a generalization cannot be done with (12), but with (13):

$$S(\Delta t)N_A(t_0) = \sum_{k\geq 0} p_k(\Delta t)N_A(t_0-T_k(\Delta t)), \tag{14}$$

where $p_k(\Delta t)$ is given by (11), with

$$G_0(\Delta t) = \{x \in G \,|\, \tau_r(x) \geq T_1(\Delta t)\}\,, \tag{15}$$

and, for all $k \in \mathbb{N}^*$,

$$G_k(\Delta t) = \{x \in G \,|\, \tau_r(x) < T_1(\Delta t),\ \tau_e(x) = T_k(\Delta t)\}\,. \tag{16}$$

This formula is to be linked with expression (8) in [10].

We assume by now that $\Delta t \mapsto p_0(\Delta t)$ has a right limit at 0, denoted $p_0(0^+)$. Now we specify values of $p_k(\Delta t)$ and $T_k(\Delta t)$ in the case of chaotic Hamiltonian systems.

4. Dynamical Traps and Anomalous Diffusion

Zaslavsky studies in [33] the general shape of chaotic Hamiltonian phase spaces. In chapter 12, he introduces the notion of *dynamical trap* so as to describe the behavior of trajectories near KAM tori. This area possesses a self-similar structure: it is composed of imbricated subsets P_k which verify

$$m(P_{k+1}) = \lambda_M\, m(P_k), \text{ with } \lambda_M < 1.$$

Moreover, the trapping times T_k associated also have a self-similar property:

$$T_{k+1} = \lambda_T\, T_k, \text{ with } \lambda_T > 1.$$

Trapping times are hence all the longer as traps are small. This analysis can also be found in [32, 34].

This kind of structure has "macroscopic" consequences: when one studies diffusion of particules through a probabilistic description of the system, the moment of order 2 is ruled by the following law:

$$\langle x^2 \rangle \propto t^{\mu}.$$

The classical case (normal diffusion) corresponds to $\mu = 1$. The terms subdiffusion and superdiffusion are respectively used for $\mu < 1$ and $\mu > 1$. See [33, part.3] and [28] for more details.

Those anomalous diffusion phenomena can be described with the introduction of fractional derivatives into some specific partial derivatives equations [20, 31, 25], [33, chap. 16].

One of the equations proposed by Zaslavsky [33, p.249], a simplified fractional Fokker-Planck-Kolmogorov equation, is given by

$$\frac{\partial^\beta}{\partial t^\beta} P(x,t) = \frac{\partial^\alpha}{\partial x^\alpha}\left(\mathcal{A}(x) P(x,t)\right),\ 0 < \beta \leq 1,\ 0 < \alpha \leq 2, \tag{17}$$

where $P(x,t)$ is the probability to find the particule at position x at time t.

If we assume $\mathcal{A}$ constant, this equation leads to the following transport equation [33, p.251]:

$$\langle x^\alpha \rangle \propto t^{\beta}.$$

The classical case corresponds to $\beta = 1$ and $\alpha = 2$. The transport exponent [33, p.192] is defined by

$$\mu = \frac{2\beta}{\alpha}. \tag{18}$$

According to Zaslavsky [33, p.251, p.263], the influence of the dynamical traps appears through the following relation:

$$\mu = \frac{|\ln(\lambda_M)|}{\ln(\lambda_T)}. \tag{19}$$

Equality between (18) and (19) provides a connection between the fractal structure of the phase space and the fractional derivatives of (17). However, the justification for the introduction of those derivatives in equations (16.3) and (16.4) of [33] is not clear. An approach based on Continuous Time Random Walks (CTRW) [18, 20, 30] leads to such derivatives, but those probabilistic models do not rely on the "microscopic" dynamics of the trajectories.

We propose here to link the emergence of fractional operators with the self-similar structure of the phase space described above. To do so, we simplify the original dynamics by only considering the dynamical traps which appear in typical chaotic Hamiltonian systems (see figure 4), with both of their fractal properties. In that case, we may use the model presented in section 3.

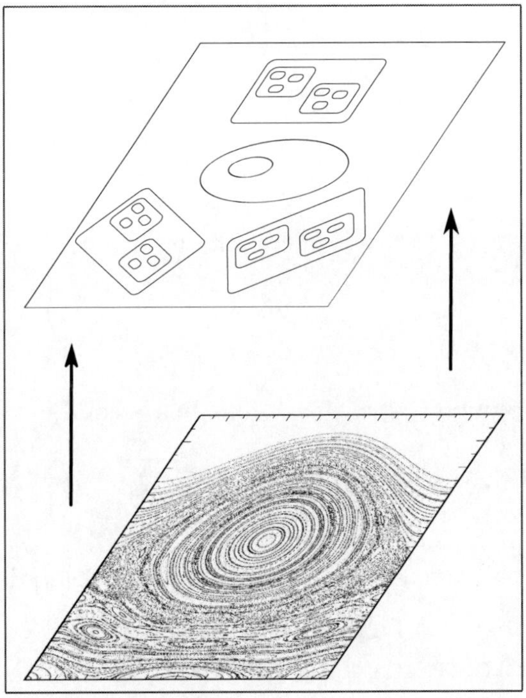

Figure 4. Simplification of the original dynamics.

Coefficients λ_M and λ_T *a priori* depend on Δt. Because of the dynamical definition of traps P_k, the subsets $G_k(\Delta t)$ also verify, for $k \geq 1$,

$$m'(G_{k+1}(\Delta t)) = \lambda_M(\Delta t)\, m'(G_k(\Delta t)).$$

Concerning the characteristic times, we have, for all $k \geq 1$,

$$T_k(\Delta t) = T_1(\Delta t)\, \lambda_T(\Delta t)^{k-1}.$$

We would like traps structure to become thiner when $\Delta t \to 0$, while remaining self-similar. This leads us to assume

$$\lambda_M(\Delta t) = (\lambda_M)^{\Delta t} \text{ and } \lambda_T(\Delta t) = (\lambda_T)^{\Delta t}, \tag{20}$$

where by sake of lisibility, λ_M and λ_T are now two real numbers such that $0 < \lambda_M < 1$ and $\lambda_T > 1$.

Remark 2. *The transport exponent remains unchanged with definition* (20)*:*

$$\forall \Delta t > 0,\ \mu = \frac{|\ln(\lambda_M(\Delta t))|}{\ln(\lambda_T(\Delta t))} = \frac{|\ln(\lambda_M)|}{\ln(\lambda_T)}.$$

Consequently, for all $k \geq 1$,

$$m'(G_k(\Delta t)) = (\lambda_M)^{(k-1)\Delta t}\, m'(G_1(\Delta t)),$$

and

$$T_k(\Delta t) = T_1(\Delta t)\, \lambda_T^{(k-1)\Delta t}. \tag{21}$$

In order to obtain smaller characteristic times when $\Delta t \to 0$, we suppose that

$$\lim_{\Delta t \to 0^+} T_1(\Delta t) = 0. \tag{22}$$

Using relation $\sum_{k \geq 0} p_k(\Delta t) = 1$, we find that for all $k \geq 1$,

$$p_k(\Delta t) = (1 - p_0(\Delta t))(1 - \lambda_M^{\Delta t})\lambda_M^{(k-1)\Delta t}. \tag{23}$$

The infinitesimal evolution (14) of the system thus becomes

$$S(\Delta t)N_A(t_0) = p_0(\Delta t)N_A(t_0) + (1 - p_0(\Delta t))\,(1 - \lambda_M^{\Delta t}) \sum_{k \geq 0} \lambda_M^{k\Delta t} N_A\left(t_0 - T_1(\Delta t)\lambda_T^{k\Delta t}\right). \tag{24}$$

The infinitesimal generators related to (24) can now be determined.

5. Fractional Infinitesimal Generator

Hölder conditions on N_A appear in this part, so we need the following definitions. Let $\Omega \subset \mathbb{R}$, $f : \Omega \to \mathbb{R}$ and $\alpha \in (0, 1]$.

Definition 1. *Let $x \in \Omega$. The function f satisfies the Hölder condition of order α at x if*

$$\exists c > 0,\ \exists \eta > 0,\ \forall y \in \Omega,\ |x - y| \leq \eta \Rightarrow |f(x) - f(y)| \leq c|x - y|^{\alpha}.$$

Definition 2. *The function f locally satisfies the Hölder condition of order α if for all $x \in \Omega$, f satisfies the Hölder condition of order α at x.*

Definition 3. *The function f satisfies the Hölder condition of order α if*

$$\exists c > 0,\ \forall x,\ y \in \Omega,\ |f(x) - f(y)| \leq c|x - y|^{\alpha}.$$

If $\alpha = 1$, f is called Lipschitz continuous.

Now we go back to our problem and we begin to show that $S(\Delta t)$ still fulfills (6).

Lemma 1. *If N_A satisfies the Hölder condition of order β, with $\beta < \mu$, then*

$$\lim_{\Delta t \to 0^+} S(\Delta t) N_A(t_0) = N_A(t_0).$$

Proof. The difference $S(\Delta t)N_A(t_0) - N_A(t_0)$ verifies

$$S(\Delta t)N_A(t_0) - N_A(t_0) = (1 - p_0(\Delta t))\,(1 - \lambda_M^{\Delta t}) \sum_{k \geq 0} \lambda_M^{k\Delta t} \left[N_A\left(t_0 - T_1(\Delta t)\lambda_T^{k\Delta t}\right) - N_A(t_0) \right].$$

We remark that $\lambda_M \lambda_T^{\beta} < 1$ if and only if $\beta < \mu$. Given that N_A satisfies the Hölder condition of order β, we obtain

$$\begin{aligned} |S(\Delta t)N_A(t_0) - N_A(t_0)| &\leq (1 - p_0(\Delta t))\,(1 - \lambda_M^{\Delta t}) \sum_{k \geq 0} \lambda_M^{k\Delta t} \left(T_1(\Delta t)\lambda_T^{k\Delta t} \right)^{\beta} \\ &\leq (1 - p_0(\Delta t))\, T_1(\Delta t)^{\beta} \frac{1 - \lambda_M^{\Delta t}}{1 - \left(\lambda_M \lambda_T^{\beta}\right)^{\Delta t}} \end{aligned}$$

On the one hand,

$$\lim_{\Delta t \to 0^+} (1 - p_0(\Delta t)) \frac{1 - \lambda_M^{\Delta t}}{1 - \left(\lambda_M \lambda_T^{\beta}\right)^{\Delta t}} = (1 - p_0(0^+)) \frac{\ln(\lambda_M)}{\ln(\lambda_M \lambda_T^{\beta})},$$

and on the other hand, $\lim_{\Delta t \to 0^+} T_1(\Delta t) = 0$ from assumption (22).
Consequently, $\lim_{\Delta t \to 0^+} (S(\Delta t)N_A(t_0) - N_A(t_0)) = 0$.

□

As it has already be seen, definition (4) cannot be used to check property (7). So we just assume that (7) is fulfilled.

We recall that the infinitesimal generator $\mathcal{G}$ associated to this semi-group verifies

$$\mathcal{G} N_A(t_0) = \lim_{\Delta t \to 0^+} \frac{S(\Delta t)N_A(t_0) - N_A(t_0)}{\Delta t}.$$

In [14, 10, 13, 11], Hilfer shows that fractional derivatives may appear as infinitesimal generators of renormalized evolution operators. A similar result will now be obtained, but without using any renormalization.

For all $\Delta t > 0$, we note

$$\mathcal{G}(\Delta t)N_A(t_0) = \frac{S(\Delta t)N_A(t_0) - N_A(t_0)}{\Delta t}.$$

We also introduce the function f defined by

$$\begin{array}{rccl} f: & \mathbb{R}^+ \times \mathbb{R}^+ & \longrightarrow & \mathbb{R} \\ & (\Delta t, y) & \longmapsto & \lambda_M^y \left[N_A\left(t_0 - T_1(\Delta t)\lambda_T^y\right) - N_A(t_0)\right]. \end{array}$$

Consequently,

$$\mathcal{G}(\Delta t)N_A(t_0) = (1 - p_0(\Delta t)) \frac{1 - \lambda_M^{\Delta t}}{\Delta t} \sum_{k \geq 0} f(\Delta t, k\Delta t).$$

For all $k \in \mathbb{N}$, we note

$$\begin{aligned} I_k(\Delta t) &= f(\Delta t, k\Delta t) = \int_k^{k+1} f(\Delta t, k\Delta t)\, dx, \\ J_k(\Delta t) &= \int_k^{k+1} f(\Delta t, x\Delta t)\, dx. \end{aligned}$$

Hence $\mathcal{G}(\Delta t)N_A(t_0)$ can be written as

$$\mathcal{G}(\Delta t)N_A(t_0) = (1 - p_0(\Delta t)) \frac{1 - \lambda_M^{\Delta t}}{\Delta t} \sum_{k \geq 0} I_k(\Delta t).$$

5.1. Case $\mu > 1$

In that case, $\lambda_M \lambda_T < 1$.

Theorem 2. *If N_A is differentiable and Lipschitz continuous on $\mathbb{R}$, and if T_1 is differentiable at 0, then*

$$\mathcal{G}N_A(t_0) = -\gamma \frac{d}{dt} N_A(t_0),$$

where $\gamma = (1 - p_0(0^+)) \dfrac{\mu}{\mu - 1} T_1'(0)$.

Proof. First we prove that $\lim\limits_{\Delta t \to 0^+} \sum\limits_{k \geq 0} (I_k(\Delta t) - J_k(\Delta t)) = 0$.

The function N_A is differentiable, so is $y \mapsto f(\Delta t, y)$ for all $\Delta t \geq 0$, and

$$\partial_2 f(\Delta t, y) = \ln(\lambda_M) f(\Delta t, y) - T_1(\Delta t) \left(\lambda_M \lambda_T\right)^y N_A'(t_0 - T_1(\Delta t)\lambda_T^y).$$

If we note c the Lipschitz constant, we have

$$|\partial_2 f(\Delta t, y)| \leq \ln(\lambda_M) c T_1(\Delta t) \left(\lambda_M \lambda_T\right)^y + c T_1(\Delta t) \left(\lambda_M \lambda_T\right)^y.$$

By setting $c' = c(1 + \ln(\lambda_T))$, we obtain

$$|\partial_2 f(\Delta t, y)| \leq c' T_1(\Delta t) \left(\lambda_M \lambda_T\right)^y.$$

Let $k \in \mathbb{N}$. Then

$$\begin{aligned}|I_k(\Delta t) - J_k(\Delta t)| &\leq \int_k^{k+1} |f(\Delta t, x\Delta t) - f(\Delta t, k\Delta t)|\, dx,\\ &\leq \Delta t \sup_{[k,k+1]} |\partial_2 f(\Delta t, x\Delta t)|,\\ &\leq c'\Delta t T_1(\Delta t)\,(\lambda_M\lambda_T)^{k\Delta t}.\end{aligned}$$

Given that $\lim_{\Delta t\to 0^+} \Delta t \sum_{k\geq 0} (\lambda_M\lambda_T)^{k\Delta t} = \frac{1}{|\ln(\lambda_M\lambda_T)|}$ and $\lim_{\Delta t\to 0^+} T_1(\Delta t) = 0$, we infer that

$$\lim_{\Delta t\to 0^+} \sum_{k\geq 0} (I_k(\Delta t) - J_k(\Delta t)) = 0.$$

Consequently,

$$\mathcal{G}(\Delta t)N_A(t_0) \underset{\Delta t\to 0^+}{\sim} -\left(1 - p_0(0^+)\right)\ln(\lambda_M) \sum_{k\geq 0} J_k(\Delta t). \tag{25}$$

Now we can evaluate $\lim_{\Delta t\to 0^+} \sum_{k\geq 0} J_k(\Delta t)$. Firstly,

$$\sum_{k\geq 0} J_k(\Delta t) = \int_0^\infty \lambda_M^{x\Delta t}\left[N_A\left(t_0 - T_1(\Delta t)\lambda_T^{x\Delta t}\right) - N_A(t_0)\right] dx.$$

With substitution $t = \lambda_T^{x\Delta t}$, we obtain

$$\begin{aligned}\sum_{k\geq 0} J_k(\Delta t) &= \frac{1}{\Delta t \ln(\lambda_T)} \int_1^\infty t^{-(1+\mu)}\left[N_A(t_0 - tT_1(\Delta t)) - N_A(t_0)\right] dt, \qquad (26)\\ &= \frac{1}{\ln(\lambda_T)} \frac{T_1(\Delta t)}{\Delta t} \int_1^\infty t^{-\mu} \frac{N_A(t_0 - tT_1(\Delta t)) - N_A(t_0)}{tT_1(\Delta t)}\, dt.\end{aligned}$$

Secondly, for all $t \geq 1$, $\left|t^{-\mu}\frac{N_A(t_0 - tT_1(\Delta t)) - N_A(t_0)}{tT_1(\Delta t)}\right| \leq ct^{-\mu}$, and $t \mapsto ct^{-\mu}$ is integrable on $[1, +\infty)$. Moreover,

$$\lim_{\Delta t\to 0^+} \frac{N_A(t_0 - tT_1(\Delta t)) - N_A(t_0)}{tT_1(\Delta t)} = -N_A'(t_0).$$

Consequently, by dominated convergence,

$$\begin{aligned}\lim_{\Delta t\to 0^+} \sum_{k\geq 0} J_k(\Delta t) &= \frac{1}{\ln(\lambda_T)} T_1'(0)(-N_A'(t_0)) \int_1^\infty t^{-\mu}\, dt,\\ &= \frac{1}{(1-\mu)\ln(\lambda_T)} T_1'(0) N_A'(t_0).\end{aligned}$$

Finally, from (25),

$$\lim_{\Delta t\to 0^+} \mathcal{G}(\Delta t)N_A(t_0) = -\left(1-p_0(0^+)\right)\frac{\ln(\lambda_M)}{(1-\mu)\ln(\lambda_T)}T_1'(0)N_A'(t_0),$$
$$= -\left(1-p_0(0^+)\right)\frac{\mu}{\mu-1}T_1'(0)N_A'(t_0).$$

□

Remark 3. *If N_A verifies assumptions of theorem 2, then N_A satisfies the Hölder condition of order 1 and consequently fulfills conditions of lemma 1.*

5.2. Case $\mu < 1$

We can try to estimate $\lim_{\Delta t\to 0^+}\sum_{k\geq 0} J_k(\Delta t)$, assuming that N_A is smooth enough and rapidly decreasing in $-\infty$, in order that all the following quantities are well-defined. We integrate by parts (26):

$$\sum_{k\geq 0} J_k(\Delta t) = \frac{-T_1(\Delta t)}{\mu\Delta t\ln(\lambda_T)}\int_1^\infty t^{-\mu}N_A'(t_0 - tT_1(\Delta t))\,dt + \frac{N_A(t_0 - T_1(\Delta t)) - N_A(t_0)}{\mu\Delta t\ln(\lambda_T)}.$$

Substitution $u = T_1(\Delta t)t$ leads to

$$\sum_{k\geq 0} J_k(\Delta t) = \frac{-T_1(\Delta t)^\mu}{\mu\Delta t\ln(\lambda_T)}\int_{T_1(\Delta t)}^\infty u^{-\mu}N_A'(t_0 - u)\,du + \frac{N_A(t_0 - T_1(\Delta t)) - N_A(t_0)}{\mu\Delta t\ln(\lambda_T)}.$$

The integral is not problematic:

$$\lim_{\Delta t\to 0^+}\int_{T_1(\Delta t)}^\infty u^{-\mu}N_A'(t_0-u)\,du = \int_0^\infty u^{-\mu}N_A'(t_0-u)\,du.$$

If T_1 is differentiable at 0, then

$$\lim_{\Delta t\to 0^+}\frac{N_A(t_0 - T_1(\Delta t)) - N_A(t_0)}{\mu\Delta t\ln(\lambda_T)} = -\frac{T_1'(0)}{\mu\ln(\lambda_T)}N_A'(t_0).$$

Conversely, $\lim_{\Delta t\to 0^+}\dfrac{T_1(\Delta t)^\mu}{\Delta t} = +\infty$. So we cannot find any infinitesimal generator. The assumption on the differentiability of T_1 at 0 should hence be replaced.

In order that $\dfrac{T_1(\Delta t)^\mu}{\Delta t}$ has a finite limit, we suppose that there exists $b > 0$ such that

$$T_1(\Delta t) \underset{\Delta t\to 0^+}{\sim} b(\Delta t)^{1/\mu}.$$

From a physical point of view, $T_1(\Delta t)$ and Δt are homogeneous to time, so we introduce a constant of time τ such that $b = \tau^{1-1/\mu}$:

$$T_1(\Delta t) \underset{\Delta t\to 0^+}{\sim} \tau^{1-1/\mu}(\Delta t)^{1/\mu}. \tag{27}$$

Under this assumption on T_1, a fractional derivative defined as follows will appear.

Definition 4. *Let $f : \mathbb{R} \to \mathbb{R}$ and $\alpha \in (0,1)$. The Marchaud fractional derivative of order α is defined as*

$$\mathcal{D}_+^\alpha f(t) = \frac{\alpha}{\Gamma(1-\alpha)} \int_0^\infty u^{-(1+\alpha)} \left[f(t) - f(t-u) \right] du,$$

where Γ is the Gamma function.

This derivative is well-defined if f is bounded and locally satisfies the Hölder condition of order δ, with $\delta > \alpha$. See [26, p.109] for more details.

Now we can enunciate the main result of the article.

Theorem 3. *If N_A satisfies the Hölder condition of order β and locally satisfies the Hölder condition of order ν, with $\beta < \mu < \nu$, then*

$$\mathcal{G} N_A(t_0) = -\tilde{\gamma}\, \tau^{\mu-1}\, \mathcal{D}_+^\mu N_A(t_0), \tag{28}$$

where $\tilde{\gamma} = \Gamma(1-\mu)(1-p_0(0^+))$.

Proof. As previously, we firstly prove that $\lim_{\Delta t \to 0^+} \sum_{k \geq 0} (I_k(\Delta t) - J_k(\Delta t)) = 0$.

Let $k \in \mathbb{N}$. For all $x \in [k, k+1]$,

$$\begin{aligned}
f(\Delta t, x\Delta t) - f(\Delta t, k\Delta t) &= \lambda_M^{x\Delta t} \left[(N_A(t_0 - T_1(\Delta t)\lambda_T^{x\Delta t}) - N_A(t_0)) \right. \\
&\quad \left. - \left(N_A(t_0 - T_1(\Delta t)\lambda_T^{k\Delta t}) - N_A(t_0) \right) \right] \\
&\quad + \left[\lambda_M^{x\Delta t} - \lambda_M^{k\Delta t} \right] \left(N_A(t_0 - T_1(\Delta t)\lambda_T^{k\Delta t}) - N_A(t_0) \right), \\
&= \lambda_M^{x} \left(N_A(t_0 - T_1(\Delta t)\lambda_T^{x\Delta t}) - N_A(t_0 - T_1(\Delta t)\lambda_T^{k\Delta t}) \right) \\
&\quad + \left[\lambda_M^{x\Delta t} - \lambda_M^{k\Delta t} \right] \left(N_A(t_0 - T_1(\Delta t)\lambda_T^{k\Delta t}) - N_A(t_0) \right).
\end{aligned}$$

Concerning the first right-hand member, we obtain the following inequality:

$$\begin{aligned}
&\left| \int_k^{k+1} \lambda_M^{x\Delta t} \left(N_A(t_0 - T_1(\Delta t)\lambda_T^{x\Delta t}) - N_A(t_0 - T_1(\Delta t)\lambda_T^{k\Delta t}) \right) dx \right| \\
&\quad \leq \lambda_M^{k\Delta t} \int_k^{k+1} \left| N_A(t_0 - T_1(\Delta t)\lambda_T^{x\Delta t}) - N_A(t_0 - T_1(\Delta t)\lambda_T^{k\Delta t}) \right| dx, \\
&\quad \leq \lambda_M^{k\Delta t} \int_k^{k+1} \left| T_1(\Delta t) \left(\lambda_T^{x\Delta t} - \lambda_T^{k\Delta t} \right) \right|^\beta dx, \\
&\quad \leq \lambda_M^{k\Delta t} T_1(\Delta t)^\beta \left(\lambda_T^{(k+1)\Delta t} - \lambda_T^{k\Delta t} \right)^\beta, \\
&\quad \leq T_1(\Delta t)^\beta \left(\lambda_T^{\Delta t} - 1 \right) \left(\lambda_M \lambda_T^\beta \right)^{k\Delta t}.
\end{aligned}$$

For the second one, we have

$$
\begin{aligned}
|\int_k^{k+1} \left[\lambda_M^{x\Delta t} - \lambda_M^{k\Delta t}\right] \left(N_A(t_0 - T_1(\Delta t)\lambda_T^{k\Delta t}) - N_A(t_0)\right)\, dx| \\
&\leq T_1(\Delta t)^\beta \lambda_T^{k\beta\Delta t} \int_k^{k+1} \left|\lambda_M^{x\Delta t} - \lambda_M^{k\Delta t}\right| dx, \\
&\leq T_1(\Delta t)^\beta \lambda_T^{k\beta\Delta t} \left(\lambda_M^{(k+1)\Delta t} - \lambda_M^{k\Delta t}\right), \\
&\leq T_1(\Delta t)^\beta \left(\lambda_M^{\Delta t} - 1\right) \left(\lambda_M \lambda_T^\beta\right)^{k\Delta t}.
\end{aligned}
$$

Consequently,

$$
|I_k(\Delta t) - J_k(\Delta t)| \leq T_1(\Delta t)^\beta \left(\lambda_T^{\Delta t} - 1 + \lambda_M^{\Delta t} - 1\right) \left(\lambda_M \lambda_T^\beta\right)^{k\Delta t}.
$$

Since $\beta < \mu$, $\lambda_M \lambda_T^\beta < 1$. Then,

$$
\sum_{k\geq 0} |I_k(\Delta t) - J_k(\Delta t)| \leq T_1(\Delta t)^\beta \frac{\lambda_T^{\Delta t} - 1 + \lambda_M^{\Delta t} - 1}{1 - \lambda_M \lambda_T^\beta}.
$$

Given that $\lim_{\Delta t\to 0^+} \frac{\lambda_T^{\Delta t} - 1 + \lambda_M^{\Delta t} - 1}{1 - \lambda_M \lambda_T^\beta} = -\frac{\ln(\lambda_T) + \ln(\lambda_M)}{\ln\left(\lambda_M \lambda_T^\beta\right)}$ and $\lim_{\Delta t\to 0^+} T_1(\Delta t)^\beta = 0$, we infer that

$$
\lim_{\Delta t\to 0^+} \sum_{k\geq 0} (I_k(\Delta t) - J_k(\Delta t)) = 0.
$$

Relation (25) is hence still valid here. Furthermore, (26) holds for $\mu < 1$. Substitution $u = tT_1(\Delta t)$ leads to

$$
\sum_{k\geq 0} J_k(\Delta t) = \frac{T_1(\Delta t)^\mu}{\Delta t \ln(\lambda_T)} \int_{T_1(\Delta t)}^{\infty} u^{-(1+\mu)} \left[N_A(t_0 - u) - N_A(t_0)\right] du.
$$

By definition, $0 \leq N_A(t) \leq m'(G)$ for all $t \in \mathbb{R}$. Since we have also assumed that N_A locally satisfies the Hölder condition of order $\nu > \mu$, its Marchaud fractional derivative of order μ is well-defined.

As a consequence,

$$
\lim_{\Delta t\to 0^+} \int_{T_1(\Delta t)}^{\infty} u^{-(1+\mu)} \left[N_A(t_0 - u) - N_A(t_0)\right] du = -\frac{\Gamma(1-\mu)}{\mu} \mathcal{D}_+^\mu N_A(t_0).
$$

With relation (27), we obtain

$$
\lim_{\Delta t\to 0^+} \sum_{k\geq 0} J_k(\Delta t) = -\frac{\tau^{\mu-1}}{\ln(\lambda_T)} \frac{\Gamma(1-\mu)}{\mu} \mathcal{D}_+^\mu N_A(t_0).
$$

Finally,

$$
\begin{aligned}
\mathcal{G} N_A(t_0) &= (1 - p_0(0^+)) \frac{\tau^{\mu-1}}{\ln(\lambda_T)} \frac{\ln(\lambda_M)\Gamma(1-\mu)}{\mu} \mathcal{D}_+^\mu N_A(t_0) \\
&= -(1 - p_0(0^+)) \tau^{\mu-1} \Gamma(1-\mu)\, \mathcal{D}_+^\mu N_A(t_0).
\end{aligned}
$$

□

We have deliberately let the constant τ appear in (28) for reasons of dimensional homogeneity [16]: the relevant derivative is not $\mathcal{D}_+^\mu$, but $\tau^{\mu-1}\mathcal{D}_+^\mu$, in order to be homogeneous to the inverse of a time.

Remark 4. *So as to respect dimensional homogeneity, a constant of time τ' should have been introduced for the traps constants:*

$$\lambda_T(\Delta t) = (\lambda_T)^{\Delta t/\tau'}, \quad \lambda_M(\Delta t) = (\lambda_M)^{\Delta t/\tau'}.$$

However, from remark 2, the transport exponent should have eventually been unchanged.

6. Discussion

6.1. Characterization of μ

From distribution $(p_k(\Delta t))_{k\geq 0}$ and characteristic times $(T_k(\Delta t))_{k\geq 0}$ (with $T_0(\Delta t) = 0$), we can evaluate moments $\langle T^\alpha\rangle_{\Delta t}$ with $\alpha > 0$, defined by

$$\langle T^\alpha\rangle_{\Delta t} = \sum_{k\geq 0} p_k(\Delta t) T_k(\Delta t)^\alpha.$$

In the case of dynamical traps described by (21) and (23), we obtain:

$$\langle T^\alpha\rangle_{\Delta t} = \begin{cases} (1 - p_0(\Delta t)) \dfrac{1 - \lambda_M^{\Delta t}}{1 - (\lambda_M \lambda_T^\alpha)^{\Delta t}} T_1(\Delta t)^\alpha & \text{if } \alpha < \mu, \\ +\infty & \text{if } \alpha \geq \mu. \end{cases}$$

Consequently, if we note $\langle T^\alpha\rangle = \lim_{\Delta t \to 0^+} \langle T^\alpha\rangle_{\Delta t}$, parameter μ appears as a critical point:

$$\langle T^\alpha\rangle = \begin{cases} 0 & \text{if } \alpha < \mu, \\ +\infty & \text{if } \alpha \geq \mu. \end{cases}$$

However, if $m(G) > 0$ and $\mu \leq 1$, $\langle T\rangle = \infty$, which does not respect the Kac lemma [17, 19]. Then we should assume $m(G) = 0$. This remark is closely akin to the approach of Hilfer [11], where the fractional infinitesimal generator only appears for sets of measure 0.

Remark 5. *If G is a section transverse to the Hamiltonian flow (a Poincaré section for instance), then $m(G) = 0$, and since all the trajectories cross G, $\tau_r(x) = 0$ for all $x \in G$. Consequently, $G_0(\Delta t) = \emptyset$ and $p_0(\Delta t) = 0$, for all $\Delta t > 0$. Hence $p_0(0^+) = 0$, which leads to $\mathcal{G}N_A(t_0) = -\Gamma(1-\mu)\tau^{\mu-1}\mathcal{D}_+^\mu N_A(t_0)$.*

6.2. Fractional Kinetic Equation

The model presented here do not explain the emergence of fractional derivatives in equations such as (17). However, the fractional exponent we have obtained is exactly the transport coefficient (19). This result is compatible with relation (18) which involves the fractional exponents of Zaslavsky.

Indeed, let us assume that $S(\Delta t)$ could be applied to $P(x,t)$ in order to describe "a generalized shift of $P(x,t)$ along t by Δt" [33, p.246]. Then the temporal derivative associated to the temporal evolution of $P(x,t)$ is the infinitesimal generator $\mathcal{G}$. In the case of anomalous diffusion, exponents α and β in (17) become completely determined.

- If $\mu > 1$ (superdiffusion), then the temporal derivative is classic: $\beta = 1$. Superdiffusion is exclusively taken into account by the spatial derivative of order $\alpha = \frac{2}{\mu}$. Equation (17) thus becomes

$$\frac{\partial P(x,t)}{\partial t} = \frac{\partial^{2/\mu}}{\partial x^{2/\mu}} \left(\mathcal{A}(x) P(x,t) \right).$$

- If $\mu < 1$ (subdiffusion), $\beta = \mu$ so $\alpha = 2$: the temporal derivative is the only one to be fractional. Consequently, (17) turns into

$$\frac{\partial^{\mu}}{\partial t^{\mu}} P(x,t) = \frac{\partial^2}{\partial x^2} \left(\mathcal{A}(x) P(x,t) \right).$$

In particular, if our model applies to $P(x,t)$, fractional derivatives in space and time cannot coexist.

7. Conclusion

The model which has been described in this article attempts to explain, from a dynamical view point, the emergence of fractional derivatives in chaotic Hamiltonian systems. It seems simplier than the formalism of Hilfer, in particular because no renormalization appears. Moreover, it strongly relies on fractal properties of the phase space. Our approach is obviously perfectible on several aspects. It does not explain why $T_1(\Delta t)$ should fulfill (27), and condition $m(G) = 0$ imposed by the Kac lemma should be clarified. So as to test the validity of the model, other systems should also be considered, in particular strongly chaotic systems, where the distribution of recurrence times is similar to an exponential law [3, 15]. Finally, we believe that there are still enough freedom degrees in our model for allowing us to enhance it in forthcoming studies.

References

[1] O.P. Agrawal. Formulation of Euler-Lagrange equations for fractional variational problems. *J. Math. Anal. Appl.*, 272:368–379, 2002.

[2] D. Baleanu and S.I. Muslih. Lagrangian formulation of classical fields within Riemann-Liouville fractional derivatives. *Physica Scripta*, 72:119–121, 2005.

[3] N. Buric, A. Rampioni, G. Turchetti, and S. Vaienti. Weak chaos and poincaré recurrences for area preserving maps. *J. Phys. A*, 36, 2003.

[4] A. Compte. Stochastic foundations of fractional dynamics. *Physical Review E*, 53(4):4191–4193, 1996.

[5] J. Cresson. Fractional embedding of differential operators and Lagrangian systems. *Journal of Mathematical Physics*, 48(3):033504, 2007.

[6] J. Cresson. Inverse problem of fractional calculus of variations for partial differential equations. *Commun Nonlinear Sci Numer Simulat*, 2009.

[7] J. Cresson and P. Inizan. About fractional Hamiltonian systems. *Physica Scripta*, T136:014007, 2009.

[8] Feller. *An Introduction to Probability Theory and Its Applications, Volume II*. Wiley, New York, 1971.

[9] R. Gorenflo, F. Mainardi, D. Moretti, and P. Paradisi. Time fractional diffusion: A discrete randomw walk approach. *Nonlinear Dynamics*, 29:129–143, 2002.

[10] R. Hilfer. Foundations of fractional dynamics. *Fractals*, 3(3):549–556, 1995.

[11] R. Hilfer. Fractional dynamics, irreversibility and ergodicity breaking. *Chaos, Solition & Fractals*, 5(8):1475–1484, 1995.

[12] R. Hilfer. *Applications of Fractional Calculus in Physics*. World Scientific Publ. Co., Singapore, 2000.

[13] R. Hilfer. Fractional evolution equations and irreversibility. In D. Helbing, H. Herrmann, M. Schreckenberg, and D. Wolf, editors, *Traffic and Granular Flow '99*. Springer, 2000.

[14] R. Hilfer. Remarks on fractional time. In L. Castell and O. Ischebeck, editors, *Time, Quantum and Information*, page 235. Springer-Verlag, Berlin, 2003.

[15] M. Hirata, B. Saussol, and S. Vaienti. Statistics of return times: A general framework and new applications. *Commun. Math. Phys.*, 206:33–55, 1999.

[16] P. Inizan. Homogeneous fractional embeddings. *Journal of Mathematical Physics*, 49(8):082901, 2008.

[17] M. Kac. *Probability and Related Topics in Physical Sciences*. Interscience, New York, 1957.

[18] M.M. Meerschaert and H.-P. Scheffler. Limit theorems for continuous-time random walks with infinite mean waiting times. *J. Appl. Prob.*, 41:623–638, 2004.

[19] J.D. Meiss. Average exit time for volume-preserving maps. *Chaos*, 7(1):139–147, 1997.

[20] R. Metzler and J. Klafter. The random walk's guide to anomalous diffusion: a fractional dynamics approach. *Pysics Reports*, 339:1–77, 2000.

[21] K.B. Oldham and J. Spanier. *The Fractional Calculus*. Academic Press, New York and London, 1974.

[22] I. Podlubny. *Fractional Differential Equations*. Academic Press, 1999.

[23] F. Riewe. Nonconservative Lagrangian and Hamiltonian mechanics. *Physical Review E*, 53(2):1890, 1996.

[24] J. Sabatier, O.P. Agrawal, and J.A. Tenreiro Machado. *Advances in fractional calculus*. Springer, 2007.

[25] A.I. Saichev and G.M. Zaslavsky. Fractional kinetic equations: solutions and applications. *Chaos*, 7(4):753–764, 1997.

[26] S.G. Samko, A.A. Kilbas, and O.I. Marichev. *Fractional integrals ans derivatives: theory and applications*. Gordon and Breach, 1993.

[27] W.R. Schneider and W. Wyss. Fractional diffusion and wave equations. *Journal of Mathematical Physics*, 30(1), 1989.

[28] M.F. Shlesinger, G.M. Zaslavsky, and J. Klafter. Strange kinetics. *Nature*, 363:31–37, 1993.

[29] A.A. Stanislavsky. Probability interpretation of the integral of fractional order. *Theoritical and Mathematical Physics*, 138(3), 2004.

[30] A.A. Stanislavsky. Hamiltonian formalism of fractional systems. *Eur. Phys. J. B*, 49:93–101, 2006.

[31] G.M. Zaslavsky. Renormalization group theory of anomalous transport in systems with hamiltonian chaos. *Chaos*, 4(1):25–33, 1994.

[32] G.M. Zaslavsky. Chaotic dynamics and the origin of statistical laws. *Physics Today*, August 1999.

[33] G.M. Zaslavsky. *Hamiltonian Chaos & Fractional Dynamics*. Oxford University Press, Oxford, 2005.

[34] G.M. Zaslavsky, M. Edelman, and B.A. Niyazov. Self-similarity, renormalization, and phase space nonuniformity of hamiltonian chaotic dynamics. *Chaos*, 7(1):159–181, 1997.

In: Fractional Calculus in Analysis, Dynamics ... ISBN: 978-1-62948-635-2
Editor: Jacky Cresson, pp. 201-239

Chapter 5

Numerical Approximations to Fractional Problems of the Calculus of Variations and Optimal Control

Shakoor Pooseh,* Ricardo Almeida† and Delfim F. M. Torres‡
Center for Research and Development in Mathematics and Applications (CIDMA)
Department of Mathematics, University of Aveiro, 3810-193 Aveiro, Portugal

Keywords: fractional calculus of variations, fractional optimal control, numerical methods, direct methods, indirect methods

AMS Subject Classification: 49K05, 49M25, 26A33

1. Introduction

A fractional problem of the calculus of variations and optimal control consists in the study of an optimization problem in which the objective functional or constraints depend on derivatives and integrals of arbitrary, real or complex, orders. This is a generalization of the classical theory, where derivatives and integrals can only appear in integer orders.

1.1. Preliminaries

Integer order derivatives and integrals have a unified meaning in the literature. In contrast, there are several different approaches and definitions in fractional calculus for derivatives and integrals of arbitrary order. The following definitions and notations will be used throughout this chapter. See [19].

Definition 1.1 (Gamma function). *The Euler integral of the second kind*

$$\Gamma(z) = \int_0^\infty t^{z-1}e^{-t}dt, \qquad Re(z) > 0,$$

is called the gamma function.

*E-mail address: spooseh@ua.pt
†E-mail address: ricardo.almeida@ua.pt
‡E-mail address: delfim@ua.pt

The gamma function has an important property, $\Gamma(z+1) = z\Gamma(z)$, and hence $\Gamma(z) = (z-1)!$ for $z \in \mathbb{N}$, which allows to extend the notion of factorial to real numbers. Other properties of this special function can be found in [5].

Definition 1.2 (Mittag–Leffler function). *Let $\alpha > 0$. The function E_α defined by*

$$E_\alpha(z) = \sum_{j=0}^{\infty} \frac{z^j}{\Gamma(\alpha j + 1)},$$

whenever the series converges, is called the one parameter Mittag–Leffler function. The two-parameter Mittag–Leffler function with parameters $\alpha, \beta > 0$ is defined by

$$E_{\alpha,\beta}(z) = \sum_{j=0}^{\infty} \frac{z^j}{\Gamma(\alpha j + \beta)}. \tag{1}$$

Definition 1.3 (Grünwald–Letnikov derivative). *Let $0 < \alpha < 1$ and $\binom{\alpha}{k}$ be the generalization of binomial coefficients to real numbers.*

- *The left Grünwald–Letnikov fractional derivative is defined as*

$$ {}_{a}^{GL}D_t^\alpha x(t) = \lim_{h \to 0^+} \frac{1}{h^\alpha} \sum_{k=0}^{\infty} (-1)^k \binom{\alpha}{k} x(t - kh). \tag{2}$$

- *The right Grünwald–Letnikov derivative is*

$$ {}_{t}^{GL}D_b^\alpha x(t) = \lim_{h \to 0^+} \frac{1}{h^\alpha} \sum_{k=0}^{\infty} (-1)^k \binom{\alpha}{k} x(t + kh). \tag{3}$$

In the above mentioned definitions, $\binom{\alpha}{k}$ is the generalization of binomial coefficients to real numbers, defined by

$$\binom{\alpha}{k} = \frac{\Gamma(\alpha + 1)}{\Gamma(k+1)\Gamma(\alpha - k + 1)}.$$

In this relation, k and α can be any integer, real or complex number, except that $\alpha \notin \{-1, -2, -3, \ldots\}$.

Definition 1.4 (Riemann–Liouville fractional integral). *Let $x(\cdot)$ be an integrable function in $[a, b]$ and $\alpha > 0$.*

- *The left Riemann–Liouville fractional integral of order α is given by*

$$ {}_aI_t^\alpha x(t) = \frac{1}{\Gamma(\alpha)} \int_a^t (t - \tau)^{\alpha - 1} x(\tau) d\tau, \quad t \in [a, b].$$

- *The right Riemann–Liouville fractional integral of order α is given by*

$$ {}_tI_b^\alpha x(t) = \frac{1}{\Gamma(\alpha)} \int_t^b (\tau - t)^{\alpha - 1} x(\tau) d\tau, \quad t \in [a, b].$$

Definition 1.5 (Riemann–Liouville fractional derivative). *Let $x(\cdot)$ be an absolutely continuous function in $[a,b]$, $x(\cdot) \in AC[a,b]$, and $0 \leq \alpha < 1$.*

- *The left Riemann–Liouville fractional derivative of order α is given by*

$$_aD_t^\alpha x(t) = \frac{1}{\Gamma(1-\alpha)} \frac{d}{dt} \int_a^t (t-\tau)^{-\alpha} x(\tau) d\tau, \quad t \in [a,b].$$

- *The right Riemann–Liouville fractional derivative of order α is given by*

$$_tD_b^\alpha x(t) = \frac{1}{\Gamma(1-\alpha)} \left(-\frac{d}{dt}\right) \int_t^b (\tau-t)^{-\alpha} x(\tau) d\tau, \quad t \in [a,b].$$

Another type of fractional derivatives, introduced by Caputo, is closely related to the Riemann–Liouville definitions.

Definition 1.6 (Caputo's fractional derivative). *For a function $x(\cdot) \in AC[a,b]$ with $0 \leq \alpha < 1$:*

- *The left Caputo fractional derivative of order α is given by*

$${}_a^C D_t^\alpha x(t) = \frac{1}{\Gamma(1-\alpha)} \int_a^t (t-\tau)^{-\alpha} \dot{x}(\tau) d\tau, \quad t \in [a,b].$$

- *The right Caputo fractional derivative of order α is given by*

$${}_t^C D_b^\alpha x(t) = \frac{-1}{\Gamma(1-\alpha)} \int_t^b (\tau-t)^{-\alpha} \dot{x}(\tau) d\tau, \quad t \in [a,b].$$

Definition 1.7 (Hadamard fractional integral). *Let $x : [a,b] \to \mathbb{R}$.*

- *The left Hadamard fractional integral of order $\alpha > 0$ is defined by*

$$_a\mathcal{I}_t^\alpha x(t) = \frac{1}{\Gamma(\alpha)} \int_a^t \left(\ln \frac{t}{\tau}\right)^{\alpha-1} \frac{x(\tau)}{\tau} d\tau, \quad t \in]a,b[.$$

- *The right Hadamard fractional integral of order $\alpha > 0$ is defined by*

$$_t\mathcal{I}_b^\alpha x(t) = \frac{1}{\Gamma(\alpha)} \int_t^b \left(\ln \frac{\tau}{t}\right)^{\alpha-1} \frac{x(\tau)}{\tau} d\tau, \quad t \in]a,b[.$$

When $\alpha = m$ is an integer, these fractional integrals are *m-fold* integrals:

$$_a\mathcal{I}_t^m x(t) = \int_a^t \frac{d\tau_1}{\tau_1} \int_a^{\tau_1} \frac{d\tau_2}{\tau_2} \dots \int_a^{\tau_{m-1}} \frac{x(\tau_m)}{\tau_m} d\tau_m,$$

and

$$_t\mathcal{I}_b^m x(t) = \int_t^b \frac{d\tau_1}{\tau_1} \int_{\tau_1}^b \frac{d\tau_2}{\tau_2} \dots \int_{\tau_{m-1}}^b \frac{x(\tau_m)}{\tau_m} d\tau_m.$$

Definition 1.8 (Hadamard fractional derivative). *For $\alpha > 0$ and $n = [\alpha] + 1$,*

- *The left Hadamard fractional derivative of order α is defined by*

$$ {}_a\mathcal{D}_t^\alpha x(t) = \left(t\frac{d}{dt}\right)^n \frac{1}{\Gamma(n-\alpha)} \int_a^t \left(\ln\frac{t}{\tau}\right)^{n-\alpha-1} \frac{x(\tau)}{\tau} d\tau, \quad t \in]a,b[. $$

- *The right Hadamard fractional derivative of order α is defined by*

$$ {}_t\mathcal{D}_b^\alpha x(t) = \left(-t\frac{d}{dt}\right)^n \frac{1}{\Gamma(n-\alpha)} \int_t^b \left(\ln\frac{\tau}{t}\right)^{n-\alpha-1} \frac{x(\tau)}{\tau} d\tau, \quad t \in]a,b[. $$

When $\alpha = m$ is an integer, we have

$$ {}_a\mathcal{D}_t^m x(t) = \left(t\frac{d}{dt}\right)^m x(t) \text{ and } {}_t\mathcal{D}_b^m x(t) = \left(-t\frac{d}{dt}\right)^m x(t). $$

1.2. Fractional Calculus of Variations and Optimal Control

Many generalizations to the classical calculus of variations and optimal control have been made to extend the theory to cover fractional variational and fractional optimal control problems. A simple fractional variational problem, for example, consists in finding a function $x(\cdot)$ that minimizes the functional

$$ J[x(\cdot)] = \int_a^b L(t, x(t), {}_aD_t^\alpha x(t))dt, \tag{4} $$

where ${}_aD_t^\alpha$ is the left Riemann–Liouville fractional derivative. Typically, some boundary conditions are prescribed as $x(a) = x_a$ and/or $x(b) = x_b$. Classical techniques have been adopted to solve such problems. The Euler–Lagrange equation for a Lagrangian of the form $L(t, x(t), {}_aD_t^\alpha x(t))$ has been derived firstly in [30, 31]. Many variants of necessary conditions of optimality have been studied. A generalization of the problem to include fractional integrals, i.e., $L = L(t, {}_aI_t^{1-\alpha}x(t), {}_aD_t^\alpha x(t))$, the transversality conditions of fractional variational problems and many other aspects can be found in the literature of recent years. See [1, 4, 6] and references therein. Furthermore, it has been shown that a variational problem with fractional derivatives can be reduced to a classical problem using an approximation of the Riemann–Liouville fractional derivatives in terms of a finite sum, where only derivatives of integer order are present [6].

On the other hand, fractional optimal control problems usually appear in the form of

$$ J[x(\cdot)] = \int_a^b L(t, x(t), u(t))dt \to \min $$

$$ s.t. \begin{cases} {}_aD_t^\alpha x(t) = f(t, x(t), u(t)) \\ x(a) = x_a, \; x(b) = x_b, \end{cases} $$

where an optimal control $u(\cdot)$ together with an optimal trajectory $x(\cdot)$ are required to follow a fractional dynamic and, at the same time, optimize an objective functional. Again, classical techniques are generalized to derive necessary optimality conditions. Euler–Lagrange equations have been introduced, e.g., in [2]. A Hamiltonian formalism for fractional optimal control problems can be found in [9] that exactly follows the same procedure of the regular optimal control theory, i.e., those with only integer-order derivatives.

Due to the growing number of applications of fractional calculus in science and engineering (see, e.g., [11, 12, 33, 34]), numerical methods are being developed to provide tools for solving such problems. Using the Grünwald–Letnikov approach, it is convenient to approximate the fractional differentiation operator, D^α, by generalized finite differences. In [25] some problems have been solved by this approximation. In [13] a predictor-corrector method is presented that converts an initial value problem into an equivalent Volterra integral equation, while [20] shows the use of numerical methods to solve such integral equations. A good survey on numerical methods for fractional differential equations can be found in [16].

A numerical scheme to solve fractional differential equations has been introduced in [7, 8], and [17], making an adaptation, uses this technique to solve fractional optimal control problems. The scheme is based on an expansion formula to approximate the Riemann–Liouville fractional derivative. The approximations transform fractional derivatives into finite sums containing only derivatives of integer order.

In this chapter, we try to analyze problems for which an analytic solution is available. This approach gives us the ability of measuring the accuracy of each method. To this end, we need to measure how close we get to the exact solutions. We use the L^2-norm and define the error function $E[x(\cdot), \tilde{x}(\cdot)]$ by

$$E = \|x(\cdot) - \tilde{x}(\cdot)\|_2 = \left(\int_a^b [x(t) - \tilde{x}(t)]^2 dt \right)^{\frac{1}{2}},$$

where $x(\cdot)$ is defined on $[a, b]$.

1.3. A General Formulation

The appearance of fractional terms of different types, derivatives and integrals, and the fact that there are several definitions for such operators, makes it difficult to present a typical problem to represent all possibilities. Nevertheless, one can consider the optimization of functionals of the form

$$J[\mathbf{x}(\cdot)] = \int_a^b L(t, \mathbf{x}(t), D^{\boldsymbol{\alpha}}\mathbf{x}(t))dt \tag{5}$$

that depends on a fractional derivative, $D^{\boldsymbol{\alpha}}$, in which $\mathbf{x} = (x_1, x_2, \ldots, x_n)$, $\boldsymbol{\alpha} = (\alpha_1, \alpha_2, \ldots, \alpha_n)$ and α_i, $i = 1, 2, \ldots, n$, are arbitrary real positive numbers. The problem can be with or without boundary conditions. Many settings of fractional variational and optimal control problems can be transformed to the optimization of (5). Constraints that usually appear in the calculus of variations and are always present in optimal control problems can be included in the functional using Lagrange multipliers. More precisely, in presence of dynamic constraints as fractional differential equations, we assume that it is possible to transform such equations to a vector fractional differential equation of the form

$$D^{\boldsymbol{\alpha}}\mathbf{x}(t) = f(t, \mathbf{x}(t)).$$

In this stage, we introduce a new variable $\boldsymbol{\lambda} = (\lambda_1, \lambda_2, \dots, \lambda_n)$ and consider the optimization of

$$J[\mathbf{x}(\cdot)] = \int_a^b \left[L(t, \mathbf{x}(t), D^{\alpha}\mathbf{x}(t)) + \boldsymbol{\lambda}(t) D^{\alpha}\mathbf{x}(t) - \boldsymbol{\lambda}(t) f(t, \mathbf{x}(t))\right] dt.$$

When the problem depends on fractional integrals, I^{α}, a new variable can be defined as $z(t) = I^{\alpha}x(t)$. Recall that $D^{\alpha}I^{\alpha}x = x$ (see, e.g., [19]). The equation

$$D^{\alpha}z(t) = D^{\alpha}I^{\alpha}x(t) = x(t)$$

can be regarded as an extra constraint to be added to the original problem. However, problems containing fractional integrals can be treated directly to avoid the complexity of adding an extra variable to the original problem. Interested readers are addressed to [4, 28].

Throughout this chapter, by a fractional variational problem, we mainly consider the following one-variable problem with given boundary conditions:

$$J[x(\cdot)] = \int_a^b L(t, x(t), D^{\alpha}x(t))dt \to \min$$

$$s.t. \begin{cases} x(a) = x_a, \\ x(b) = x_b. \end{cases}$$

In this setting D^{α} can be replaced by any fractional operator that is available in the literature, say, Riemann–Liouville, Caputo, Grünwald–Letnikov, Hadamard and so forth. The inclusion of constraints is done by Lagrange multipliers. The transition from this problem to the general one, equation (5), is straightforward and is not discussed here.

1.4. Solution Methods

There are two main approaches to solve variational, including optimal control, problems. On the one hand, there are direct methods. In a branch of direct methods, the problem is discretized over a mesh on the interested time interval. Discrete values of the unknown function on mesh points, finite differences for derivatives, and, finally, a quadrature rule for the integral, are used. This procedure reduces the variational problem, a continuous dynamic optimization problem, to static multi-variable optimization. Better accuracies are achieved by refining the underlying mesh size. Another class of direct methods uses function approximation through a linear combination of the elements of a certain basis, e.g., power series. The problem is then transformed into the determination of the unknown coefficients. To get better results in this sense, is the matter of using more adequate or higher order function approximations.

On the other hand, there are indirect methods that reduce a variational problem to the solution of a differential equation by applying some necessary optimality conditions. Euler–Lagrange equations and Pontryagin's maximum principle are used, in this context, to make the transformation process. Once we solve the resulting differential equation, an extremal for the original problem is reached. Therefore, to reach better results using indirect methods, one has to employ powerful integrators. It is worth, however, to mention here that numerical methods are usually used to solve practical problems.

These two methods have been generalized to cover fractional problems, which is the essential subject of this chapter.

2. Expansion Formulas to Approximate Fractional Derivatives

This section is devoted to present two approximations for the Riemann–Liouville, Caputo and Hadamard derivatives that are referred as fractional operators afterwards. We introduce the expansions of fractional operators in terms of infinite sums involving only integer order derivatives. These expansions are then used to approximate fractional operators in problems of the fractional calculus of variations and fractional optimal control. In this way, one can transform such problems into classical variational or optimal control problems. Hereafter, a suitable method, that can be found in the classical literature, is employed to find an approximated solution for the original fractional problem. Here we focus mainly on the left derivatives and the details of extracting corresponding expansions for right derivatives are given whenever it is needed to apply new techniques.

2.1. Riemann–Liouville Derivative

2.1.1. Approximation by a Sum of Integer Order Derivatives

Recall the definition of the left Riemann–Liouville derivative for $\alpha \in (0,1)$,

$$ {}_aD_t^\alpha x(t) = \frac{1}{\Gamma(1-\alpha)} \frac{d}{dt} \int_a^t (t-\tau)^{-\alpha} x(\tau) d\tau. \tag{6} $$

The following theorem holds for any function $x(\cdot)$ that is analytic in an interval $(c,d) \supset [a,b]$. See [6] for a more detailed discussion and [32] for a different proof.

Theorem 2.1. *Let (c,d), $-\infty < c < d < +\infty$, be an open interval in $\mathbb{R}$, and $[a,b] \subset (c,d)$ be such that for each $t \in [a,b]$ the closed ball $B_{b-a}(t)$, with center at t and radius $b-a$, lies in (c,d). If $x(\cdot)$ is analytic in (c,d), then*

$$ {}_aD_t^\alpha x(t) = \sum_{k=0}^{\infty} \frac{(-1)^{k-1} \alpha x^{(k)}(t)}{k!(k-\alpha)\Gamma(1-\alpha)} (t-a)^{k-\alpha}. \tag{7} $$

Proof. Since $x(t)$ is analytic in (c,d), and $B_{b-a}(t) \subset (c,d)$ for any $\tau \in (a,t)$ with $t \in (a,b)$, the Taylor expansion of $x(\tau)$ at t is a convergent power series, i.e.,

$$ x(\tau) = x(t-(t-\tau)) = \sum_{k=0}^{\infty} \frac{(-1)^k x^{(k)}(t)}{k!} (t-\tau)^k, $$

and then, by (6),

$$ {}_aD_t^\alpha x(t) = \frac{1}{\Gamma(1-\alpha)} \frac{d}{dt} \int_a^t \left((t-\tau)^{-\alpha} \sum_{k=0}^{\infty} \frac{(-1)^k x^{(k)}(t)}{k!} (t-\tau)^k \right) d\tau. \tag{8} $$

Since $(t-\tau)^{k-\alpha}x^{(k)}(t)$ is analytic, we can interchange integration with summation, so

$$\begin{aligned}
{}_aD_t^\alpha x(t) &= \frac{1}{\Gamma(1-\alpha)}\frac{d}{dt}\left(\sum_{k=0}^{\infty}\frac{(-1)^k x^{(k)}(t)}{k!}\int_a^t (t-\tau)^{k-\alpha}d\tau\right)\\
&= \frac{1}{\Gamma(1-\alpha)}\frac{d}{dt}\sum_{k=0}^{\infty}\left(\frac{(-1)^k x^{(k)}(t)}{k!(k+1-\alpha)}(t-a)^{k+1-\alpha}\right)\\
&= \frac{1}{\Gamma(1-\alpha)}\sum_{k=0}^{\infty}\left(\frac{(-1)^k x^{(k+1)}(t)}{k!(k+1-\alpha)}(t-a)^{k+1-\alpha}+\frac{(-1)^k x^{(k)}(t)}{k!}(t-a)^{k-\alpha}\right)\\
&= \frac{x(t)}{\Gamma(1-\alpha)}(t-a)^{-\alpha}\\
&\quad+\frac{1}{\Gamma(1-\alpha)}\sum_{k=1}^{\infty}\left(\frac{(-1)^{k-1}}{(k-\alpha)(k-1)!}+\frac{(-1)^k}{k!}\right)x^{(k)}(t)(t-a)^{k-\alpha}.
\end{aligned}$$

Observe that

$$\begin{aligned}
\frac{(-1)^{k-1}}{(k-\alpha)(k-1)!}+\frac{(-1)^k}{k!} &= \frac{k(-1)^{k-1}+k(-1)^k-\alpha(-1)^k}{(k-\alpha)k!}\\
&= \frac{(-1)^{k-1}\alpha}{(k-\alpha)k!},
\end{aligned}$$

since for any $k=0,1,2,\ldots$ we have $k(-1)^{k-1}+k(-1)^k=0$. Therefore, the expansion formula is reached as required. □

For numerical purposes, a finite number of terms in (7) is used and one has

$$ {}_aD_t^\alpha x(t) \approx \sum_{k=0}^{N}\frac{(-1)^{k-1}\alpha x^{(k)}(t)}{k!(k-\alpha)\Gamma(1-\alpha)}(t-a)^{k-\alpha}. \tag{9}$$

Remark 2.2. *With the same assumptions of Theorem 2.1, we can expand $x(\tau)$ at t,*

$$x(\tau)=x(t+(\tau-t))=\sum_{k=0}^{\infty}\frac{x^{(k)}(t)}{k!}(\tau-t)^k,$$

where $\tau\in(t,b)$. Similar calculations result in the following approximation for the right Riemann–Liouville derivative:

$$ {}_tD_b^\alpha x(t) \approx \sum_{k=0}^{N}\frac{-\alpha x^{(k)}(t)}{k!(k-\alpha)\Gamma(1-\alpha)}(b-t)^{k-\alpha}.$$

A proof for this expansion is available at [32] that uses a similar relation for fractional integrals. The proof discussed here, however, allows to extract an error term for this expansion easily.

2.1.2. Approximation Using Moments of a Function

By moments of a function, we have no physical or distributive sense in mind. The naming comes from the fact that, during expansion, the terms of the form

$$V_p(t) := V_p(x(t)) = (1-p)\int_a^t (\tau - a)^{p-2} x(\tau)d\tau, \quad p \in \mathbb{N}, \quad \tau \geq a, \tag{10}$$

resemble the formulas of central moments (cf. [8]). We assume that $V_p(x(\cdot))$, $p \in \mathbb{N}$, denotes the $(p-2)$th moment of a function $x(\cdot) \in AC^2[a,b]$.

The following lemma, that is given here without a proof, is the key relation to extract an expansion formula for Riemann–Liouville derivatives.

Lemma 2.3 (cf. Lemma 2.12 of [12]). *Let $x(\cdot) \in AC[a,b]$ and $0 < \alpha < 1$. Then the left Riemann–Liouville fractional derivative, ${}_aD_t^\alpha x(\cdot)$, exists almost everywhere in $[a,b]$. Moreover, ${}_aD_t^\alpha x(\cdot) \in L_p[a,b]$ for $1 \leq p < \frac{1}{\alpha}$ and*

$${}_aD_t^\alpha x(t) = \frac{1}{\Gamma(1-\alpha)}\left[\frac{x(a)}{(t-a)^\alpha} + \int_a^t (t-\tau)^{-\alpha}\dot{x}(\tau)d\tau\right], \qquad t \in (a,b). \tag{11}$$

The same argument is valid for the right Riemann–Liouville derivative and

$${}_tD_b^\alpha x(t) = \frac{1}{\Gamma(1-\alpha)}\left[\frac{x(b)}{(b-t)^\alpha} - \int_t^b (\tau-t)^{-\alpha}\dot{x}(\tau)d\tau\right], \qquad t \in (a,b).$$

Theorem 2.4 (cf. [7]). *With the same assumptions of Lemma 2.3, the left Riemann–Liouville derivative can be expanded as*

$$\begin{aligned}{}_aD_t^\alpha x(t) &= \frac{(t-a)^{-\alpha}}{\Gamma(1-\alpha)}x(t) + B(\alpha)(t-a)^{1-\alpha}\dot{x}(t)\\ &\quad - \sum_{p=2}^{\infty}\left[C(\alpha,p)(t-a)^{1-p-\alpha}V_p(t) - \frac{\Gamma(p-1+\alpha)}{\Gamma(\alpha)\Gamma(1-\alpha)(p-1)!}(t-a)^{-\alpha}x(t)\right],\end{aligned} \tag{12}$$

where $V_p(t)$ is defined by (10) *and*

$$\begin{aligned} B(\alpha) &= \frac{1}{\Gamma(2-\alpha)}\left(1 + \sum_{p=1}^{\infty}\frac{\Gamma(p-1+\alpha)}{\Gamma(\alpha-1)p!}\right),\\ C(\alpha,p) &= \frac{1}{\Gamma(2-\alpha)\Gamma(\alpha-1)}\frac{\Gamma(p-1+\alpha)}{(p-1)!}.\end{aligned}$$

Proof. Integration by parts on the right-hand-side of (11) gives

$$\begin{aligned}{}_aD_t^\alpha x(t) &= \frac{x(a)}{\Gamma(1-\alpha)}(t-a)^{-\alpha} + \frac{\dot{x}(a)}{\Gamma(2-\alpha)}(t-a)^{1-\alpha}\\ &\quad + \frac{1}{\Gamma(2-\alpha)}\int_a^t (t-\tau)^{1-\alpha}\ddot{x}(\tau)d\tau.\end{aligned} \tag{13}$$

Since $(\tau - a) \leq (t - a)$,

$$(t-\tau)^{1-\alpha} = (t-a)^{1-\alpha}\left(1 - \frac{\tau - a}{t - a}\right)^{1-\alpha}.$$

Using the binomial theorem, we have

$$\left(1 - \frac{\tau - a}{t - a}\right)^{1-\alpha} = \sum_{p=0}^{\infty} \frac{\Gamma(p-1+\alpha)}{\Gamma(\alpha-1)p!}\left(\frac{\tau - a}{t - a}\right)^{p},$$

in which the infinite series converges. Replacing for $(t-\tau)^{1-\alpha}$ in (13) gives

$$\begin{aligned}
{}_aD_t^\alpha x(t) &= \frac{x(a)}{\Gamma(1-\alpha)}(t-a)^{-\alpha} + \frac{\dot{x}(a)}{\Gamma(2-\alpha)}(t-a)^{1-\alpha} \\
&+ \frac{(t-a)^{1-\alpha}}{\Gamma(2-\alpha)} \int_a^t \left(\sum_{p=0}^{\infty} \frac{\Gamma(p-1+\alpha)}{\Gamma(\alpha-1)p!}\left(\frac{\tau - a}{t - a}\right)^{p}\right) \ddot{x}(\tau)d\tau, \quad t > a.
\end{aligned}$$

Interchanging the summation and integration operations is possible, and yields

$$\begin{aligned}
{}_aD_t^\alpha x(t) &= \frac{x(a)}{\Gamma(1-\alpha)}(t-a)^{-\alpha} + \frac{\dot{x}(a)}{\Gamma(2-\alpha)}(t-a)^{1-\alpha} \\
&+ \frac{(t-a)^{1-\alpha}}{\Gamma(2-\alpha)} \sum_{p=0}^{\infty} \frac{\Gamma(p-1+\alpha)}{\Gamma(\alpha-1)p!(t-a)^p} \int_a^t (\tau-a)^p \ddot{x}(\tau)d\tau, \quad t > a.
\end{aligned}$$

Decomposing the infinite sum, integrating, and doing another integration by parts, allow us to write

$$\begin{aligned}
{}_aD_t^\alpha x(t) &= \frac{x(a)}{\Gamma(1-\alpha)}(t-a)^{-\alpha} + \frac{\dot{x}(a)}{\Gamma(2-\alpha)}(t-a)^{1-\alpha} + \frac{(t-a)^{1-\alpha}}{\Gamma(2-\alpha)} \int_a^t \ddot{x}(\tau)d\tau \\
&+ \frac{(t-a)^{1-\alpha}}{\Gamma(2-\alpha)} \sum_{p=1}^{\infty} \frac{\gamma(\alpha,p)}{p!(t-a)^p} \left[(t-a)^p \dot{x}(t) - p\int_a^t (\tau-a)^{p-1}\dot{x}(\tau)d\tau\right] \\
&= \frac{x(a)}{\Gamma(1-\alpha)}(t-a)^{-\alpha} + \frac{\dot{x}(t)}{\Gamma(2-\alpha)}(t-a)^{1-\alpha} + \frac{(t-a)^{1-\alpha}}{\Gamma(2-\alpha)} \sum_{p=1}^{\infty} \frac{\gamma(\alpha,p)}{p!}\dot{x}(t) \\
&+ \frac{(t-a)^{1-\alpha}}{\Gamma(2-\alpha)} \sum_{p=1}^{\infty} \frac{\gamma(\alpha,p)}{(p-1)!(t-a)^p} \int_a^t (\tau-a)^{p-1}\dot{x}(\tau)d\tau,
\end{aligned}$$

where

$$\gamma(\alpha,p) = \frac{\Gamma(p-1+\alpha)}{\Gamma(\alpha-1)}.$$

Repeating this procedure again, and simplifying the results, ends the proof. □

The moments $V_p(t)$, $p = 2, 3, \ldots$, can be regarded as the solutions to the following system of differential equations:

$$\begin{cases} \dot{V}_p(t) = (1-p)(t-a)^{p-2}x(t) \\ V_p(a) = 0, \qquad p = 2, 3, \ldots \end{cases} \tag{14}$$

As before, a numerical approximation is achieved by taking only a finite number of terms in the series (12). We approximate the fractional derivative as

$$ {}_aD_t^\alpha x(t) \approx A(t-a)^{-\alpha}x(t) + B(t-a)^{1-\alpha}\dot{x}(t) - \sum_{p=2}^{N} C(\alpha,p)(t-a)^{1-p-\alpha}V_p(t), \quad (15) $$

where $A = A(\alpha, N)$ and $B = B(\alpha, N)$ are given by

$$ A(\alpha,N) = \frac{1}{\Gamma(1-\alpha)}\left(1+\sum_{p=2}^{N}\frac{\Gamma(p-1+\alpha)}{\Gamma(\alpha)(p-1)!}\right), \quad (16) $$

$$ B(\alpha,N) = \frac{1}{\Gamma(2-\alpha)}\left(1+\sum_{p=1}^{N}\frac{\Gamma(p-1+\alpha)}{\Gamma(\alpha-1)p!}\right). \quad (17) $$

Remark 2.5. *This expansion has been proposed in [14] and a simplification has been made in [8], which uses the fact that the infinite series* $\sum_{p=1}^{\infty}\frac{\Gamma(p-1+\alpha)}{\Gamma(\alpha-1)p!}$ *tends to* -1*, and concludes that* $B(\alpha) = 0$*, and thus*

$$ {}_aD_t^\alpha x(t) \approx A(\alpha,N)t^{-\alpha}x(t) - \sum_{p=2}^{N} C(\alpha,p)t^{1-p-\alpha}V_p(t). \quad (18) $$

In practice, however, we only use a finite number of terms in series. Therefore,

$$ 1+\sum_{p=1}^{N}\frac{\Gamma(p-1+\alpha)}{\Gamma(\alpha-1)p!} \neq 0, $$

and we keep here the approximation in the form of equation (15)*, [3]. To be more precise, the values of* $B(\alpha, N)$*, for different choices of* N *and* α*, are given in Table 1. It shows that even for a large* N*, when* α *tends to one,* $B(\alpha, N)$ *cannot be ignored.*

Table 1. $B(\alpha, N)$ **for different values of** α **and** N

N	4	7	15	30	70	120	170
$B(0.1, N)$	0.0310	0.0188	0.0095	0.0051	0.0024	0.0015	0.0011
$B(0.3, N)$	0.1357	0.0928	0.0549	0.0339	0.0188	0.0129	0.0101
$B(0.5, N)$	0.3085	0.2364	0.1630	0.1157	0.0760	0.0581	0.0488
$B(0.7, N)$	0.5519	0.4717	0.3783	0.3083	0.2396	0.2040	0.1838
$B(0.9, N)$	0.8470	0.8046	0.7481	0.6990	0.6428	0.6092	0.5884
$B(0.99, N)$	0.9849	0.9799	0.9728	0.9662	0.9582	0.9531	0.9498

Remark 2.6. *Similar computations give rise to an expansion formula for* ${}_tD_b^\alpha$*, the right Riemann–Liouville fractional derivative:*

$$ {}_tD_b^\alpha x(t) \approx A(b-t)^{-\alpha}x(t) - B(b-t)^{1-\alpha}\dot{x}(t) - \sum_{p=2}^{N} C(\alpha,p)(b-t)^{1-p-\alpha}W_p(t), $$

where

$$W_p(t) = (1-p)\int_t^b (b-\tau)^{p-2}x(\tau)d\tau.$$

The coefficients $A = A(\alpha, N)$ *and* $B = B(\alpha, N)$ *are the same as* (16) *and* (17) *respectively, and* $C(\alpha, p)$ *is as before.*

Remark 2.7. *As stated before, Caputo derivatives are closely related to those of Riemann–Liouville. For any function,* $x(\cdot)$*, and for* $\alpha \in (0,1)$ *for which these two kind of fractional derivatives, left and right, exist, we have*

$${}_a^C D_t^\alpha x(t) = {}_a D_t^\alpha x(t) - \frac{x(a)}{(t-a)^\alpha},$$

and

$${}_t^C D_b^\alpha x(t) = {}_t D_b^\alpha x(t) - \frac{x(b)}{(b-t)^\alpha}.$$

Using these relations, we can easily construct approximation formulas for the left and right Caputo fractional derivatives, e.g.,

$$\begin{aligned} {}_a^C D_t^\alpha x(t) \approx\ & A(\alpha, N)(t-a)^{-\alpha}x(t) + B(\alpha, N)(t-a)^{1-\alpha}\dot{x}(t) \\ & - \sum_{p=2}^{N} C(\alpha, p)(t-a)^{1-p-\alpha}V_p(t) - \frac{x(a)}{(t-a)^\alpha}. \end{aligned}$$

2.1.3. Examples

To examine the approximations provided so far, we take some test functions, and apply (9) and (15) to evaluate their fractional derivatives. We compute ${}_a D_t^\alpha x(t)$, with $\alpha = \frac{1}{2}$, for $x(t) = t^4$ and $x(t) = e^{2t}$. The exact formulas for the fractional derivatives of polynomials are derived from

$${}_0 D_t^{0.5}(t^n) = \frac{\Gamma(n+1)}{\Gamma(n+1-0.5)} t^{n-0.5},$$

and for the exponential function one has

$${}_0 D_t^{0.5}(e^{\lambda t}) = t^{-0.5} E_{1,1-0.5}(\lambda t),$$

where $E_{\alpha,\beta}$ is the two parameter Mittag–Leffler function (1).

Figure 1 shows the results using approximation (9). As we can see, the third approximations are reasonably accurate for both cases. Indeed, for $x(t) = t^4$, the approximation with $N = 4$ coincides with the exact solution because the derivatives of order five and more vanish.

Now we use approximation (15) to evaluate fractional derivatives of the same test functions. In this case, for a given function $x(\cdot)$, we can compute V_p by definition, equation (10). One can also integrate the system (14) analytically, if possible, or use any numerical integrator. It is clearly seen in Figure 2 that one can get better results by using larger values of N. Comparing Figures 1 and 2, we find out that the approximation (9) shows a faster convergence. Observe that both functions are analytic and it is easy to compute higher-order derivatives.

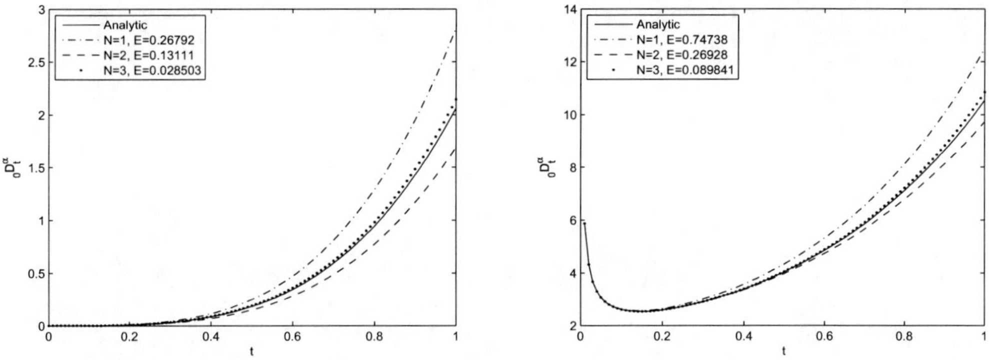

(a) ${}_0D_t^{0.5}(t^4)$ (b) ${}_0D_t^{0.5}(e^{2t})$

Figure 1. Analytic (solid line) versus numerical approximation (9).

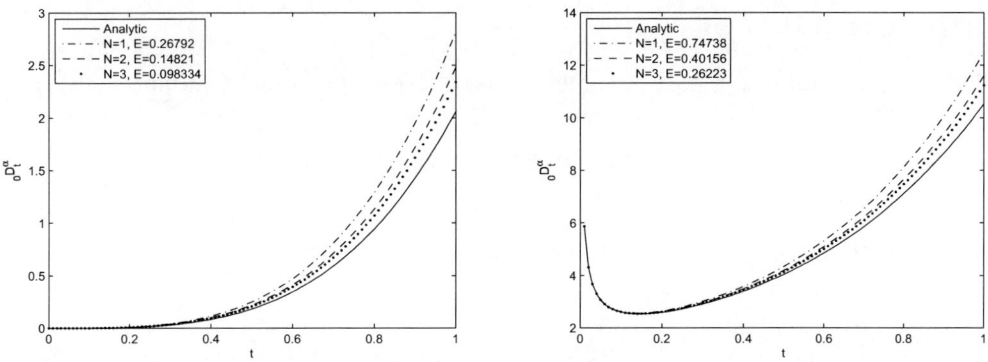

(a) ${}_0D_t^{0.5}(t^4)$ (b) ${}_0D_t^{0.5}(e^{2t})$

Figure 2. Analytic (solid line) versus numerical approximation (15).

Remark 2.8. *A closer look to* (9) *and* (15) *reveals that in both cases the approximations are not computable at* a *and* b *for the left and right fractional derivatives, respectively. At these points we assume that it is possible to extend them continuously to the closed interval* $[a, b]$.

Following Remark 2.5, we show here that neglecting the first derivative in the expansion (15) can cause a considerable loss of accuracy in computation. Once again, we compute the fractional derivatives of $x(t) = t^4$ and $x(t) = e^{2t}$, but this time we use the approximation given by (18). Figure 3 summarizes the results. Approximation (15) gives a more realistic approximation using quite small N, 3 in this case.

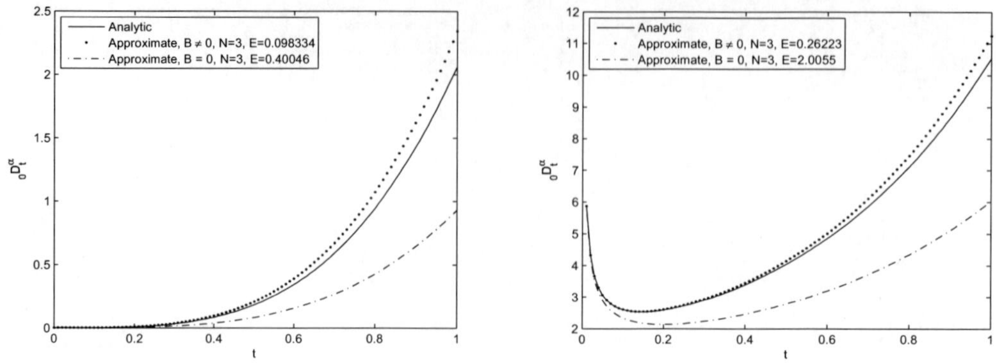

(a) ${}_0D_t^{0.5}(t^4)$ (b) ${}_0D_t^{0.5}(e^{2t})$

Figure 3. Comparison of approximation (15) and approximation (18) of [8].

2.2. Hadamard Derivatives

For Hadamard derivatives, the expansions can be obtained in a quiet similar way [27].

2.2.1. Approximation by a Sum of Integer Order Derivatives

Assume that a function $x(\cdot)$ admits derivatives of any order, then expansion formulas for the Hadamard fractional integrals and derivatives of x, in terms of its integer-order derivatives, are given in [10, Theorem 17]:

$$ {}_0\mathcal{I}_t^{\alpha} x(t) = \sum_{k=0}^{\infty} S(-\alpha, k) t^k x^{(k)}(t) $$

and

$$ {}_0\mathcal{D}_t^{\alpha} x(t) = \sum_{k=0}^{\infty} S(\alpha, k) t^k x^{(k)}(t), $$

where

$$ S(\alpha, k) = \frac{1}{k!} \sum_{j=1}^{k} (-1)^{k-j} \binom{k}{j} j^{\alpha} $$

is the Stirling function.

As approximations, we truncate infinite sums at an appropriate order N and get the following formulas:

$$ {}_0\mathcal{I}_t^{\alpha} x(t) \approx \sum_{k=0}^{N} S(-\alpha, k) t^k x^{(k)}(t), $$

and

$$ {}_0\mathcal{D}_t^{\alpha} x(t) \approx \sum_{k=0}^{N} S(\alpha, k) t^k x^{(k)}(t). $$

2.2.2. Approximation Using Moments of a Function

The same idea of expanding Riemann–Liouville derivatives, with slightly different techniques, is used to derive expansion formulas for left and right Hadamard derivatives. The following lemma is a basis for these new relations.

Lemma 2.9. *Let $\alpha \in (0,1)$ and $x(\cdot)$ be an absolutely continuous function on $[a,b]$. Then the Hadamard fractional derivatives may be expressed by*

$$ {}_a\mathcal{D}_t^\alpha x(t) = \frac{x(a)}{\Gamma(1-\alpha)}\left(\ln\frac{t}{a}\right)^{-\alpha} + \frac{1}{\Gamma(1-\alpha)}\int_a^t \left(\ln\frac{t}{\tau}\right)^{-\alpha} \dot{x}(\tau)d\tau \tag{19}$$

and

$$ {}_t\mathcal{D}_b^\alpha x(t) = \frac{x(b)}{\Gamma(1-\alpha)}\left(\ln\frac{b}{t}\right)^{-\alpha} - \frac{1}{\Gamma(1-\alpha)}\int_t^b \left(\ln\frac{\tau}{t}\right)^{-\alpha} \dot{x}(\tau)d\tau.$$

A proof of this lemma, for an arbitrary $\alpha > 0$, can be found in [18, Theorem 3.2].

Theorem 2.10. *Let $0 < a < b$ and $x : [a,b] \to \mathbb{R}$ be an absolutely continuous function. Then*

$$ \begin{aligned} {}_a\mathcal{D}_t^\alpha x(t) &= \frac{1}{\Gamma(1-\alpha)}\left(\ln\frac{t}{a}\right)^{-\alpha} x(t) + B(\alpha)\left(\ln\frac{t}{a}\right)^{1-\alpha} t\dot{x}(t) \\ &- \sum_{p=2}^{\infty}\left[C(\alpha,p)\left(\ln\frac{t}{a}\right)^{1-\alpha-p} V_p(t) - \frac{\Gamma(p+\alpha-1)}{\Gamma(\alpha)\Gamma(1-\alpha)(p-1)!}\left(\ln\frac{t}{a}\right)^{-\alpha} x(t)\right] \end{aligned}$$

with

$$ \begin{aligned} B(\alpha) &= \frac{1}{\Gamma(2-\alpha)}\left(1+\sum_{p=1}^{\infty}\frac{\Gamma(p+\alpha-1)}{\Gamma(\alpha\ \ 1)p!}\right), \\ C(\alpha,p) &= \frac{\Gamma(p+\alpha-1)}{\Gamma(-\alpha)\Gamma(1+\alpha)(p-1)!}, \\ V_p(t) &= (1-p)\int_a^t \left(\ln\frac{\tau}{a}\right)^{p-2}\frac{x(\tau)}{\tau}d\tau. \end{aligned}$$

Proof. We rewrite (19) as

$$ {}_a\mathcal{D}_t^\alpha x(t) = \frac{x(a)}{\Gamma(1-\alpha)}\left(\ln\frac{t}{a}\right)^{-\alpha} + \frac{1}{\Gamma(1-\alpha)}\int_a^t \frac{1}{\tau}\left(\ln\frac{t}{\tau}\right)^{-\alpha} \tau\dot{x}(\tau)d\tau$$

and then integrating by parts gives

$$ \begin{aligned} {}_a\mathcal{D}_t^\alpha x(t) &= \frac{x(a)}{\Gamma(1-\alpha)}\left(\ln\frac{t}{a}\right)^{-\alpha} + \frac{a\dot{x}(a)}{\Gamma(2-\alpha)}\left(\ln\frac{t}{a}\right)^{1-\alpha} \\ &+ \frac{1}{\Gamma(2-\alpha)}\int_a^t \left(\ln\frac{t}{\tau}\right)^{1-\alpha}[\dot{x}(\tau)+\tau\ddot{x}(\tau)]d\tau. \end{aligned}$$

Now we use the following expansion for $\left(\ln\frac{t}{\tau}\right)^{1-\alpha}$, using the binomial theorem,

$$\begin{aligned}\left(\ln\frac{t}{\tau}\right)^{1-\alpha} &= \left(\ln\frac{t}{a}\right)^{1-\alpha}\left(1-\frac{\ln\frac{\tau}{a}}{\ln\frac{t}{a}}\right)^{1-\alpha}\\ &= \left(\ln\frac{t}{a}\right)^{1-\alpha}\sum_{p=0}^{\infty}\frac{\Gamma(p-1+\alpha)}{\Gamma(\alpha-1)p!}\cdot\frac{(\ln\frac{\tau}{a})^p}{(\ln\frac{t}{a})^p}.\end{aligned}$$

This implies that

$$\begin{aligned}{}_a\mathcal{D}_t^{\alpha}x(t) &= \frac{x(a)}{\Gamma(1-\alpha)}\left(\ln\frac{t}{a}\right)^{-\alpha}+\frac{a\dot{x}(a)}{\Gamma(2-\alpha)}\left(\ln\frac{t}{a}\right)^{1-\alpha}+\frac{1}{\Gamma(2-\alpha)}\left(\ln\frac{t}{a}\right)^{1-\alpha}\\ &\quad\times\sum_{p=0}^{\infty}\frac{\Gamma(p-1+\alpha)}{\Gamma(\alpha-1)p!}\left(\ln\frac{t}{a}\right)^{-p}\int_a^t\left(\ln\frac{\tau}{a}\right)^p[\dot{x}(\tau)+\tau\ddot{x}(\tau)]d\tau.\end{aligned}$$

Extracting the first term of the infinite sum, simplifications and another integration by parts using $u=\left(\ln\frac{\tau}{a}\right)^p$, $du=(p)\frac{1}{\tau}\left(\ln\frac{\tau}{a}\right)^{p-1}$ and $dv=[\dot{x}(\tau)+\tau\ddot{x}(\tau)]d\tau$, $v=\tau\dot{x}(\tau)$ yields

$$\begin{aligned}{}_a\mathcal{D}_t^{\alpha}x(t) &= \frac{x(a)}{\Gamma(1-\alpha)}\left(\ln\frac{t}{a}\right)^{-\alpha}+B(\alpha)\left(\ln\frac{t}{a}\right)^{1-\alpha}t\dot{x}(t)-\frac{1}{\Gamma(2-\alpha)}\left(\ln\frac{t}{a}\right)^{1-\alpha}\\ &\quad\times\sum_{p=1}^{\infty}\frac{\Gamma(p-1+\alpha)}{\Gamma(\alpha-1)(p-1)!}\left(\ln\frac{t}{a}\right)^{-p}\int_a^t\left(\ln\frac{\tau}{a}\right)^{p-1}\dot{x}(\tau)d\tau.\end{aligned}$$

A final step of extracting the first term in the sum and integration by parts finishes the proof. □

For practical purposes, finite sums up to order N are considered and the approximation becomes

$$\begin{aligned}{}_a\mathcal{D}_t^{\alpha}x(t) &\approx A(\alpha,N)\left(\ln\frac{t}{a}\right)^{-\alpha}x(t)+B(\alpha,N)\left(\ln\frac{t}{a}\right)^{1-\alpha}t\dot{x}(t)\\ &\quad+\sum_{p=2}^{N}C(\alpha,p)\left(\ln\frac{t}{a}\right)^{1-\alpha-p}V_p(t)\end{aligned}\tag{20}$$

with

$$\begin{aligned}A(\alpha,N) &= \frac{1}{\Gamma(1-\alpha)}\left(1+\sum_{p=2}^{N}\frac{\Gamma(p+\alpha-1)}{\Gamma(\alpha)(p-1)!}\right),\\ B(\alpha,N) &= \frac{1}{\Gamma(2-\alpha)}\left(1+\sum_{p=1}^{N}\frac{\Gamma(p+\alpha-1)}{\Gamma(\alpha-1)p!}\right).\end{aligned}$$

Remark 2.11. *The right Hadamard fractional derivative can be expanded in the same way. This gives the following approximation:*

$$
\begin{aligned}
{}_t\mathcal{D}_b^\alpha x(t) \quad \approx \quad & A(\alpha, N)\left(\ln \frac{b}{t}\right)^{-\alpha} x(t) - B(\alpha, N)\left(\ln \frac{b}{t}\right)^{1-\alpha} t\dot{x}(t) \\
& - \sum_{p=2}^{N} C(\alpha, p)\left(\ln \frac{b}{t}\right)^{1-\alpha-p} W_p(t)
\end{aligned}
$$

with

$$
W_p(t) = (1-p)\int_t^b \left(\ln \frac{b}{\tau}\right)^{p-2} \frac{x(\tau)}{\tau} d\tau.
$$

2.2.3. Examples

In this section we apply (20) to compute fractional derivatives, of order $\alpha = \frac{1}{2}$, for $x(t) = t^4$ and $x(t) = \ln(t)$. The exact Hadamard fractional derivative is available for $x(t) = t^4$ and we have

$$
{}_1\mathcal{D}_t^{0.5}(t^4) = \frac{\sqrt{\ln t}}{\Gamma(1.5)}.
$$

For $x(t) = \ln(t)$, only an approximation of the Hadamard fractional derivative is found in the literature:

$$
{}_1\mathcal{D}_t^{0.5} \ln(t) \approx \frac{1}{\Gamma(0.5)\sqrt{\ln t}} + \frac{0.5908179503}{\Gamma(0.5)} 9t^9 \operatorname{erf}(3\sqrt{\ln t}).
$$

The results of applying (20) to evaluate fractional derivatives are depicted in Figure 4.

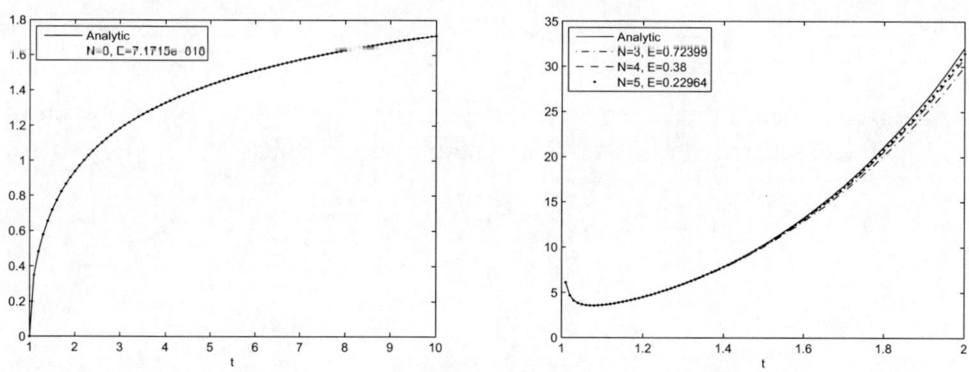

(a) ${}_1\mathcal{D}_t^{0.5}(\ln t)$ (b) ${}_1\mathcal{D}_t^{0.5}(t^4)$

Figure 4. Analytic versus numerical approximation (20).

2.2.4. Error Analysis

When we approximate an infinite series by a finite sum, the choice of the order of approximation is a key question. Having an estimate knowledge of truncation errors, one can choose properly up to which order the approximations should be made to suit the accuracy requirements. In this section we study the errors of the approximations presented so far.

Separation of an error term in (8) concludes in

$$
\begin{aligned}
{}_aD_t^\alpha x(t) &= \frac{1}{\Gamma(1-\alpha)}\frac{d}{dt}\int_a^t \left((t-\tau)^{-\alpha}\sum_{k=0}^{N}\frac{(-1)^k x^{(k)}(t)}{k!}(t-\tau)^k\right)d\tau \\
&\quad + \frac{1}{\Gamma(1-\alpha)}\frac{d}{dt}\int_a^t \left((t-\tau)^{-\alpha}\sum_{k=N+1}^{\infty}\frac{(-1)^k x^{(k)}(t)}{k!}(t-\tau)^k\right)d\tau. \qquad (21)
\end{aligned}
$$

The first term in (21) gives (9) directly and the second term is the error caused by truncation. The next step is to give a local upper bound for this error, $E_{tr}(t)$.

The series

$$
\sum_{k=N+1}^{\infty}\frac{(-1)^k x^{(k)}(t)}{k!}(t-\tau)^k, \quad \tau\in(a,t), \quad t\in(a,b),
$$

is the remainder of the Taylor expansion of $x(\tau)$ and thus bounded by $\left|\frac{M}{(N+1)!}(t-\tau)^{N+1}\right|$ in which

$$
M = \max_{\tau\in[a,t]}|x^{(N+1)}(\tau)|.
$$

Then,

$$
E_{tr}(t) \le \left|\frac{M}{\Gamma(1-\alpha)(N+1)!}\frac{d}{dt}\int_a^t (t-\tau)^{N+1-\alpha}d\tau\right| = \frac{M}{\Gamma(1-\alpha)(N+1)!}(t-a)^{N+1-\alpha}.
$$

In order to estimate a truncation error for approximation (15), the expansion procedure is carried out with separation of N terms in binomial expansion as

$$
\begin{aligned}
\left(1-\frac{\tau-a}{t-a}\right)^{1-\alpha} &= \sum_{p=0}^{\infty}\frac{\Gamma(p-1+\alpha)}{\Gamma(\alpha-1)p!}\left(\frac{\tau-a}{t-a}\right)^p \\
&= \sum_{p=0}^{N}\frac{\Gamma(p-1+\alpha)}{\Gamma(\alpha-1)p!}\left(\frac{\tau-a}{t-a}\right)^p + R_N(\tau), \qquad (22)
\end{aligned}
$$

where

$$
R_N(\tau) = \sum_{p=N+1}^{\infty}\frac{\Gamma(p-1+\alpha)}{\Gamma(\alpha-1)p!}\left(\frac{\tau-a}{t-a}\right)^p.
$$

Substituting (22) into (13), we get

$$\begin{aligned}
{}_aD_t^\alpha x(t) &= \frac{x(a)}{\Gamma(1-\alpha)}(t-a)^{-\alpha} + \frac{\dot{x}(a)}{\Gamma(2-\alpha)}(t-a)^{1-\alpha} \\
&+\frac{(t-a)^{1-\alpha}}{\Gamma(2-\alpha)}\int_a^t \left(\sum_{p=0}^{N}\frac{\Gamma(p-1+\alpha)}{\Gamma(\alpha-1)p!}\left(\frac{\tau-a}{t-a}\right)^p + R_N(\tau)\right)\ddot{x}(\tau)d\tau \\
&= \frac{x(a)}{\Gamma(1-\alpha)}(t-a)^{-\alpha} + \frac{\dot{x}(a)}{\Gamma(2-\alpha)}(t-a)^{1-\alpha} \\
&+\frac{(t-a)^{1-\alpha}}{\Gamma(2-\alpha)}\int_a^t \left(\sum_{p=0}^{N}\frac{\Gamma(p-1+\alpha)}{\Gamma(\alpha-1)p!}\left(\frac{\tau-a}{t-a}\right)^p\right)\ddot{x}(\tau)d\tau \\
&+\frac{(t-a)^{1-\alpha}}{\Gamma(2-\alpha)}\int_a^t R_N(\tau)\ddot{x}(\tau)d\tau.
\end{aligned}$$

At this point, we apply the techniques of [8] to the first three terms with finite sums. Then, we receive (15) with an extra term of truncation error:

$$E_{tr}(t) = \frac{(t-a)^{1-\alpha}}{\Gamma(2-\alpha)}\int_a^t R_N(\tau)\ddot{x}(\tau)d\tau.$$

Since $0 \leq \frac{\tau-a}{t-a} \leq 1$ for $\tau \in [a,t]$, one has

$$\begin{aligned}
|R_N(\tau)| &\leq \sum_{p=N+1}^{\infty}\left|\frac{\Gamma(p-1+\alpha)}{\Gamma(\alpha-1)p!}\right| = \sum_{p=N+1}^{\infty}\left|\binom{1-\alpha}{p}\right| \leq \sum_{p=N+1}^{\infty}\frac{e^{(1-\alpha)^2+1-\alpha}}{p^{2-\alpha}} \\
&\leq \int_{p=N}^{\infty}\frac{e^{(1-\alpha)^2+1-\alpha}}{p^{2-\alpha}}dp = \frac{e^{(1-\alpha)^2+1-\alpha}}{(1-\alpha)N^{1-\alpha}}.
\end{aligned}$$

Finally, assuming $L_2 = \max_{\tau\in[a,t]}\left|x^{(2)}(\tau)\right|$, we conclude that

$$|E_{tr}(t)| \leq L_2\frac{e^{(1-\alpha)^2+1-\alpha}}{\Gamma(2-\alpha)(1-\alpha)N^{1-\alpha}}(t-a)^{2-\alpha}.$$

Remark 2.12. *Following similar techniques, one can extract an error bound for the approximations of Hadamard derivatives. When we consider finite sums in* (20), *the error is bounded by*

$$|E_{tr}(t)| \leq L(t)\frac{e^{(1-\alpha)^2+1-\alpha}}{\Gamma(2-\alpha)(1-\alpha)N^{1-\alpha}}\left(\ln\frac{t}{a}\right)^{1-\alpha}(t-a),$$

where

$$L(t) = \max_{\tau\in[a,t]}|\dot{x}(\tau)+\tau\ddot{x}(\tau)|.$$

3. Direct Methods

There are two main classes of direct methods in the classical calculus of variations and optimal control. On the one hand, we specify a discretization scheme by choosing a set of mesh points on the horizon of interest, say $a = t_0, t_1, \ldots, t_n = b$ for $[a, b]$. Then we use some approximations for derivatives in terms of unknown function values at t_i and, using an appropriate quadrature, the problem is transformed to a finite dimensional optimization problem. This method is known as Euler's method in the literature [15]. Regarding Figure 5, the solid line is the function that we are looking for, nevertheless, the method gives the polygonal dashed line as an approximate solution.

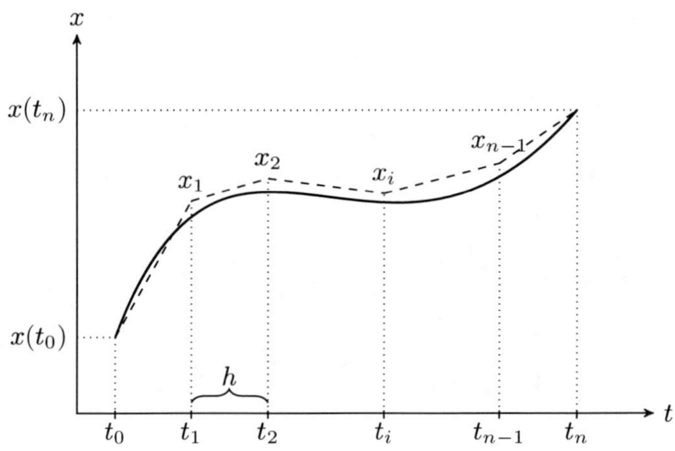

Figure 5. Euler's finite differences method.

On the other hand, there is the Ritz method, that has an extension to functionals of several independent variables which is called Kantorovich's method. We assume that the admissible functions can be expanded in some kind of series, e.g., power or Fourier's series, of the form

$$x(t) = \sum_{k=0}^{\infty} a_k \phi_k(t).$$

Using a finite number of terms in the sum as an approximation, and some sort of quadrature again, the original problem can be transformed to an equivalent optimization problem for a_k, $k = 0, 1, \ldots, n$.

In the presence of fractional operators, the same ideas are applied to discretize a problem. Many works can be found in the literature that use different types of basis functions to establish Ritz-like methods for fractional calculus of variations and optimal control.

3.1. Euler-like Methods

The Euler method in the classical theory of the calculus of variations uses finite differences approximations for derivatives and is referred also as the method of finite differences. The basic idea of this method is that instead of considering the values of a functional

$$J[x(\cdot)] = \int_a^b L(t, x(t), \dot{x}(t))dt$$

with boundary conditions $x(a) = x_a$ and $x(b) = x_b$, on arbitrary admissible curves, we only track the values at an $n+1$ grid points, t_i, $i = 0, \ldots, n$, of the interested time interval [29]. The functional $J[x(\cdot)]$ is then transformed into a function $\Psi(x(t_1), x(t_2), \ldots, x(t_{n-1}))$ of the values of unknown function on mesh points. Assuming $h = t_i - t_{i-1}$, $x(t_i) = x_i$ and $\dot{x}_i \approx \frac{x_i - x_{i-1}}{h}$, one has

$$\begin{aligned} J[x(\cdot)] &\approx \Psi(x_1, x_2, \ldots, x_{n-1}) = h \sum_{i=1}^{n} L\left(t_i, x_i, \frac{x_i - x_{i-1}}{h}\right), \\ x_0 &= x_a, \quad x_n = x_b. \end{aligned}$$

The desired values of x_i, $i = 1, \ldots, n-1$, are the extremum of the multi-variable function Ψ which is the solution to the system

$$\frac{\partial \Psi}{\partial x_i} = 0, \quad i = 1, \ldots, n-1.$$

The fact that only two terms in the sum, $(i-1)$th and ith, depend on x_i, makes it rather easy to find the extremum of Ψ solving a system of algebraic equations. For each n, we obtain a polygonal line which is an approximate solution of the original problem. It has been shown that passing to the limit as $h \to 0$, the linear system corresponding to finding the extremum of Ψ is equivalent to the Euler–Lagrange equation of the problem.

3.1.1. Finite Differences for Fractional Derivatives

In classical theory, given a derivative of a certain order, $x^{(n)}$, there is a finite difference approximation of the form

$$x^{(n)}(t) = \lim_{h \to 0^+} \frac{1}{h^n} \sum_{k=0}^{n} (-1)^k \binom{n}{k} x(t - kh),$$

where $\binom{n}{k}$ is the binomial coefficient and

$$\binom{n}{k} = \frac{n(n-1)(n-2)\cdots(n-k+1)}{k!}, \quad n, k \in \mathbb{N}.$$

The Grünwald–Letnikov definition of fractional derivative is a generalization of this formula to derivatives of arbitrary order.

The series in (2) and (3), the Grünwald–Letnikov definitions, converge absolutely and uniformly if $x(\cdot)$ is bounded. The infinite sums, backward differences for the left and forward differences for the right derivative in the Grünwald–Letnikov definitions for fractional derivatives, reveals that the arbitrary order derivative of a function at a time t depends on all values of that function in $(-\infty, t]$ and $[t, \infty)$, for left and right derivatives respectively. This is due to the non-local property of fractional derivatives.

Remark 3.1. *Equations* (2) *and* (3) *need to be consistent in closed time intervals and we need the values of* $x(t)$ *outside the interval* $[a, b]$. *To overcome this difficulty, we can take*

$$x^*(t) = \begin{cases} x(t) & t \in [a, b], \\ 0 & t \notin [a, b]. \end{cases}$$

Then we assume ${}^{GL}_{a}D^{\alpha}_{t}x(t) = {}^{GL}_{a}D^{\alpha}_{t}x^*(t)$ *and* ${}^{GL}_{t}D^{\alpha}_{b}x(t) = {}^{GL}_{t}D^{\alpha}_{b}x^*(t)$ *for* $t \in [a, b]$.

This definition coincides with Riemann–Liouville and Caputo derivatives. The latter is believed to be more applicable in practical fields such as engineering and physics.

Proposition 3.2 (See [25])**.** *Let* $0 < \alpha < n$, $n \in \mathbb{N}$ *and* $x(\cdot) \in C^{n-1}[a, b]$. *Suppose also that* $x^{(n)}(\cdot)$ *is integrable on* $[a, b]$. *Then, for every* α, *the Riemann–Liouville derivative exists and coincides with the Grünwald–Letnikov derivative and the following holds:*

$$\begin{aligned} {}_{a}D^{\alpha}_{t}x(t) &= \sum_{i=0}^{n-1} \frac{x^{(i)}(a)(t-a)^{i-\alpha}}{\Gamma(1+i-\alpha)} + \frac{1}{\Gamma(n-\alpha)} \int_a^t (t-\tau)^{n-1-\alpha} x^{(n)}(\tau) d\tau \\ &= {}^{GL}_{a}D^{\alpha}_{t}x(t). \end{aligned}$$

Remark 3.3. *For numerical purposes we need a finite series in* (2). *Given a grid on* $[a, b]$ *as* $a = t_0, t_1, \ldots, t_n = b$, *where* $t_i = t_0 + ih$ *for some* $h > 0$, *we approximate the left Riemann–Liouville derivative as*

$$ {}_{a}D^{\alpha}_{t}x(t_i) \approx \frac{1}{h^\alpha} \sum_{k=0}^{i} (\omega^\alpha_k)\, x(t_i - kh), \tag{23}$$

where $(\omega^\alpha_k) = (-1)^k \binom{\alpha}{k} = \frac{\Gamma(k-\alpha)}{\Gamma(-\alpha)\Gamma(k+1)}$.

Similarly, one can approximate the right Riemann–Liouville derivative by

$$ {}_{t}D^{\alpha}_{b}x(t_i) \approx \frac{1}{h^\alpha} \sum_{k=0}^{n-i} (\omega^\alpha_k)\, x(t_i + kh). \tag{24}$$

Remark 3.4. *The Grünwald–Letnikov approximation of Riemann–Liouville is a first order approximation [25], i.e.,*

$$ {}_{a}D^{\alpha}_{t}x(t_i) = \frac{1}{h^\alpha} \sum_{k=0}^{i} (\omega^\alpha_k)\, x(t_i - kh) + \mathcal{O}(h).$$

Remark 3.5. *It has been shown that the implicit Euler method solution to a certain fractional partial differential equation based on the Grünwald–Letnikov approximation to the fractional derivative, is unstable [23]. Therefore, discretizing fractional derivatives, shifted Grünwald–Letnikov derivatives are used and, despite the slight difference, they exhibit a stable performance, at least for certain cases. The shifted Grünwald–Letnikov derivative is defined by*

$$ {}^{sGL}_{a}D^{\alpha}_{t}x(t_i) \approx \frac{1}{h^\alpha} \sum_{k=0}^{i} (\omega^\alpha_k)\, x(t_i - (k-1)h).$$

Other finite difference approximations can be found in the literature. We refer here to the Diethelm backward finite difference formula for Caputo's fractional derivative, with $0 < \alpha < 2$ and $\alpha \neq 1$, which is an approximation of order $\mathcal{O}(h^{2-\alpha})$ [16]:

$$ {}_a^C D_t^\alpha x(t_i) \approx \frac{h^{-\alpha}}{\Gamma(2-\alpha)} \sum_{j=0}^{i} a_{i,j} \left(x_{i-j} - \sum_{k=0}^{\lfloor \alpha \rfloor} \frac{(i-j)^k h^k}{k!} x^{(k)}(a) \right), $$

where

$$ a_{i,j} = \begin{cases} 1, & \text{if } i = 0, \\ (j+1)^{1-\alpha} - 2j^{1-\alpha} + (j-1)^{1-\alpha}, & \text{if } 0 < j < i, \\ (1-\alpha)i^{-\alpha} - i^{1-\alpha} + (i-1)^{1-\alpha}, & \text{if } j = i. \end{cases} $$

3.1.2. Euler-like Direct Method for Fractional Variational Problems

As mentioned earlier, we consider a simple version of fractional variational problems where the fractional term has a Riemann–Liouville form on a finite time interval $[a, b]$. The boundary conditions are given and we approximate the problem using the Grünwald–Letnikov approximation given by (23). In this context, we discretize the functional in (4) using a simple quadrature rule on the mesh points, $a = t_0, t_1, , \ldots, t_n = b$, with $h = \frac{b-a}{n}$. The goal is to find the values $x_1, x_2, \ldots, x_{n-1}$ of the unknown function $x(\cdot)$ at points t_i, $i = 1, \ldots, n-1$. The values of x_0 and x_n are given. Applying the quadrature rule gives

$$ J[x(\cdot)] = \sum_{i=1}^{n} \int_{t_{i-1}}^{t_i} L(t_i, x_i, {}_aD_t^\alpha x_i) dt \approx \sum_{i=1}^{n} hL(t_i, x_i, {}_aD_t^\alpha x_i) $$

and by approximating the fractional derivatives at mesh points using (23) we have

$$ J[x(\cdot)] \approx \sum_{i=1}^{n} hL \left(t_i, x_i, \frac{1}{h^\alpha} \sum_{k=0}^{i} (\omega_k^\alpha) \, x_{i-k} \right). \tag{25} $$

Hereafter the procedure is the same as in the classical case. The right-hand-side of (25) can be regarded as a function Ψ of $n-1$ unknowns $\mathbf{x} = (x_1, x_2, \ldots, x_{n-1})$,

$$ \Psi(\mathbf{x}) = \sum_{i=1}^{n} hL \left(t_i, x_i, \frac{1}{h^\alpha} \sum_{k=0}^{i} (\omega_k^\alpha) \, x_{i-k} \right). \tag{26} $$

To find an extremum for Ψ, one has to solve the following system of algebraic equations:

$$ \frac{\partial \Psi}{\partial x_i} = 0, \qquad i = 1, \ldots, n-1. \tag{27} $$

Unlike the classical case, all terms, starting from the ith term in (26), depend on x_i and we have

$$ \frac{\partial \Psi}{\partial x_i} = h \frac{\partial L}{\partial x}(t_i, x_i, {}_aD_t^\alpha x_i) + h \sum_{k=0}^{n-i} \frac{1}{h^\alpha} (\omega_k^\alpha) \frac{\partial L}{\partial \, {}_aD_t^\alpha x}(t_{i+k}, x_{i+k}, {}_aD_t^\alpha x_{i+k}). \tag{28} $$

Equating the right-hand-side of (28) with zero, one has

$$\frac{\partial L}{\partial x}(t_i, x_i, {}_aD_t^\alpha x_i) + \frac{1}{h^\alpha}\sum_{k=0}^{n-i}(\omega_k^\alpha)\frac{\partial L}{\partial\, {}_aD_t^\alpha x}(t_{i+k}, x_{i+k}, {}_aD_t^\alpha x_{i+k}) = 0.$$

Passing to the limit, and considering the approximation formula for the right Riemann–Liouville derivative, equation (24), it is straightforward to verify that:

Theorem 3.6. *The Euler-like method for a fractional variational problem of the form* (4) *is equivalent to the fractional Euler–Lagrange equation*

$$\frac{\partial L}{\partial x} + {}_tD_b^\alpha \frac{\partial L}{\partial\, {}_aD_t^\alpha x} = 0,$$

as the mesh size, h, tends to zero.

Proof. Consider a minimizer $(x_1, \ldots, x_{n-1})$ of Ψ, a variation function $\eta \in C[a,b]$ with $\eta(a) = \eta(b) = 0$ and define $\eta_i = \eta(t_i)$, for $i = 0, \ldots, n$. We remark that $\eta_0 = \eta_n = 0$ and that $(x_1 + \epsilon\eta_1, \ldots, x_{n-1} + \epsilon\eta_{n-1})$ is a variation of $(x_1, \ldots, x_{n-1})$, with $|\epsilon| < r$, for some fixed $r > 0$. Therefore, since $(x_1, \ldots, x_{n-1})$ is a minimizer for Ψ, proceeding with Taylor's expansion, we deduce that

$$\begin{aligned} 0 &\leq \Psi(x_1 + \epsilon\eta_1, \ldots, x_{n-1} + \epsilon\eta_{n-1}) - \Psi(x_1, \ldots, x_{n-1}) \\ &= \epsilon\sum_{i=1}^{n} h\left[\frac{\partial L}{\partial x}[i]\eta_i + \frac{\partial L}{\partial_a D_t^\alpha}[i]\frac{1}{h^\alpha}\sum_{k=0}^{i}(\omega_k^\alpha)\eta_{i-k}\right] + \mathcal{O}(\epsilon), \end{aligned}$$

where

$$[i] = \left(t_i, x_i, \frac{1}{h^\alpha}\sum_{k=0}^{i}(\omega_k^\alpha)x_{i-k}\right).$$

Since ϵ takes any value, it follows that

$$\sum_{i=1}^{n} h\left[\frac{\partial L}{\partial x}[i]\eta_i + \frac{\partial L}{\partial_a D_t^\alpha}[i]\frac{1}{h^\alpha}\sum_{k=0}^{i}(\omega_k^\alpha)\eta_{i-k}\right] = 0. \tag{29}$$

On the other hand, since $\eta_0 = 0$, reordering the terms of the sum, it follows immediately that

$$\sum_{i=1}^{n}\frac{\partial L}{\partial_a D_t^\alpha}[i]\sum_{k=0}^{i}(\omega_k^\alpha)\eta_{i-k} = \sum_{i=1}^{n}\eta_i\sum_{k=0}^{n-i}(\omega_k^\alpha)\frac{\partial L}{\partial_a D_t^\alpha}[i+k].$$

Substituting this relation into equation (29), we obtain

$$\sum_{i=1}^{n}\eta_i h\left[\frac{\partial L}{\partial x}[i] + \frac{1}{h^\alpha}\sum_{k=0}^{n-i}(\omega_k^\alpha)\frac{\partial L}{\partial_a D_t^\alpha}[i+k]\right] = 0.$$

Since η_i is arbitrary, for $i = 1, \ldots, n-1$, we deduce that

$$\frac{\partial L}{\partial x}[i] + \frac{1}{h^\alpha}\sum_{k=0}^{n-i}(\omega_k^\alpha)\frac{\partial L}{\partial_a D_t^\alpha}[i+k] = 0, \quad \text{for } i = 1, \ldots, n-1.$$

Let us study the case when n goes to infinity. Let $\overline{t} \in]a,b[$ and $i \in \{1,\ldots,n\}$ such that $t_{i-1} < \overline{t} \leq t_i$. First observe that, in such case, we also have $i \to \infty$ and $n - i \to \infty$. In fact, let $i \in \{1,\ldots,n\}$ be such that

$$a + (i-1)h < \overline{t} \leq a + ih.$$

So, $i < (\overline{t} - a)/h + 1$, which implies that

$$n - i > n\frac{b - \overline{t}}{b - a} - 1.$$

Then

$$\lim_{n\to\infty, i\to\infty} t_i = \overline{t}.$$

Assume that there exists a function $\overline{x} \in C[a,b]$ satisfying

$$\forall \epsilon > 0 \, \exists N \, \forall n \geq N : |x_i - \overline{x}(t_i)| < \epsilon, \quad \forall i = 1,\ldots,n-1.$$

As $\overline{x}$ is uniformly continuous, we have

$$\forall \epsilon > 0 \, \exists N \, \forall n \geq N : |x_i - \overline{x}(\overline{t})| < \epsilon, \quad \forall i = 1,\ldots,n-1.$$

By the continuity assumption of $\overline{x}$, we deduce that

$$\lim_{n\to\infty, i\to\infty} \frac{1}{h^\alpha} \sum_{k=0}^{n-i} (\omega_k^\alpha) \frac{\partial L}{\partial_a D_t^\alpha}[i+k] = {}_tD_b^\alpha \frac{\partial L}{\partial_a D_t^\alpha}(\overline{t}, \overline{x}(\overline{t}), {}_aD_{\overline{t}}^\alpha \overline{x}(\overline{t})).$$

For n sufficiently large (and therefore i also sufficiently large),

$$\lim_{n\to\infty, i\to\infty} \frac{\partial L}{\partial x}[i] = \frac{\partial L}{\partial x}(\overline{t}, \overline{x}(\overline{t}), {}_aD_{\overline{t}}^\alpha \overline{x}(\overline{t})).$$

In conclusion,

$$\frac{\partial L}{\partial x}(\overline{t}, \overline{x}(\overline{t}), {}_aD_{\overline{t}}^\alpha \overline{x}(\overline{t})) + {}_tD_b^\alpha \frac{\partial L}{\partial_a D_t^\alpha}(\overline{t}, \overline{x}(\overline{t}), {}_aD_{\overline{t}}^\alpha \overline{x}(\overline{t})) = 0. \tag{30}$$

Using the continuity condition, we prove that the fractional Euler–Lagrange equation (30) holds for all values on the closed interval $a \leq t \leq b$. □

3.1.3. Examples

Now we apply the Euler-like direct method to some test problems for which the exact solutions are known. Although we propose problems for the interval $[0,1]$, moving to arbitrary intervals is only a matter of more computations. To measure the errors related to approximations, different norms can be used. Since a direct method seeks for the function values at certain points, we use the maximum norm to determine how close we can get to the exact value at that point. Assume that the exact value of the function $x(\cdot)$, at the point t_i, is $x(t_i)$ and it is approximated by x_i. The error is defined as

$$E = \max\{|x(t_i) - x_i|, \; i = 1, \cdots, n-1\}.$$

Example 3.7. *Our goal here is to minimize a quadratic Lagrangian on* $[0,1]$ *with fixed boundary conditions. Consider the following minimization problem:*

$$\begin{cases} J[x(\cdot)] = \int_0^1 \left({}_0D_t^{0.5} x(t) - \frac{2}{\Gamma(2.5)} t^{1.5} \right)^2 dt \to \min \\ x(0) = 0, \ x(1) = 1. \end{cases} \tag{31}$$

Since the Lagrangian is always positive, problem (31) *attains its minimum when*

$$ {}_0D_t^{0.5} x(t) - \frac{2}{\Gamma(2.5)} t^{1.5} = 0$$

and has the obvious solution of the form $x(t) = t^2$ *because* ${}_0D_t^{0.5} t^2 = \frac{2}{\Gamma(2.5)} t^{1.5}$.

To begin with, we approximate the fractional derivative by

$$ {}_0D_t^{0.5} x(t_i) \approx \frac{1}{h^{0.5}} \sum_{k=0}^{i} \left(\omega_k^{0.5}\right) x(t_i - kh)$$

for a fixed $h > 0$. The functional is now transformed into

$$J[x(\cdot)] = \int_0^1 \left(\frac{1}{h^{0.5}} \sum_{k=0}^{i} \left(\omega_k^{0.5}\right) x_{i-k} - \frac{2}{\Gamma(2.5)} t^{1.5} \right)^2 dt.$$

Finally, we approximate the integral by a rectangular rule and end with the discrete problem

$$\Psi(\mathbf{x}) = \sum_{i=1}^{n} h \left(\frac{1}{h^{0.5}} \sum_{k=0}^{i} \left(\omega_k^{0.5}\right) x_{i-k} - \frac{2}{\Gamma(2.5)} t_i^{1.5} \right)^2.$$

Since the Lagrangian in this example is quadratic, system (27) has a linear form and therefore is easy to solve. Other problems may end with a system of nonlinear equations. Simple calculations lead to the system

$$\mathbf{A}\mathbf{x} = \mathbf{b}, \tag{32}$$

in which

$$\mathbf{A} = \begin{bmatrix} \sum_{i=0}^{n-1} A_i^2 & \sum_{i=1}^{n-1} A_i A_{i-1} & \cdots & \sum_{i=n-2}^{n-1} A_i A_{i-(n-2)} \\ \sum_{i=0}^{n-2} A_i A_{i+1} & \sum_{i=1}^{n-2} A_i^2 & \cdots & \sum_{i=n-3}^{n-2} A_i A_{i-(n-3)} \\ \sum_{i=0}^{n-3} A_i A_{i+2} & \sum_{i=1}^{n-3} A_i A_{i+1} & \cdots & \sum_{i=n-4}^{n-3} A_i A_{i-(n-4)} \\ \vdots & \vdots & \ddots & \vdots \\ \sum_{i=0}^{1} A_i A_{i+n-2} & \sum_{i=0}^{1} A_i A_{i+n-3} & \cdots & \sum_{i=0}^{1} A_i^2 \end{bmatrix},$$

where $A_i = (-1)^i h^{1.5} \binom{0.5}{i}$ and $\mathbf{b} = (b_1, b_2, \cdots, b_{n-1})$ with

$$b_i = \sum_{k=0}^{n-i} \frac{2h^2 A_k}{\Gamma(2.5)} t_{k+i}^{1.5} - A_{n-i} A_0 - \left(\sum_{k=0}^{n-i} A_k A_{k+i} \right).$$

Since system (32) is linear, it is easily solved for different values of n. As indicated in Figure 6, by increasing the value of n we get better solutions.

Let us now move to another example for which the solution is obtained by the fractional Euler–Lagrange equation.

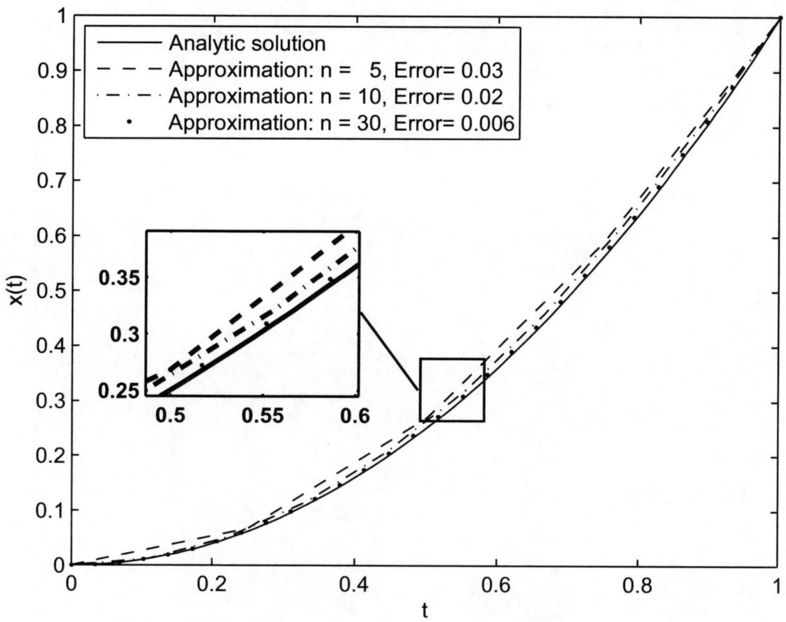

Figure 6. Analytic and approximate solutions of Example 3.7.

Example 3.8. *Consider the following minimization problem:*

$$\begin{cases} J[x(\cdot)] = \int_0^1 \left({}_0D_t^{0.5}x(t) - \dot{x}^2(t)\right) dt \to \min \\ x(0) = 0, \; x(1) = 1. \end{cases} \tag{33}$$

In this case the only way to get a solution is by use of Euler–Lagrange equations. The Lagrangian depends not only on the fractional derivative, but also on the first order derivative of the function. The Euler–Lagrange equation for this setting becomes

$$\frac{\partial L}{\partial x} + {}_tD_b^\alpha \frac{\partial L}{\partial_a D_t^\alpha} - \frac{d}{dt}\left(\frac{\partial L}{\partial \dot{x}}\right) = 0,$$

and by direct computations a necessary condition for $x(\cdot)$ *to be a minimizer of* (33) *is*

$${}_tD_1^\alpha 1 + 2\ddot{x}(t) = 0 \;\; or \;\; \ddot{x}(t) = \frac{1}{2\Gamma(1-\alpha)}(1-t)^{-\alpha}.$$

Subject to the given boundary conditions, the above second order ordinary differential equation has the solution

$$x(t) = -\frac{1}{2\Gamma(3-\alpha)}(1-t)^{2-\alpha} + \left(1 - \frac{1}{2\Gamma(3-\alpha)}\right)t + \frac{1}{2\Gamma(3-\alpha)}. \tag{34}$$

Discretizing problem (33) with the same assumptions of Example 3.7 ends in a linear

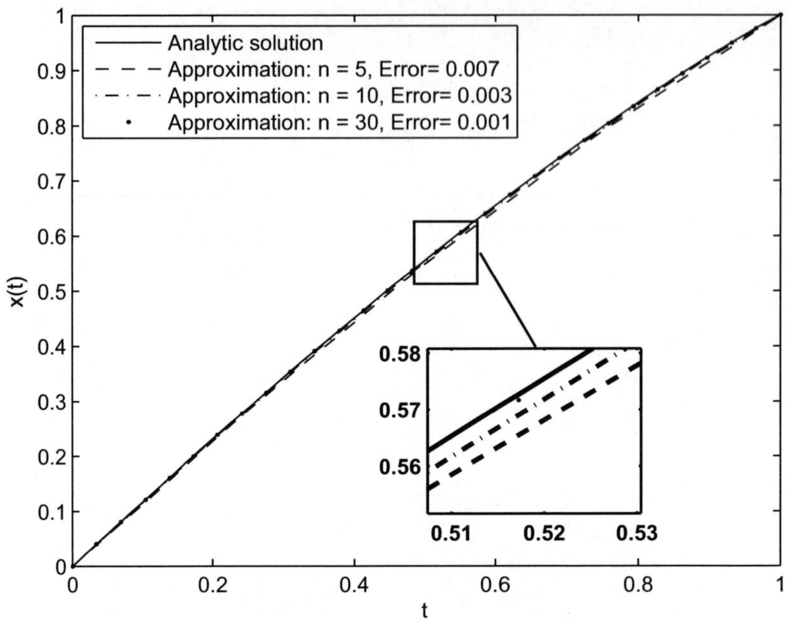

Figure 7. Analytic and approximate solutions of Example 3.8.

system of the form

$$\begin{bmatrix} 2 & -1 & 0 & 0 & \cdots & 0 & 0 \\ -1 & 2 & -1 & 0 & \cdots & 0 & 0 \\ 0 & -1 & 2 & -1 & \cdots & 0 & 0 \\ \vdots & \vdots & \vdots & \vdots & \ddots & \vdots & \vdots \\ 0 & 0 & 0 & 0 & \cdots & -1 & 2 \end{bmatrix} \begin{bmatrix} x_1 \\ x_2 \\ x_3 \\ \vdots \\ x_{n-1} \end{bmatrix} = \begin{bmatrix} b_1 \\ b_2 \\ b_3 \\ \vdots \\ b_{n-1} \end{bmatrix}, \tag{35}$$

where

$$b_i = \frac{h}{2} \sum_{k=0}^{n-i-1} (-1)^k h^{0.5} \binom{0.5}{k}, \qquad i = 1, 2, \ldots, n-2,$$

and

$$b_{n-1} = \frac{h}{2} \sum_{k=0}^{1} \left((-1)^k h^{0.5} \binom{0.5}{k} \right) + x_n.$$

System (35) is linear and can be solved for any n to reach the desired accuracy. The analytic solution together with some approximated solutions are shown in Figure 7.

Both examples above end with linear systems and their solvability is simply dependant to the matrix of coefficients. Now we try this method on a more complicated problem, yet analytically solvable, with an oscillating solution.

Example 3.9. *Consider the problem of minimizing $\int_0^1 L dt$ subject to the boundary conditions $x(0) = 0$ and $x(1) = 1$, where the Lagrangian L is given by*

$$L = \left({}_0D_t^{0.5} x(t) - \frac{16\Gamma(6)}{\Gamma(5.5)} t^{4.5} + \frac{20\Gamma(4)}{\Gamma(3.5)} t^{2.5} - \frac{5}{\Gamma(1.5)} t^{0.5} \right)^4 .$$

This example has an obvious solution too. Since L is positive, the minimizer is

$$x(t) = 16t^5 - 20t^3 + 5t.$$

Note that ${}_aD_t^{\alpha}(t-a)^{\nu} = \frac{\Gamma(\nu+1)}{\Gamma(\nu+\alpha)} t^{\nu-\alpha}$.

The appearance of a fourth power in the Lagrangian, results in a nonlinear system as we apply the Euler-like direct method to this problem. For $j = 1, \cdots, n-1$ we have

$$\sum_{i=j}^{n} \left(\omega_{i-j}^{0.5}\right) \left(\frac{1}{h^{0.5}} \sum_{k=0}^{i} \left(\omega_k^{0.5}\right) x_{i-k} - \phi(t_i) \right)^3 = 0, \tag{36}$$

where

$$\phi(t) = \frac{16\Gamma(6)}{\Gamma(5.5)} t^{4.5} + \frac{20\Gamma(4)}{\Gamma(3.5)} t^{2.5} - \frac{5}{\Gamma(1.5)} t^{0.5} .$$

System (36) is solved for different values of n and the results are depicted in Figure 8.

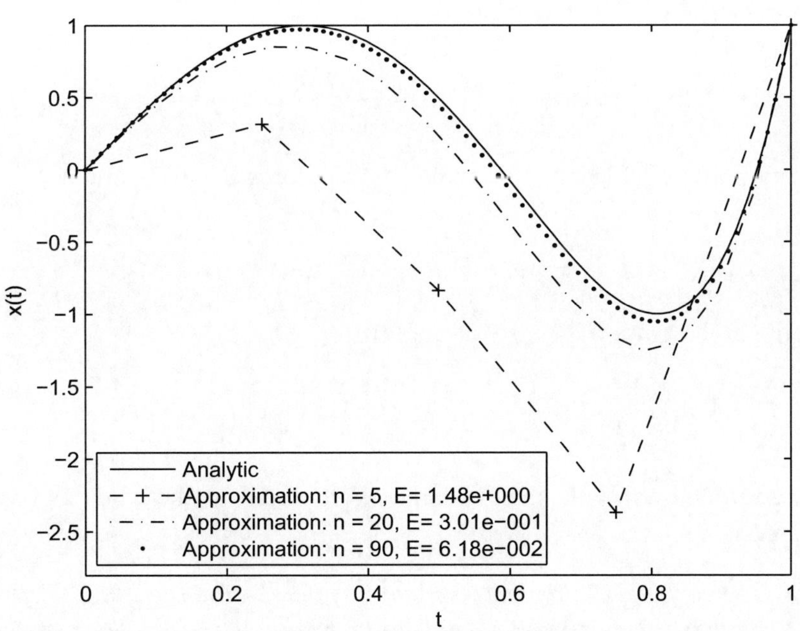

Figure 8. Analytic and approximate solutions of Example 3.9.

4. Indirect Methods

As in the classical case, indirect methods in fractional sense provide the necessary conditions of optimality using the first variation. Fractional Euler–Lagrange equations are now a well-known and well-studied subject in fractional calculus. For a simple problem of the form (4), following [1], a necessary condition implies that the solution must satisfy a fractional boundary value differential equation.

Theorem 4.1 (cf. [1]). *Let $x(\cdot)$ have a continuous left Riemann–Liouville derivative of order α and J be a functional of the form*

$$J[x(\cdot)] = \int_a^b L(t, x(t), {}_aD_t^\alpha x(t))dt \tag{37}$$

subject to the boundary conditions $x(a) = x_a$ and $x(b) = x_b$. Then a necessary condition for J to have an extremum for a function $x(\cdot)$ is that $x(\cdot)$ satisfies the following Euler-Lagrange equation:

$$\begin{cases} \frac{\partial L}{\partial x} + {}_tD_b^\alpha \frac{\partial L}{\partial\, {}_aD_t^\alpha x} = 0, \\ x(a) = x_a, \quad x(b) = x_b, \end{cases} \tag{38}$$

which is called the fractional Euler–Lagrange equation.

Proof. Assume that $x^*(t)$ is the desired function and let $x(t) = x^*(t) + \epsilon\eta(t)$ be a family of curves that satisfy boundary conditions, i.e., $\eta(a) = \eta(b) = 0$. Since ${}_aD_t^\alpha$ is a linear operator, for any $x(\cdot)$, the functional becomes

$$J[x(\cdot)] = \int_a^b L(t, x^*(t) + \epsilon\eta(t), {}_aD_t^\alpha x^*(t) + \epsilon {}_aD_t^\alpha \eta(t))dt,$$

which is a function of ϵ, $J[\epsilon]$. Since J assumes its extremum at $\epsilon = 0$, one has $\frac{dJ}{d\epsilon}\big|_{\epsilon=0} = 0$, i.e.,

$$\int_a^b \left[\frac{\partial L}{\partial x}\eta + \frac{\partial L}{\partial_a D_t^\alpha x} {}_aD_t^\alpha \eta\right] dt = 0.$$

Using the fractional integration by parts of the form

$$\int_a^b g(t) {}_aD_t^\alpha f(t)dt = \int_a^b f(t) {}_tD_b^\alpha g(t)dt$$

on the second term and applying the fundamental theorem of the calculus of variations completes the proof. □

Remark 4.2. *Many variants of this theorem can be found in the literature. Different types of fractional terms have been embedded in the Lagrangian and appropriate versions of Euler–Lagrange equations have been derived using proper integration by parts formulas. See [1, 3, 6, 22, 24] for details.*

For fractional optimal control problems, a so-called Hamiltonian system is constructed using Lagrange multipliers. For example, cf. [9], assume that we are required to minimize a functional of the form

$$J[x(\cdot), u(\cdot)] = \int_a^b L(t, x(t), u(t))dt$$

such that $x(a) = x_a$, $x(b) = x_b$ and ${}_aD_t^\alpha x(t) = f(t, x(t), u(t))$. Similar to the classical methods, one can introduce a Hamiltonian

$$H = L(t, x(t), u(t)) + \lambda(t) f(t, x(t), u(t)),$$

where $\lambda(t)$ is considered as a Lagrange multiplier. In this case we define the augmented functional as

$$J[x(\cdot), u(\cdot), \lambda(\cdot)] = \int_a^b [H(t, x(t), u(t), \lambda(t)) - \lambda(t)_a D_t^\alpha x(t)]dt.$$

Optimizing the latter functional results in the following necessary optimality conditions:

$$\begin{cases} {}_aD_t^\alpha x(t) = \frac{\partial H}{\partial \lambda} \\ {}_tD_b^\alpha \lambda(t) = \frac{\partial H}{\partial x} \\ \frac{\partial H}{\partial u} = 0. \end{cases} \tag{39}$$

Together with the prescribed boundary conditions, this makes a two point fractional boundary value problem.

These arguments reveal that, like the classical case, fractional variational problems end with fractional boundary value problems. To reach an optimal solution, one needs to deal with a fractional differential equation or a system of fractional differential equations.

The classical theory of differential equations is furnished with several solution methods, theoretical and numerical. Nevertheless, solving a fractional differential equation is a rather tough task [12]. To benefit those methods, especially all solvers that are available to solve an integer order differential equation numerically, we can either approximate a fractional variational problem by an equivalent integer-order one or approximate the necessary optimality conditions (38) and (39). The rest of this section discusses two types of approximations that are used to transform a fractional problem to one in which only integer order derivatives are present; i.e., we approximate the original problem by substituting a fractional term by its corresponding expansion formulas. This is mainly done by case studies on certain examples. The examples are chosen so that either they have a trivial solution or it is possible to get an analytic solution using fractional Euler–Lagrange equations.

By substituting the approximations (9) or (15) for the fractional derivative in (37), the problem is transformed to

$$\begin{aligned} J[x(\cdot)] &= \int_a^b L\left(t, x(t), \sum_{k=0}^{N} \frac{(-1)^{k-1}\alpha x^{(k)}(t)}{k!(k-\alpha)\Gamma(1-\alpha)}(t-a)^{k-\alpha}\right) dt \\ &= \int_a^b L'\left(t, x(t), \dot{x}(t), \ldots, x^{(N)}(t)\right) dt \end{aligned}$$

or

$$
\begin{aligned}
J[x(\cdot)] &= \int_a^b L\left(t, x(t), \frac{Ax(t)}{(t-a)^\alpha} + \frac{B\dot{x}(t)}{(t-a)^{\alpha-1}} - \sum_{p=2}^{N} \frac{C(\alpha,p)V_p(t)}{(t-a)^{p+\alpha-1}}\right) dt \\
&= \int_a^b L'\left(t, x(t), \dot{x}(t), V_2(t), \ldots, V_N(t)\right) dt \\
&\begin{cases} \dot{V}_p(t) = (1-p)(t-a)^{p-2}x(t) \\ V_p(a) = 0, \qquad p = 2, 3, \ldots \end{cases}
\end{aligned}
$$

The former problem is a classical variational problem containing higher order derivatives. The latter is a multi-variable problem, subject to some ordinary differential equation constraint. Together with the boundary conditions, both above problems belong to classes of well studied variational problems.

To accomplish a detailed study, as test problems, we consider here Example 3.8,

$$
\begin{cases} J[x(\cdot)] = \int_0^1 \left({}_0D_t^{0.5}x(t) - \dot{x}^2(t)\right) dt \to \min \\ x(0) = 0, \; x(1) = 1, \end{cases} \tag{40}
$$

and the following example.

Example 4.3. *Given* $\alpha \in (0,1)$, *consider the functional*

$$
J[x(\cdot)] = \int_0^1 \left({}_aD_t^\alpha x(t) - 1\right)^2 dt \tag{41}
$$

to be minimized subject to the boundary conditions $x(0) = 0$ *and* $x(1) = \frac{1}{\Gamma(\alpha+1)}$. *Since the integrand in* (41) *is non-negative, the functional attains its minimum when* ${}_aD_t^\alpha x(t) = 1$, *i.e., for* $x(t) = \frac{t^\alpha}{\Gamma(\alpha+1)}$.

We illustrate the use of the two different expansions separately.

4.1. Expansion to Integer Orders

Using approximation (9) for the fractional derivative in (40), we get the approximated problem

$$
\begin{aligned}
\min \quad \tilde{J}[x(\cdot)] &= \int_0^1 \left[\sum_{n=0}^{N} C(n,\alpha) t^{n-\alpha} x^{(n)}(t) - \dot{x}^2(t)\right] dt \\
x(0) = 0, &\quad x(1) = 1,
\end{aligned} \tag{42}
$$

which is a classical higher-order problem of the calculus of variations that depends on derivatives up to order N. The corresponding necessary optimality condition is a well-known result.

Theorem 4.4 (cf., e.g., [21]). *Suppose that* $x(\cdot) \in C^{2N}[a,b]$ *minimizes*

$$
\int_a^b L(t, x(t), x^{(1)}(t), x^{(2)}(t), \ldots, x^{(N)}(t)) dt
$$

with given boundary conditions

$$\begin{aligned} x(a) &= a_0, & x(b) &= b_0, \\ x^{(1)}(a) &= a_1, & x^{(1)}(b) &= b_1, \\ &\vdots \\ x^{(N-1)}(a) &= a_{N-1}, & x^{(N-1)}(b) &= b_{N-1}. \end{aligned}$$

Then $x(\cdot)$ satisfies the Euler–Lagrange equation

$$\frac{\partial L}{\partial x} - \frac{d}{dt}\left(\frac{\partial L}{\partial x^{(1)}}\right) + \frac{d^2}{dt^2}\left(\frac{\partial L}{\partial x^{(2)}}\right) - \cdots + (-1)^N \frac{d^N}{dt^N}\left(\frac{\partial L}{\partial x^{(N)}}\right) = 0. \tag{43}$$

In general (43) is an ODE of order $2N$, depending on the order N of the approximation we choose, and the method leaves $2N - 2$ parameters unknown. In our example, however, the Lagrangian in (42) is linear with respect to all derivatives of order higher than two. The resulting Euler–Lagrange equation is the second order ODE

$$\sum_{n=0}^{N} (-1)^n C(n,\alpha) \frac{d^n}{dt^n}(t^{n-\alpha}) - \frac{d}{dt}[-2\dot{x}(t)] = 0$$

that has the solution

$$x(t) = M_1(\alpha, N)t^{2-\alpha} + M_2(\alpha, N)t,$$

where

$$\begin{aligned} M_1(\alpha, N) &= -\frac{1}{2\Gamma(3-\alpha)}\left[\sum_{n=0}^{N}(-1)^n\Gamma(n+1-\alpha)C(n,\alpha)\right], \\ M_2(\alpha, N) &= \left[1 + \frac{1}{2\Gamma(3-\alpha)}\sum_{n=0}^{N}(-1)^n\Gamma(n+1-\alpha)C(n,\alpha)\right]. \end{aligned}$$

Figure 9 shows the analytic solution together with several approximations. It reveals that by increasing N, approximate solutions do not converge to the analytic one. The reason is the fact that the solution (34) to Example 3.8 is not an analytic function. We conclude that (9) may not be a good choice to approximate fractional variational problems. In contrast, as we shall see, the approximation (15) leads to good results.

To solve Example 3.8 using (9) as an approximation for the fractional derivative, the problem becomes

$$\begin{gathered} \min \quad \tilde{J}[x(\cdot)] = \int_0^1 \left(\sum_{n=0}^{N} C(n,\alpha)t^{n-\alpha}x^{(n)}(t) - 1\right)^2 dt, \\ x(0) = 0, \quad x(1) = \frac{1}{\Gamma(\alpha+1)}. \end{gathered}$$

The Euler–Lagrange equation (43) gives a $2N$ order ODE. For $N \geq 2$ this approach is inappropriate since the two given boundary conditions $x(0) = 0$ and $x(1) = \frac{1}{\Gamma(\alpha+1)}$ are not enough to determine the $2N$ constants of integration.

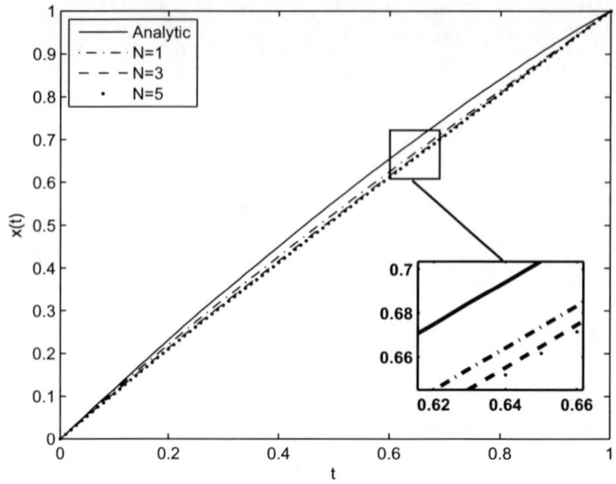

Figure 9. Analytic versus approximate solutions to Example 3.8 using approximation (9) with $\alpha = 0.5$.

4.2. Expansion through the Moments of a Function

If we use (15) to approximate the optimization problem (40), with $A = A(\alpha, N)$, $B = B(\alpha, N)$ and $C_p = C(\alpha, p)$, we have

$$
\begin{aligned}
&\tilde{J}[x(\cdot)] = \int_0^1 \left[At^{-\alpha}x(t) + Bt^{1-\alpha}\dot{x}(t) - \sum_{p=2}^{N} C_p t^{1-p-\alpha} V_p(t) - \dot{x}^2(t) \right] dt, \\
&\dot{V}_p(t) = (1-p)t^{p-2}x(t), \quad p = 2, \ldots, N, \\
&V_p(0) = 0, \quad p = 2, \ldots, N, \\
&x(0) = 0, \quad x(1) = 1.
\end{aligned} \tag{44}
$$

Problem (44) is constrained with a set of ordinary differential equations and is natural to look to it as an optimal control problem [26]. For that we introduce the control variable $u(t) = \dot{x}(t)$. Then, using the Lagrange multipliers $\lambda_1, \lambda_2, \ldots, \lambda_N$, and the Hamiltonian system, one can reduce (44) to the study of the two point boundary value problem

$$
\begin{cases}
\dot{x}(t) &= \frac{1}{2}Bt^{1-\alpha} - \frac{1}{2}\lambda_1(t), \\
\dot{V}_p(t) &= (1-p)t^{p-2}x(t), \quad p = 2, \ldots, N, \\
\dot{\lambda}_1(t) &= At^{-\alpha} - \sum_{p=2}^{N}(1-p)t^{p-2}\lambda_p(t), \\
\dot{\lambda}_p(t) &= -C_p t^{(1-p-\alpha)}, \quad p = 2, \ldots, N,
\end{cases} \tag{45}
$$

with boundary conditions

$$
\begin{cases}
x(0) = 0, \\
V_p(0) = 0, \quad p = 2, \ldots, N,
\end{cases}
\qquad
\begin{cases}
x(1) = 1, \\
\lambda_p(1) = 0, \quad p = 2, \ldots, N,
\end{cases}
$$

where $x(0) = 0$ and $x(1) = 1$ are given. We have $V_p(0) = 0$, $p = 2, \ldots, N$, due to (14) and $\lambda_p(1) = 0$, $p = 2, \ldots, N$, because V_p is free at final time for $p = 2, \ldots, N$ [26]. In

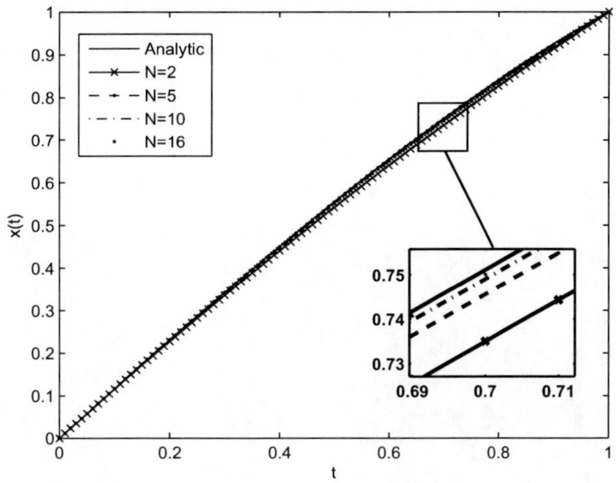

Figure 10. Analytic versus approximate solutions to Example 3.8 using approximation (15) with $\alpha = 0.5$.

general, the Hamiltonian system is a nonlinear, hard to solve, two point boundary value problem that needs special numerical methods. In this case, however, (45) is a non-coupled system of ordinary differential equations and is easily solved to give

$$x(t) = M(\alpha, N)t^{2-\alpha} - \sum_{p=2}^{N} \frac{C(\alpha, p)}{2p(2-p-\alpha)} t^p + \left[1 - M(\alpha, N) + \sum_{p=2}^{N} \frac{C(\alpha, p)}{2p(2-p-\alpha)}\right] t,$$

where

$$M(\alpha, N) = \frac{1}{2(2-\alpha)} \left[B(\alpha, N) - \frac{A(\alpha, N)}{1-\alpha} - \sum_{p=2}^{N} \frac{C(\alpha, p)(1-p)}{(1-\alpha)(2-p-\alpha)}\right].$$

Figure 10 shows the graph of $x(\cdot)$ for different values of N.

Let us now approximate Example 4.3 using (15). The resulting minimization problem has the following form:

$$\begin{aligned} \min \quad & \tilde{J}[x(\cdot)] = \int_0^1 \left[At^{-\alpha}x(t) + Bt^{1-\alpha}\dot{x}(t) - \sum_{p=2}^{N} C_p t^{1-p-\alpha} V_p(t) - 1\right]^2 dt, \\ & \dot{V}_p(t) = (1-p)t^{p-2}x(t), \quad p = 2, \ldots, N, \\ & V_p(0) = 0, \quad p = 2, \ldots, N, \\ & x(0) = 0, \quad x(1) = \frac{1}{\Gamma(\alpha+1)}. \end{aligned} \tag{46}$$

Following the classical optimal control approach of Pontryagin [26], this time with

$$u(t) = At^{-\alpha}x(t) + Bt^{1-\alpha}\dot{x}(t) - \sum_{p=2}^{N} C_p t^{1-p-\alpha} V_p(t),$$

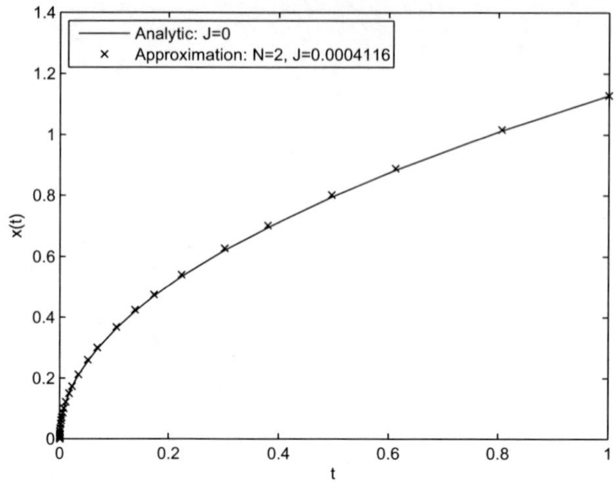

Figure 11. Analytic versus approximate solution to Example 4.3 using approximation (15) with $\alpha = 0.5$.

we conclude that the solution to (46) satisfies the system of differential equations

$$\begin{cases} \dot{x}(t) &= -AB^{-1}t^{-1}x(t) + \sum_{p=2}^{N} B^{-1}C_p t^{-p} V_p(t) + \frac{1}{2}B^{-2}t^{2\alpha-2}\lambda_1(t) + B^{-1}t^{\alpha-1}, \\ \dot{V}_p(t) &= (1-p)t^{p-2}x(t), \quad p = 2, \ldots, N, \\ \dot{\lambda}_1(t) &= AB^{-1}t^{-1}\lambda_1 - \sum_{p=2}^{N}(1-p)t^{p-2}\lambda_p(t), \\ \dot{\lambda}_p(t) &= -B^{-1}C_p t^{-p}\lambda_1, \quad p = 2, \ldots, N, \end{cases} \tag{47}$$

where $A = A(\alpha, N)$, $B = B(\alpha, N)$ and $C_p = C(\alpha, p)$ are defined according to Section 2.1.2, subject to the boundary conditions

$$\begin{cases} x(0) = 0, \\ V_p(0) = 0, \quad p = 2, \ldots, N, \end{cases} \qquad \begin{cases} x(1) = \frac{1}{\Gamma(\alpha+1)}, \\ \lambda_p(1) = 0, \quad p = 2, \ldots, N. \end{cases} \tag{48}$$

The solution to system (47)–(48), with $N = 2$, is shown in Figure 11.

5. Conclusion

The realm of numerical methods in scientific fields is vastly growing due to the very fast progresses in computational sciences and technologies. Nevertheless, the intrinsic complexity of fractional calculus, caused partially by non-local properties of fractional derivatives and integrals, makes it rather difficult to find efficient numerical methods in this field. It seems enough to mention here that, up to the time of this manuscript, and to the best of our knowledge, there is no routine available for solving a fractional differential equation as Runge–Kutta for ordinary ones. Despite this fact, however, the literature exhibits a growing interest and improving achievements in numerical methods for fractional calculus in general and fractional variational problems specifically.

This chapter was devoted to discuss some aspects of the very well-known methods for solving variational problems. Namely, we studied the notions of direct and indirect methods in the classical calculus of variations and we also mentioned some connections to optimal control. Consequently, we introduced the generalizations of these notions to the field of fractional calculus of variations and fractional optimal control.

The method of finite differences, as discussed here, seems to be a potential first candidate to solve fractional variational problems. Although a first order approximation was used for all examples, the results are satisfactory and even though it is more complicated than in the classical case, it still inherits some sort of simplicity and an ease of implementation.

The lack of efficient numerical methods for fractional variational problems is overcome, partially, by the indirect methods of this chapter. Once we transformed the fractional variational problem to an approximated classical one, the majority of classical methods can be applied to get an approximate solution. Nevertheless, the procedure is not completely straightforward. The singularity of fractional operators is still present in the approximating formulas and it makes the solution procedure more complicated.

Acknowledgments

Part of first author's Ph.D., carried out at the University of Aveiro under the Doctoral Program in Mathematics and Applications (PDMA) of Universities of Aveiro and Minho. Work supported by *FEDER* funds through *COMPETE* — Operational Programme Factors of Competitiveness ("Programa Operacional Factores de Competitividade") and by Portuguese funds through the *Center for Research and Development in Mathematics and Applications* (University of Aveiro) and the Portuguese Foundation for Science and Technology ("FCT–Fundação para a Ciência e a Tecnologia"), within project PEst-C/MAT/UI4106/2011 with COMPETE number FCOMP-01-0124-FEDER-022690. Pooseh was also supported by the FCT Ph.D. fellowship SFRH/BD/33761/2009; Torres by EU funding under the 7th Framework Programme FP7-PEOPLE-2010-ITN, grant agreement no. 264735-SADCO.

References

[1] O. P. Agrawal, Formulation of Euler–Lagrange equations for fractional variational problems, *J. Math. Anal. Appl.* 272 (2002), no. 1, 368–379.

[2] O. P. Agrawal, A general formulation and solution scheme for fractional optimal control problems, *Nonlinear Dynam.* 38 (2004), no. 1-4, 323–337.

[3] R. Almeida, S. Pooseh and D. F. M. Torres, Fractional variational problems depending on indefinite integrals, *Nonlinear Anal.* 75 (2012), no. 3, 1009–1025.

[4] R. Almeida and D. F. M. Torres, Leitmann's direct method for fractional optimization problems, *Appl. Math. Comput.* 217 (2010), no. 3, 956–962.

[5] G. E. Andrews, R. Askey and R. Roy, *Special functions*, *Encyclopedia of Mathematics and its Applications*, 71, Cambridge Univ. Press, Cambridge, 1999.

[6] T. M. Atanacković, S. Konjik and S. Pilipović, Variational problems with fractional derivatives: Euler–Lagrange equations, *J. Phys. A* 41 (2008), no. 9, 095201, 12 pp.

[7] T. M. Atanacković and B. Stankovic, An expansion formula for fractional derivatives and its application, *Fract. Calc. Appl. Anal.* 7 (2004), no. 3, 365-378.

[8] T. M. Atanacković and B. Stankovic, On a numerical scheme for solving differential equations of fractional order, *Mech. Res. Comm.* 35 (2008), no. 7, 429–438.

[9] D. Baleanu, O. Defterli and O. P. Agrawal, A central difference numerical scheme for fractional optimal control problems, *J. Vib. Control* 15 (2009), no. 4, 583–597.

[10] P. L. Butzer, A. A. Kilbas and J. J. Trujillo, Stirling functions of the second kind in the setting of difference and fractional calculus, *Numer. Funct. Anal. Optim.* 24 (2003), no. 7-8, 673–711.

[11] S. Das, *Functional fractional calculus for system identification and controls*, Springer, Berlin, 2008.

[12] K. Diethelm, *The analysis of fractional differential equations*, *Lecture Notes in Mathematics*, 2004, Springer, Berlin, 2010.

[13] K. Diethelm, N. J. Ford and A. D. Freed, A predictor-corrector approach for the numerical solution of fractional differential equations, *Nonlinear Dynam.* 29 (2002), no. 1-4, 3–22.

[14] V. D. Djordjevic and T. M. Atanackovic, Similarity solutions to nonlinear heat conduction and Burgers/Korteweg-de Vries fractional equations, *J. Comput. Appl. Math.* 222 (2008), no. 2, 701–714.

[15] L. Elsgolts, *Differential equations and the calculus of variations*, translated from the Russian by George Yankovsky, Mir Publishers, Moscow, 1973.

[16] N. J. Ford and J. A. Connolly, Comparison of numerical methods for fractional differential equations, *Commun. Pure Appl. Anal.* 5 (2006), no. 2, 289–306.

[17] Z. D. Jelicic and N. Petrovacki, Optimality conditions and a solution scheme for fractional optimal control problems, *Struct. Multidiscip. Optim.* 38 (2009), no. 6, 571–581.

[18] A. A. Kilbas, Hadamard-type fractional calculus, *J. Korean Math. Soc.* 38 (2001), no. 6, 1191–1204.

[19] A. A. Kilbas, H. M. Srivastava and J. J. Trujillo, *Theory and applications of fractional differential equations*, North–Holland Mathematics Studies, 204, Elsevier, Amsterdam, 2006.

[20] P. Kumar and O. P. Agrawal, An approximate method for numerical solution of fractional differential equations, *Signal Process.* 86 (2006), 2602–2610.

[21] L. P. Lebedev and M. J. Cloud, *The calculus of variations and functional analysis*, World Sci. Publishing, River Edge, NJ, 2003.

[22] A. B. Malinowska and D. F. M. Torres, Generalized natural boundary conditions for fractional variational problems in terms of the Caputo derivative, *Comput. Math. Appl.* 59 (2010), no. 9, 3110–3116.

[23] M. M. Meerschaert and C. Tadjeran, Finite difference approximations for fractional advection-dispersion flow equations, *J. Comput. Appl. Math.* 172 (2004), no. 1, 65–77.

[24] T. Odzijewicz, A. B. Malinowska and D. F. M. Torres, Fractional variational calculus with classical and combined Caputo derivatives, *Nonlinear Anal.* 75 (2012), no. 3, 1507–1515.

[25] I. Podlubny, *Fractional differential equations*, *Mathematics in Science and Engineering*, 198, Academic Press, San Diego, CA, 1999.

[26] L. S. Pontryagin, V. G. Boltyanskii, R. V. Gamkrelidze and E. F. Mishchenko, *The mathematical theory of optimal processes*, Translated from the Russian by K. N. Trirogoff; edited by L. W. Neustadt Interscience Publishers John Wiley & Sons, Inc. New York, 1962.

[27] S. Pooseh, R. Almeida and D. F. M. Torres, Expansion formulas in terms of integer-order derivatives for the Hadamard fractional integral and derivative, *Numer. Funct. Anal. Optim.* 33 (2012), no. 3, 301–319.

[28] S. Pooseh, R. Almeida and D. F. M. Torres, Approximation of fractional integrals by means of derivatives, *Comput. Math. Appl.* 64 (2012), no. 10, 3090–3100.

[29] S. Pooseh, R. Almeida and D. F. M. Torres, Discrete direct methods in the fractional calculus of variations, *Comput. Math. Appl.* 66 (2013), no. 5, 668–676.

[30] F. Riewe, Nonconservative Lagrangian and Hamiltonian mechanics, *Phys. Rev. E* (3) 53 (1996), no. 2, 1890–1899.

[31] F. Riewe, Mechanics with fractional derivatives, *Phys. Rev. E* (3) 55 (1997), no. 3, part B, 3581–3592.

[32] S. G. Samko, A. A. Kilbas and O. I. Marichev, *Fractional integrals and derivatives*, translated from the 1987 Russian original, Gordon and Breach, Yverdon, 1993.

[33] J. A. Tenreiro Machado, V. Kiryakova and F. Mainardi, Recent history of fractional calculus, *Commun. Nonlinear Sci. Numer. Simul.* 16 (2011), no. 3, 1140–1153.

[34] J. A. Tenreiro Machado, M. F. Silva, R. S. Barbosa, I. S. Jesus, C. M. Reis, M. G. Marcos and A. F. Galhano, Some applications of fractional calculus in engineering, *Math. Probl. Eng.*, Art. ID 639801 (2010) 34 pp.

INDEX

A

B

C

D

E

F

G

H

I

J

K

L

M

N

O

P

Q

R

S

T

V

W